Advanced Low-Cost Separation Techniques in Interface Science

Interface Science and Technology

Series Editor: Arthur T. Hubbard

In this Series:

Vol. 1:	Clay Surfaces: Fundamentals and Applications Edited by F. Wypych and K.G. Satyanarayana
Vol. 2:	Electrokinetics in Microfluids By Dongqing Li
Vol. 3:	Radiotracer Studies of Interfaces Edited By G. Horányi
Vol. 4:	Emulsions: Structure Stability and Interactions Edited By D.N. Petsev
Vol. 5:	Inhaled Particles By Chiu-sen Wang
Vol. 6:	Heavy Metals in the Environment: Origin, Interaction and Remediation Edited By H.B. Bradl
Vol. 7:	Activated Carbon Surfaces in Environmental Remediation Edited By Teresa J. Bandosz
Vol. 8:	Tissue Engineering: Fundamentals and Applications By Yoshito Ikada
Vol. 9:	Particles at Interfaces: Interactions, Deposition, Structure By Zbigniew Adamczyk
Vol. 10:	Interface Science in Drinking Water Treatment: Theory and Applications Edited By G. Newcombe and D. Dixon
Vol. 11:	Surface Complexation Modelling Edited By Johannes Lützenkirchen
Vol. 12:	Theory of Colloid and Interfacial Electric Phenomena By Hiroyuki Ohshima
Vol. 13:	Sorbent Deformation By A.V. Tvardovskiy
Vol. 14:	Advanced Chemistry of Monolayers at Interfaces: Trends in Methodology and Technology Edited By Toyoko Imae
Vol. 15:	Macromolecules in Solution and Brownian Relativity By Stefano A. Mezzasalma
Vol. 16:	The Properties of Water and their Role in Colloidal and Biological Systems By Carel Jan van Oss
Vol. 17:	Chemistry on Modified Oxide and Phosphate Surfaces: Fundamentals and Applications By Robson Fernandes de Farias
Vol. 18:	Interface Science and Composites By Soo-Jin Park and Min-Kang Seo
Vol. 19:	Nanoparticle Technologies: From Lab to Market By Farid Bensebaa
Vol. 20:	Particles at Interfaces By Zbigniew Adamczyk
Vol. 21:	Self-Assembly Processes at Interfaces: Multiscale Phenomena By Vincent Ball
Vol. 22:	Rheology of Emulsions: Electrohydrodynamics Principles By Aleksandar M. Spasic
Vol. 23:	Tailored Thin Coatings for Corrosion Inhibition using a Molecular Approach By Simo Olavi Pehkonen and Shaojun Yuan
Vol. 24:	Charge and Energy Storage in Electrical Double Layers By Silvia Ahualli and Ángel V. Delgado
Vol. 25:	Stimuli Responsive Polymeric Membranes By Mihir Kumar Purkait, Manish Kumar Sinha, Piyal Mondal, and Randeep Singh
Vol. 26:	Theory of Electrophoresis and Diffusiophoresis of Highly Charged Colloidal Particles By Eric Lee
Vol. 27:	Graphene Surfaces: Particles and Catalysts By Karim Kakaei, Mehdi Esrafili and Ali Ehsani
Vol. 28:	An Introduction to Green Nanotechnology By Mahmoud Nasrollahzadeh, Mohammad Sajadi, Monireh Atarod, Mohadeseh Sajjadi and Zahra Isaabadi
Vol. 29:	Emerging Natural and Tailored Nanomaterials for Radioactive Waste Treatment and Environmental Remediation Edited by Changlun Chen

Interface Science and Technology
Volume 30

Advanced Low-Cost Separation Techniques in Interface Science

Edited by

George Z. Kyzas
Athanasios C. Mitropoulos
Hephaestus Advanced Laboratory, Department of Chemistry,
International Hellenic University, Kavala, Greece

ELSEVIER

ACADEMIC PRESS
An imprint of Elsevier

Academic Press is an imprint of Elsevier
125 London Wall, London EC2Y 5AS, United Kingdom
525 B Street, Suite 1650, San Diego, CA 92101, United States
50 Hampshire Street, 5th Floor, Cambridge, MA 02139, United States
The Boulevard, Langford Lane, Kidlington, Oxford OX5 1GB, United Kingdom

Library of Congress Cataloging-in-Publication Data
A catalog record for this book is available from the Library of Congress

British Library Cataloguing-in-Publication Data
A catalogue record for this book is available from the British Library

ISBN: 978-0-12-814178-6
ISSN: 1573-4285

For information on all Academic Press publications visit our website at https://www.elsevier.com/books-and-journals

Publisher: Susan Dennis
Acquisition Editor: Anita Koch
Editorial Project Manager: Emily Thomson
Production Project Manager: Prem Kumar Kaliamoorthi
Cover Designer: Alan Studholme

Typeset by TNQ Technologies

Contents

Contributors

Numbers in Parentheses indicate the pages on which the author's contributions begin.

Saeid Azizian (283), Department of Physical Chemistry, Faculty of Chemistry, Bu-Ali Sina University, Hamedan, Iran

Dimitrios V. Bandekas (1), Electrical Engineering Department, Eastern Macedonia and Thrace Institute of Technology, Kavala, Greece

George Bomis (69), Hephaestus Advanced Laboratory, Department of Chemistry, International Hellenic University, Kavala, Greece

Eleni A. Deliyanni (101), Laboratory of General and Environmental Chemistry, Department of Chemistry, Aristotle University of Thessaloniki, Greece

Eleni K. Efthimiadou (69), Institute of Nanoscience and Nanotechnology, NCSR "Demokritos", Terma Patriarchou Grigoriou and Neapoleos, Attica, Greece; Inorganic Chemistry Laboratory, Chemistry Department, National and Kapodistrian University of Athens, Zografou, Greece

Evangelos P. Favvas (69), Institute of Nanoscience and Nanotechnology, NCSR "Demokritos", Terma Patriarchou Grigoriou and Neapoleos, Attica, Greece; Hephaestus Advanced Laboratory, Department of Chemistry, International Hellenic University, Kavala, Greece

George P. Gallios (43), Division of Chemical Technology & Industrial Chemistry, Department of Chemistry, Aristotle University, Thessaloniki, Greece

Despina A. Gkika (1), Electrical Engineering Department, Eastern Macedonia and Thrace Institute of Technology, Complex Systems Lab, Kavala, Greece; Electrical Engineering Department, Eastern Macedonia and Thrace Institute of Technology, Kavala, Greece; Hephaestus Advanced Laboratory, Department of Chemistry, International Hellenic University, Kavala, Greece

Gassan Hodaifa (217, 241), Molecular Biology and Biochemical Engineering Department, Chemical Engineering Area, Faculty of Experimental Sciences, University of Pablo de Olavide Seville, Spain

Maryam Khosravi (283), Department of Physical Chemistry, Faculty of Chemistry, Bu-Ali Sina University, Hamedan, Iran

Nikolaos C. Kokkinos (17), Hephaestus Advanced Laboratory, Department of Chemistry, International Hellenic University, Kavala, Greece

Ramonna I. Kosheleva (125, 151), Hephaestus Advanced Laboratory, Department of Chemistry, International Hellenic University, Kavala, Greece

George Z. Kyzas (1, 15, 43, 69, 125, 151, 173, 199), Hephaestus Advanced Laboratory, Department of Chemistry, International Hellenic University, Kavala, Greece

Nikolaos K. Lazaridis (15), Division of Chemical Technology & Industrial Chemistry, Department of Chemistry, Aristotle University, Thessaloniki, Greece

Efstathios V. Liakos (1), Hephaestus Advanced Laboratory, Department of Chemistry, International Hellenic University, Kavala, Greece

Lykourgos Magafas (1), Electrical Engineering Department, Eastern Macedonia and Thrace Institute of Technology, Complex Systems Lab, Kavala, Greece; Electrical Engineering Department, Eastern Macedonia and Thrace Institute of Technology, Kavala, Greece

Ana Malvis (217, 241), Molecular Biology and Biochemical Engineering Department, Chemical Engineering Area, Faculty of Experimental Sciences, University of Pablo de Olavide Seville, Spain

Kostas A. Matis (15, 43), Division of Chemical Technology & Industrial Chemistry, Department of Chemistry, Aristotle University, Thessaloniki, Greece

Elisavet D. Michailidi (69), Institute of Nanoscience and Nanotechnology, NCSR "Demokritos", Terma Patriarchou Grigoriou and Neapoleos, Attica, Greece; Inorganic Chemistry Laboratory, Chemistry Department, National and Kapodistrian University of Athens, Zografou, Greece

Athanasios C. Mitropoulos (1, 69, 125, 151, 173), Hephaestus Advanced Laboratory, Department of Chemistry, International Hellenic University, Kavala, Greece

Matteo Neviani (217, 241), DIME, Department of Mechanical, Energy, Management and Transportation Engineering, University of Genoa, Genoa, Italy

Ombretta Paladino (217, 241), DICCA, Department of Civil, Chemical and Environmental Engineering, University of Genoa, Genoa, Italy

Antonios N. Papadopoulos (199), Laboratory of Wood Chemistry and Technology, Eastern Macedonia and Thrace Institute of Technology, Kavala, Greece

Likourgos V. Petridis (173), Hephaestus Advanced Laboratory, Department of Chemistry, International Hellenic University, Kavala, Greece

Mahdi Seyedsalehi (217, 241), School of Environment, Tsinghua University, Beijing, China

Athanasios Varoutoglou (69), Hephaestus Advanced Laboratory, Department of Chemistry, International Hellenic University, Kavala, Greece

Nick Vordos (1), Electrical Engineering Department, Eastern Macedonia and Thrace Institute of Technology, Kavala, Greece; Hephaestus Advanced Laboratory, Department of Chemistry, International Hellenic University, Kavala, Greece

Chapter 1

The impact of raw materials cost on the adsorption process

Despina A. Gkika,[a,b,c,*] Nick Vordos,[b,c] Efstathios V. Liakos,[c] Lykourgos Magafas,[a,b] Dimitrios V. Bandekas,[b] Athanasios C. Mitropoulos[c] and George Z. Kyzas[c]
[a]*Electrical Engineering Department, Eastern Macedonia and Thrace Institute of Technology, Complex Systems Lab, Kavala, Greece;* [b]*Electrical Engineering Department, Eastern Macedonia and Thrace Institute of Technology, Kavala, Greece;* [c]*Hephaestus Advanced Laboratory, Department of Chemistry, International Hellenic University, Kavala, Greece*
**Corresponding author*

1. Introduction

Recently, the "12 Principles of Green Chemistry" [1] have been more widely adopted and have been preferred over conventional methods. It has been used to a high extent for industrial separation and purification, ensuring that adsorbents are created in optimal conditions so that the optimal characteristics are expressed. The advantages of green synthesis include its cost-effectiveness, environment friendliness, scalability, and lack of requirement for extreme conditions like high pressure, energy or temperature, or the need to utilize toxic chemicals [2].

Organic dyes are a major type of water pollutants that may have severe environmental effects due to being highly toxic and may cause cancer [3,4]. Dyes are not easily degradable due to their stability and properties [5,6]. Multiple efforts have been made to address the issue of dye removal from wastewater systems, such as ion exchange, extraction of solvents, adsorption, use of filters, coagulation-flocculation, advanced oxidation and electrochemical methods [7–12]. Adsorption is one of the preferred methods because its operation mode is simpler and easier and it has little to none by-products compared with other methods [13]. As a result, the research on new and more effective materials has been getting more focus by scientists worldwide [13–16].

Advanced Low-Cost Separation Techniques in Interface Science. https://doi.org/10.1016/B978-0-12-814178-6.00001-7

A variety of methods have been utilized to address dye residues, including coagulation, irradiation, photocatalytic oxidation, and more. Adsorption is a cost-efficient and easy-to-apply method for the removal of existing dyes [17]. Physicochemical and biological processes are utilized for dye treatment in polluted wastewater [18]. The most common methods are coagulation-flocculation [19] and advanced oxidation process [20]. Nevertheless, such options are costly and have other operational issues, such as toxic by-products, lower removal effectiveness, and requirement for better matches to specific group of dyes [21].

The adsorption process is frequently used for the treatment of organic pollutants owing to its simplicity, performance, lack of sensitivity to toxic materials, and scalability [22]. Adsorption is considered to be better than alternative options for water reuse based on starting costs, clean design, and insensitivity to toxic substances [23,24]. Adsorption can be used either in single or combinational mode for partial or total wastewater cleaning. Recently, different, cheaper materials, such as biosorbents, industrial and agricultural waste, zeolite, silicon, and clays, have been tested as adsorbents at the third stage of waste treatment, thus replacing activated carbon (AC) [25−28].

2. Market trend for various adsorbents

Fig. 1.1 presents a comparison of the market trends and future projections for different adsorbents. There is a big increase in market size, with chitosan and activated carbon displaying the highest rise individually.

Activated carbon. The AC market might be split into different categories, including powdered, granular, and palletized, depending on the final product form. The high demand for powdered AC in applications such as decolorization and deodorization in pharmaceuticals is likely to enhance the market growth [29]. According to Statistics MRC, the global activated carbon market reached about $4.12 billion in 2017 and is anticipated to go beyond $14 billion by 2026 [30,31,32].

Chitosan. The global chitosan market report explores chitosan and its most common applications such as water treatment, cosmetics, food, and pharmaceuticals, among others. The most significant driver of the chitosan market is the abundance of the source material along with developing applications in various fields [33]. The size of the global chitosan market is expected to surpass $2500 million by 2022, compared with $1205 million in 2015 [34].

Graphene. Graphene is a carbon allotrope that contains graphite and diamond. The most common graphene types that form the market include graphene oxide; nanoplatelets; mono-, bi-, or few-layer graphene. The global graphene market is expected to achieve a noticeable growth of more than 811 million USD by 2023 [35−37].

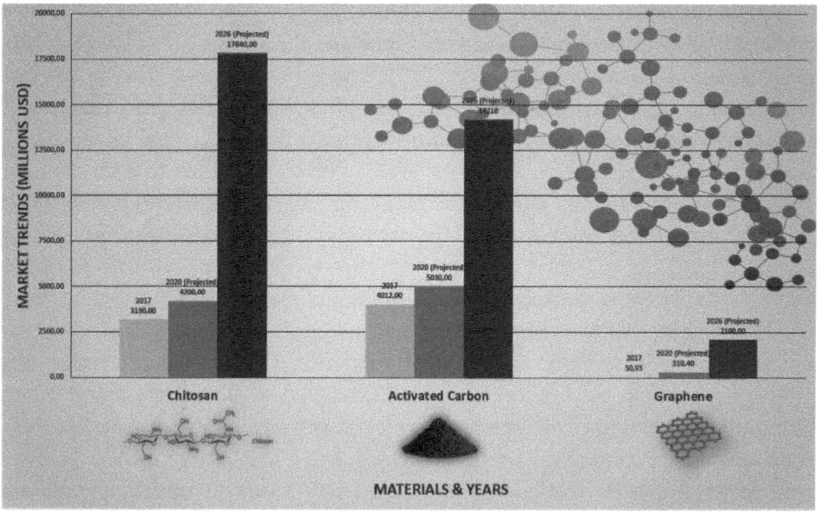

FIGURE 1.1 Trends and future projections for different adsorbents.

3. Activated carbon efficiency

ACs stem from natural materials like wood and coal, which renders their production and regeneration process quite high in price [38,39]. Many scientists have thus researched potential alternate low-cost adsorbents production processes. AC has been suggested as an adsorbent because of its structure and large surface which make it a great candidate for the removal of organic pollutants (including hard biodegradable ones) from aquatic environment. It is considered as one of the most efficient adsorbents and is used widely due to its nontoxicity for purifying drinking water and treating wastewater [22]. More specifically, AC's efficiency in removing a large variety of dyes from wastewaters has rendered it an excellent option when compared with more expensive alternatives [40].

AC is commonly retrieved from substances with high carbon content. Its adsorption ability is affected by its porous structure and surface area, which can be adjusted, thus making AC suitable for a wide application range, including pollutant removal and catalyst support. AC is not cheap, and the better the quality, the higher its cost. All methods of spent carbon regeneration, however, are costly, are impractical, or result in further waste discharge [41], despite its efficiency as adsorbent [42]. Despite its multiple applications in cleaning processes, AC is still quite costly. Commercial AC is limited due to cost, lower performance, as well as efficiency after regeneration [43].

So far, natural adsorbents, such as chitin, biomass, and chitosan, as well as industrial and agricultural waste have been for the treatment. Bioadsorbents are considered more selective, less costly, and more effective than resins and commercially activated carbon [44]. Carbon-based materials are widely used in dye removal processes [13]. During the past few years, graphene-based materials have been tested in various areas, due to attributes that make them ideal candidates, such as physicochemical, optical, and other properties [45]. Recently, they have been successfully tried as alternative adsorption options [46,47]. Graphene is, however, costly and not too easy to produce; therefore significant efforts are dedicated toward finding cheaper yet still effective ways to take advantage of graphene and related materials. Graphene oxide (GO) is one such material created by graphite oxidation [48]. Graphene oxide has shown remarkable adsorption abilities compared with other carbon-based materials and thus has an important role in wastewater treatment. Furthermore, due to its comparatively high surface area (2630 m^2/g), it has received even more attention [49]. Graphene oxide may assist in removing cationic metals and dyes due to its negative charge density. Graphene oxide sheets are made of a single-layered graphite structure connected with oxygen atoms and encompass carboxyl, ketone, epoxy, and hydroxyl groups, which transmit negative charges when in liquid solutions, in a broad pH range.

4. Biosorbents efficiency (the case of chitosan)

Biopolymers such as chitosan and chitin have emerged as effective adsorption solutions [44]. Chitosan is useful as it possesses amino and hydroxyl groups, which may act as the active sites [50] and is also nontoxic, noncorrosive, and safe to manage [51]. The main advantage of chitosan is that it is able to bind dyes in low concentrations as opposed to metal salts and is also effective in cold water. Chitosan composites are cost-effective due to their easy preparation requirement of cheaper chemical reagents. The removal of even the small amounts of pollutants is very expensive, so traditional methods are not practical [3]. Currently, multiple adsorption methods based on chitosan composites are being researched as a substitute for the pricier conventional methods [52].

5. Agricultural waste efficiency

Agricultural solid waste such as leaves, fruits, seeds, and peels, as well as forest waste materials such as sawdust and bark, have been widely used as adsorbents for dye removal [53]. Their availability reaches extremely high quantities, and their physicochemical attributes make them useful as sorbents. Several agricultural solid wastes are effective in removing both cationic and anionic dyes; however, they require activation [54,55]. The most significant factor modifying the adsorption is the pH; a high pH allows for the adsorption of cationic dyes, whereas a low pH is more effective on anionic dyes [54]. A review by Demirbas [15] provided a comprehensive list of agricultural waste and a variety of otherwise unwanted plant parts that provide a cheap and renewable supplementary source of AC. These materials have little to none economical value and would commonly be problematic to dispose, they are used in natural or modified form for dye removal. The most commonly used forms of agro-waste include pine wood, corn cob, nut shells, fruit peels, bamboo, bark [15]. Furthermore, they have desirable physicochemical attributes that render them highly efficient and easy to renew with or without being treated (they occasionally might need to be grounded, washed, and/or dried) which lowers the production and energy costs [56]. An example of agro-wastes with potential are banana and orange peels that can be used as adsorbents for the removal of copper, zinc, carbon dioxide, nickel, and lead cations from water. The residues of the peels can be treated and modified to become adsorbents due to their large surface areas, swelling capacity, strength, ease of use, and ability to adsorb hazardous pollutants.

6. Economics of adsorption

Cost is a significant factor when comparing sorbent substances. The adsorbent cost is influenced by a variety factors, such as availability, type (natural, agricultural, industrial, synthesized, by-products, etc.), conditions, recycle, stability, production country (in terms of development stage [57,58]. Due to the continual changes in adsorption, the scientific community has concentrated its efforts on developing a cheap, simple, and eco-friendly process for the removal of dyes. Traditional synthesis processes have many limitations due to their environmental risks and high costs [17]. Therefore efforts are concentrated toward establishing less expensive, more effective, and scalable separation processes as required by the industry. The feasibility of using low-cost agriculture-based waste materials has been widely studied, and biosorption technology has been found to be a sufficient alternative for the removal of heavy metals from wastewater [59]. Agricultural wastes are cheaper and often environmentally friendly due to their chemical attributes, abundancy, renewability, low cost, and efficiency for both organic and inorganic waste treatment [60].

7. Materials and methods

The information was collected from a variety of sources, primarily via in-person interviews with the managers of two Greek chemistry laboratories in Greece. The managers were chosen due to their expertise in the field of adsorption. The focal point of the interviews was the identification and analysis of the characteristics of "popular" adsorbents.

The advantage of personal interviews lies in the ability to observe the interviewee's expressions and feelings as well as listen to their responses unobtrusively. The chosen approach consisted of unstructured interviews since they are more flexible for all participants, as proposed by Gubrium and Holstein [61]. Furthermore, there is also the chance to further discuss specific "elite" issues of importance that might be brought up during interviews [62,63]. The term "elite" refers to experts on specific subjects [64]. The basic issue in such cases is to reach a balance between proper reporting of research outcomes and the potential risk of disclosing interviewees' identities, as opposed to hiding their data for less exposure but higher risk of results being disputed [65]. The confidentiality is necessary to protect the interviewees. Anonymity permits participants to express more freely [66]. For this specific research, the respondents are male professors. The interview was detailed, taking up more than 40 min. Based on the responses, we determined two important parameters affecting costs. These consist of the raw material and the adsorption energy costs.

Recipe selection: In order to further explore the adsorption phase and identify the attributes shaping the studied adsorbents, we performed a literature search to find the corresponding recipes. The most cited ones were chosen for each adsorbent, so that the cost parameters could be calculated.

Raw material price: The first cost factor to be estimated is the raw material cost. These costs refer to the necessity for the adsorption process resources. Chesbrough [67] proposes a new way of cooperation of various stakeholders in the value chain in order to gain new knowledge and technologies. The analysis included the compilation of market prices for the materials based on information of various vendors worldwide.

Energy cost: The second cost factor consists of energy costs. This refers to the energy utilized for the various stages of adsorption. Electricity costs are estimated based on the energy price in Greece for 2019 (0.194 Euro/kWh). These prices are provided by the Public Power Corporation S.A., Hellas [68]. The cost in Euros has been calculated as the energy cost in kWh multiplied by the price of energy in Greece as Euros per kWh.

Maintenance cost and labor cost: These costs are not included because corresponding data are not available for the adsorption phase.

8. Results and discussion

The effects of different parameters including solution pH, biosorbent's dosage, initial dye concentration, and contact time were studied. Table 1.1 attempts to provide an up-to-date list on adsorption properties or recycled and reused waste by-products.

A combination of multiple adsorbents (Table 1.1) has been found to yield promising results; however, this comes with the disadvantage of increased costs. Therefore, in spite of their effectiveness and suitability, certain financial considerations might limit the use of costly adsorbents.

TABLE 1.1 Synthesis recipes of the adsorption of dyes using different adsorbent (dosage: 1 g of adsorbent per 1 L of adsorbate solution).

Adsorbent	T(°C)	pH	Contact time (min)	Removal (%)	Pollutant	Ref
Graphene oxide (GO)	65	3	180	55%	Dye	[69]
GO-Chm	65	3	180	81%	Dye	[69]
Magnetic activated carbon	65	10	300	68%	Dye	[70]
CD-MIPs	25	2	225	62%	Dye	[71]
CHI-MIPs	25	2	225	62%	Dye	[71]
Chitosan biosorbent	65	12	120	97%	Metal	[72]
Chitosan biosorbent	65	4	120	85%	Metal	[72]
Coffee biosorbent	25	5	180	UCR: 70%; TCR: 77%	Metal	[73]
Coffee biosorbent	25	5	180	UCR: 60%; TCR: 62%	Metal	[73]
GO/CS	65	6	150	93%	Metal	[74]
GO/mCS	65	6	150	96%	Metal	[74]
GO	65	3	200	53%	Drug	[75]
CSA	65	3	200	68%	Drug	[75]
GO/CSA	65	3	200	93%	Drug	[75]
Activated carbon	65	3		B0: 50%; B1: 80%; B3: 82%; B4: 83%; B5: 65%	Drug	[76]
Cs	65	10		50%	Drug	[77]
CsNCB	65	10		78%	Drug	[77]
CsSLF	65	10		82%	Drug	[77]
Gho	65	2	180	PRO: 68%; ATL: 93%	Drug	[78]

ATL, atenolol; *CD-MIPs*, β-Cyclodextrin MIPs; *CHI-MIPs*, chitosan MIPs; *CSA*, chitosan grafted with poly(acrylic acid); *CsNCB*, chitosan grafted with N-(2-carboxybenzyl groups); *CsSLF*, chitosan grafted with sulfonic groups; *PRO*, propranolol; *TCR*, treated coffee residues; *UCR*, untreated coffee residues.

In general, a potential low-cost dye adsorbent should show (1) effectiveness in removing a wide range of dyes; (2) high ability and rate of adsorption; (3) differentiation in selectivity based on concentration; and (4) tolerance of a variety of wastewater attributes. Low-cost adsorbents possess certain physical and chemical characteristics such as pore size, surface area, and physical strength, as well as a range of advantages and disadvantages that affect their efficiency in treating wastewater [44].

Raw material costs are considered "hard" costs—therefore any savings in these will allow for better profits directly. The high prices of raw materials makes stakeholders nervous about the development and exploitation of new technologies and how those can be affected by fluctuations in price and availability [30].

The results from Table 1.2 reveals that only the raw material cost actually matters for the adsorption phase. The reduced consumption of raw materials could result in lowered costs and other benefits, including an improved laboratory image and working conditions.

This work presents an initial attempt to optimize the dye removal during the adsorption stage; however, an effective methodology addressing the

TABLE 1.2 Raw materials and energy costs in selected recipes.

Material produced	Raw materials	Raw materials cost per 1 g of final product (€)	Energy cost (€)	Recipe cost (€)
Polymer/graphene oxide—synthesis of magnetic nanoparticles	$KMnO_4$ (>99.0%); graphite flakes; H_2SO_4 (95%–98%); H_2O_2 (30 wt%)	5.75	5.02	10.77
Polymer/graphene oxide—synthesis of magnetic chitosan (Chm)	$FeCl_2$ $4H_2O$ (p.a. >99.0%); chitosan (high molecular weight); $FeCl_3$ 6 H_2O (reagent grade, 97%); glutaraldehyde (50 wt% in water; acetic acid solution (>99%) $KMnO_4$ (>99.0%); graphite flakes; H_2SO_4 (95%–98%); H_2O_2 (30 wt%)	2.4	3.35	5.75
Polymer—synthesis of sulfonate-grafted	Dichloroacetic acid (>99%); formamide	3.22	3.89	7.11

TABLE 1.2 Raw materials and energy costs in selected recipes.—cont'd

Material produced	Raw materials	Raw materials cost per 1 g of final product (€)	Energy cost (€)	Recipe cost (€)
chitosan adsorbent (CsSLF)	(>99.5%); chitosan (high molecular weight); glutaraldehyde (50 wt% in water); acetic acid solution (>99%)			
Polymer—synthesis of N-(2-carboxybenzyl)-grafted chitosan adsorbent (CsNCB)	Chitosan (high molecular weight); glutaraldehyde (50 wt% in water); acetic acid solution (>99%); 2-carboxybenzaldehyde (97%); sodium borohydride (>96%)	11.39	0.84	12.23
Polymer—synthesis of cross-linking chitosan adsorbents	Chitosan (high molecular weight); glutaraldehyde (50 wt% in water); acetic acid solution (>99%): 1L	2.09	4,29	6.38
Graphene oxide—synthesis of adsorbents according to a modified Hummers method	$KMnO_4$ (>99.0%); graphite flakes; H_2SO_4 (95%–98%); H_2O_2 (30 wt%)	5.57	2.47	8.04

removal of all kinds of dyes in an inexpensive manner has been found. The outcomes suggest that there are some cheaper adsorbents that allow for higher dye removal rate, as shown in Table 1.1. Such low-cost adsorbents arose as an alternative option, and they support practical solutions to toxicity-related issues.

9. Conclusions

This chapter has presented a broad range of different adsorbents that can be utilized as low-cost adsorbents. The usage of such cheap adsorbents is highly advisable due to not just their low cost but also availability, renewability, and effectiveness. The advantage of low raw material cost should be further explored in terms of regeneration, design, and confinement of the waste material for improved effectiveness and recovery rates. The outcomes of this work could help in establishing a well-founded, environmentally friendly, simple, and cost-effective process for green synthesis especially with the constant developments in its application fields.

References

[1] A. Ivanković, Review of 12 Principles of green Chemistry in practice, Int. J. Sustain. Green Energy 6 (3) (2017) 39.

[2] C.-N. Lok, et al., Silver nanoparticles: partial oxidation and antibacterial activities, JBIC J. Biol. Inorg. Chem 12 (4) (April 2007) 527–534.

[3] R.L. Droste, Theory and Practice of Water and Wastewater Treatment, J. Wiley, New York, 1997.

[4] R. Kant, Textile dyeing industry an environmental hazard, Nat. Sci. 04 (01) (2012) 22–26.

[5] V.K. Gupta, I. Ali, T.A. Saleh, A. Nayak, S. Agarwal, "Chemical treatment technologies for waste-water recycling—an overview, RSC Adv. 2 (16) (2012) 6380.

[6] Y.-C. Hsiao, T.-F. Wu, Y.-S. Wang, C.-C. Hu, C. Huang, Evaluating the sensitizing effect on the photocatalytic decoloration of dyes using anatase-TiO_2, Appl. Catal. B Environ. 148–149 (April 2014) 250–257.

[7] W. Przystaś, E. Zabłocka-Godlewska, E. Grabińska-Sota, Biological removal of azo and triphenylmethane dyes and toxicity of process by-products, Water. Air. Soil Pollut 223 (4) (May 2012) 1581–1592.

[8] A.Y. Zahrim, N. Hilal, Treatment of highly concentrated dye solution by coagulation/-flocculation–sand filtration and nanofiltration, Water Resour. Ind 3 (September 2013) 23–34.

[9] R. Brás, M. Isabel, A. Ferra, H.M. Pinheiro, I.C. Gonçalves, Batch tests for assessing decolourisation of azo dyes by methanogenic and mixed cultures, J. Biotechnol. 89 (2–3) (August 2001) 155–162.

[10] E.H. Umukoro, M.G. Peleyeju, J.C. Ngila, O.A. Arotiba, Photoelectrochemical degradation of orange II dye in wastewater at a silver–zinc oxide/reduced graphene oxide nano-composite photoanode, RSC Adv. 6 (58) (2016) 52868–52877.

[11] A. Molla, M. Sahu, S. Hussain, Under dark and visible light: fast degradation of methylene blue in the presence of Ag–In–Ni–S nanocomposites, J. Mater. Chem. 3 (30) (2015) 15616–15625.

[12] A. Molla, M. Sahu, S. Hussain, Synthesis of tunable band gap semiconductor nickel sulphide nanoparticles: rapid and round the clock degradation of organic dyes, Sci. Rep. 6 (1) (September 2016).

[13] R. Malik, D.S. Ramteke, S.R. Wate, Adsorption of malachite green on groundnut shell waste based powdered activated carbon, Waste Manag. 27 (9) (January 2007) 1129–1138.

[14] F. Perreault, A. Fonseca de Faria, M. Elimelech, Environmental applications of graphene-based nanomaterials, Chem. Soc. Rev. 44 (16) (2015) 5861–5896.

[15] A. Demirbas, Agricultural based activated carbons for the removal of dyes from aqueous solutions: a review, J. Hazard Mater. 167 (1–3) (August 2009) 1–9.

[16] P. Sharma, M.R. Das, Removal of a cationic dye from aqueous solution using graphene oxide nanosheets: investigation of adsorption parameters, J. Chem. Eng. Data 58 (1) (January 2013) 151–158.

[17] M. Danish, R. Hashim, M.N. Mohamad Ibrahim, O. Sulaiman, Response surface methodology approach for methyl orange dye removal using optimized Acacia mangium wood activated carbon, Wood Sci. Technol. 48 (5) (September 2014) 1085–1105.

[18] A. Elsagh, O. Moradi, A. Fakhri, F. Najafi, R. Alizadeh, V. Haddadi, Evaluation of the potential cationic dye removal using adsorption by graphene and carbon nanotubes as adsorbents surfaces, Arab. J. Chem. 10 (May 2017) S2862–S2869.

[19] M.S. Rahbar, E. Alipour, R.E. Sedighi, Color removal from industrial wastewater with a novel coagulant flocculant formulation, Int. J. Environ. Sci. Technol. 3 (1) (December 2006) 79–88.

[20] Z. Ai, J. Li, L. Zhang, S. Lee, Rapid decolorization of azo dyes in aqueous solution by an ultrasound-assisted electrocatalytic oxidation process, Ultrason. Sonochem. 17 (2) (February 2010) 370–375.

[21] C. Pelekani, V.L. Snoeyink, Competitive adsorption between atrazine and methylene blue on activated carbon: the importance of pore size distribution, Carbon 38 (10) (2000) 1423–1436.

[22] Z. Noorimotlagh, S.A. Mirzaee, S.S. Martinez, S. Alavi, M. Ahmadi, N. Jaafarzadeh, Adsorption of textile dye in activated carbons prepared from DVD and CD wastes modified with multi-wall carbon nanotubes: equilibrium isotherms, kinetics and thermodynamic study, Chem. Eng. Res. Des. 141 (January 2019) 290–301.

[23] J. Pal, M.K. Deb, Microwave green synthesis of PVP-stabilised gold nanoparticles and their adsorption behaviour for methyl orange, J. Exp. Nanosci. 9 (5) (May 2014) 432–443.

[24] V. Meshko, L. Markovska, M. Mincheva, A. Rodrigues, Adsorption of basic dyes on granular activated carbon and natural zeolite, Water Res. 35 (14) (October 2001) 3357–3366.

[25] G.Z. Kyzas, M. Kostoglou, A.A. Vassiliou, N.K. Lazaridis, Treatment of real effluents from dyeing reactor: experimental and modeling approach by adsorption onto chitosan, Chem. Eng. J. 168 (2) (April 2011) 577–585.

[26] G. Kyzas, A decolorization technique with spent 'Greek coffee' grounds as zero-cost adsorbents for industrial textile wastewaters, Materials 5 (11) (October 2012) 2069–2087.

[27] G.Z. Kyzas, N.K. Lazaridis, A.C. Mitropoulos, Removal of dyes from aqueous solutions with untreated coffee residues as potential low-cost adsorbents: equilibrium, reuse and thermodynamic approach, Chem. Eng. J. 189–190 (May 2012) 148–159.

[28] L.D. Fiorentin, et al., Biosorption of reactive blue 5G dye onto drying orange bagasse in batch system: kinetic and equilibrium modeling, Chem. Eng. J. 163 (1–2) (September 2010) 68–77.

[29] Prescient & Strategic (P&S) Intelligence Private Limited, activated carbon Market by product type (powdered activated carbon, granular activated carbon), by end use (water treatment, food & beverage processing, industrial, pharmaceutical & medical, automotive, air purification), by Geography (U.S., Canada, France, Germany, U.K., Italy, Spain, Japan, China, India, Brazil, Saudi Arabia, South Africa) – Global Market Size, Share, Development, Growth, and Demand Forecast, 2014–2024. Available: https://www.psmarketresearch.com/contact.

[30] Activated Carbon Market Size, Share, Report, Analysis, Trends & Forecast to 2026, Reuters Brand Feature. Available: https://www.reuters.com/brandfeatures/venture-capital/article?id=47577. .

[31] Activated Carbon Market Size, Share, Analysis - Segmented by Type (Powdered activated carbons, Granular activated carbons, Extruded or Pelletized activated carbons), Application (Gas Purification, Water Purification, Metal Extraction, Medicine), End-user Industry (Water Treatment, Food & Beverage, Healthcare, Automotive, Industrial Processing), and

Geography - Growth, Trends, and Forecast (2019–2024). Mordor Intelligence. Available: https://www.mordorintelligence.com/industry-reports/global-activated-carbon-market?gclid=CjwKCAiA45njBRBwEiwASnZT5wfuv-WnbHmwZVV2JxhEg4qgZKu0T4hfCmZOfckpQNZIeETLHTtx-BoCYr4QAvD_BwE.

[32] Activated Carbon – A Global Market Overview, Industry Experts. Available: http://industry-experts.com/verticals/chemicals-and-materials/activated-carbon-a-global-market-overview.

[33] Chitosan Market by Source (Shrimp, Crab, Squid, and Krill), Application (Water Treatment, Biomedical & Pharmaceuticals, Cosmetics, Food & beverage, and Others) – Global Opportunity Analysis and Industry Forecast, 2014–2022. Allied Market Search. [Online]. Available: https://www.alliedmarketresearch.com/chitosan-market.

[34] Chitosan Market Estimates and Trend Analysis By Application (Water Treatment, Pharmaceutical & Biomedical, Cosmetics, Food & Beverage), By Region (North America, Europe, Asia Pacific, RoW), By Country, And Segment Forecasts, 2018 – 2025, Grand View Research. Available: https://www.grandviewresearch.com/industry-analysis/global-chitosan-market.

[35] Graphene-based products market value worldwide from 2015 to 2025 (in million U.S. dollars), The Statistics Portal (January 20, 2019) [Online]. Available: https://www.statista.com/statistics/805459/graphene-based-products-global-market-value/.

[36] Graphene 2018-2023: Global Market to Reach a Value of $357.84 Million, Stimulated by the Development of Automotive and Telecom Industries," CISION PR Newswire. [Online]. Available: https://www.prnewswire.com/news-releases/graphene-2018-2023-global-market-to-reach-a-value-of-357-84-million-stimulated-by-the-development-of-automotive-and-telecom-industries-300673994.html.

[37] Graphene Market Research Report- Forecast to 2023, Market Search Future. Available: https://www.marketresearchfuture.com/reports/graphene-market-2987.

[38] J.M. Dias, M.C.M. Alvim-Ferraz, M.F. Almeida, J. Rivera-Utrilla, M. Sánchez-Polo, Waste materials for activated carbon preparation and its use in aqueous-phase treatment: a review, J. Environ. Manag. 85 (4) (December 2007) 833–846.

[39] B. Buczek, Preparation of active carbon by additional activation with potassium hydroxide and characterization of their properties, Ann. Mater. Sci. Eng. 2016 (2016) 1–4.

[40] K.R. Ramakrishna, T. Viraraghavan, Dye removal using low cost adsorbents, Water Sci. Technol. 36 (2–3) (July 1997) 189–196.

[41] Z. Aksu, Application of biosorption for the removal of organic pollutants: a review, Process Biochem. 40 (3–4) (March 2005) 997–1026.

[42] J. Pal, M.K. Deb, D.K. Deshmukh, D. Verma, Removal of methyl orange by activated carbon modified by silver nanoparticles, Appl. Water Sci. 3 (2) (June 2013) 367–374.

[43] A. Mudhoo, R.K. Gautam, M.C. Ncibi, F. Zhao, V.K. Garg, M. Sillanpää, Green synthesis, activation and functionalization of adsorbents for dye sequestration, Environ. Chem. Lett. 17 (1) (July 2018) 157–193.

[44] G. Crini, Non-conventional low-cost adsorbents for dye removal: a review, Bioresour. Technol. 97 (9) (June 2006) 1061–1085.

[45] M. Hu, Z. Yao, X. Wang, Department of chemical, biological and pharmaceutical engineering, New Jersey institute of technology, newark, NJ 07102, USA, "characterization techniques for graphene-based materials in catalysis, AIMS Mater. Sci. 4 (3) (2017) 755–788.

[46] G.K. Ramesha, A. Vijaya Kumara, H.B. Muralidhara, S. Sampath, Graphene and graphene oxide as effective adsorbents toward anionic and cationic dyes, J. Colloid Interface Sci. 361 (1) (September 2011) 270–277.
[47] N. Cai, P. Larese-Casanova, Sorption of carbamazepine by commercial graphene oxides: a comparative study with granular activated carbon and multiwalled carbon nanotubes, J. Colloid Interface Sci. 426 (July 2014) 152–161.
[48] A. kargupta, P. Dutta, K. Dutta, Review on 'Graphene and Graphene Oxide based system for the detection of synthetic azo dyes, Int. J. Adv. Eng. Technol. Manage. Appl. Sci. 5 (6) (2018) 1–8.
[49] M.R. Gandhi, S. Vasudevan, A. Shibayama, M. Yamada, Graphene and graphene-based composites: a rising star in water purification – a comprehensive overview, Chemistry 1 (15) (September 2016) 4358–4385.
[50] F.C. Wu, R.L. Tseng, R.S. Juang, Kinetic modeling of liquid-phase adsorption of reactive dyes and metal ions on chitosan, Water Res. 35 (3) (March 2001) 613–618.
[51] Y. Lee, H.-W. Kim, Y.H. Brad Kim, New route of chitosan extraction from blue crabs and shrimp shells as flocculants on soybean solutes, Food Sci. Biotechnol. 27 (2) (December 2017) 461–466.
[52] K. Li, et al., Selective adsorption of Gd^{3+} on a magnetically retrievable imprinted chitosan/carbon nanotube composite with high capacity, ACS Appl. Mater. Interfaces 7 (38) (September 2015) 21047–21055.
[53] K.S. Bharathi, S.T. Ramesh, Removal of dyes using agricultural waste as low-cost adsorbents: a review, Appl. Water Sci. 3 (4) (December 2013) 773–790.
[54] M.A.M. Salleh, D.K. Mahmoud, W.A.W.A. Karim, A. Idris, Cationic and anionic dye adsorption by agricultural solid wastes: a comprehensive review, Desalination 280 (1–3) (October 2011) 1–13.
[55] R.-L. Tseng, Physical and chemical properties and adsorption type of activated carbon prepared from plum kernels by NaOH activation, J. Hazard Mater. 147 (3) (August 2007) 1020–1027.
[56] A.S. Franca, L.S. Oliveira, M.E. Ferreira, Kinetics and equilibrium studies of methylene blue adsorption by spent coffee grounds, Desalination 249 (1) (November 2009) 267–272.
[57] V.K. Gupta, Suhas, Application of low-cost adsorbents for dye removal – a review, J. Environ. Manag. 90 (8) (June 2009) 2313–2342.
[58] S. De Gisi, G. Lofrano, M. Grassi, M. Notarnicola, Characteristics and adsorption capacities of low-cost sorbents for wastewater treatment: a review, Sustain. Mater. Technol. 9 (September 2016) 10–40.
[59] M.A. Barakat, New trends in removing heavy metals from industrial wastewater, Arab. J. Chem. 4 (4) (October 2011) 361–377.
[60] F. Obi, B. Ugwuishiwu, J. Nwakaire, Agricultural waste concept, generation, utilization and management, Niger. J. Technol. 35 (4) (September 2016) 957.
[61] J.F. Gubrium, J.A. Holstein (Eds.), Handbook of Interview Research: Context & Method, Sage Publications, Thousand Oaks, Calif, 2002.
[62] W.S. Harvey, Strategies for conducting elite interviews, Qual. Res. 11 (4) (August 2011) 431–441.
[63] R. Mikecz, Interviewing elites: addressing methodological issues, Qual. Inq. 18 (6) (July 2012) 482–493.
[64] Z.S. Morris, The truth about interviewing elites, Politics 29 (3) (October 2009) 209–217.
[65] B. Baez, Confidentiality in qualitative research: reflections on secrets, power and agency, Qual. Res. 2 (1) (April 2002) 35–58.

[66] K. Lancaster, Confidentiality, anonymity and power relations in elite interviewing: conducting qualitative policy research in a politicised domain, Int. J. Soc. Res. Methodol. 20 (1) (January 2017) 93–103.

[67] H.W. Chesbrough, Open Innovation: The New Imperative for Creating and Profiting from Technology, Harvard Business School Press, Boston, Mass, 2003.

[68] Public Power Corporation S.A.-Hellas. Available: https://www.dei.gr/en.

[69] N.A. Travlou, G.Z. Kyzas, N.K. Lazaridis, E.A. Deliyanni, Functionalization of graphite oxide with magnetic chitosan for the preparation of a nanocomposite dye adsorbent, Langmuir 29 (5) (February 2013) 1657–1668.

[70] G.Z. Kyzas, E.A. Deliyanni, N.K. Lazaridis, Magnetic modification of microporous carbon for dye adsorption, J. Colloid Interface Sci. 430 (September 2014) 166–173.

[71] G.Z. Kyzas, N.K. Lazaridis, D.N. Bikiaris, Optimization of chitosan and β-cyclodextrin molecularly imprinted polymer synthesis for dye adsorption, Carbohydr. Polym. 91 (1) (January 2013) 198–208.

[72] G.Z. Kyzas, M. Kostoglou, N.K. Lazaridis, "Copper and chromium(VI) removal by chitosan derivatives—equilibrium and kinetic studies, Chem. Eng. J. 152 (2–3) (October 2009) 440–448.

[73] G.Z. Kyzas, Commercial coffee wastes as materials for adsorption of heavy metals from aqueous solutions, Materials 5 (10) (October 2012) 1826–1840.

[74] G.Z. Kyzas, N.A. Travlou, E.A. Deliyanni, The role of chitosan as nanofiller of graphite oxide for the removal of toxic mercury ions, Colloids Surfaces B Biointerfaces 113 (January 2014) 467–476.

[75] G.Z. Kyzas, D.N. Bikiaris, M. Seredych, T.J. Bandosz, E.A. Deliyanni, Removal of dorzolamide from biomedical wastewaters with adsorption onto graphite oxide/poly(acrylic acid) grafted chitosan nanocomposite, Bioresour. Technol. 152 (January 2014) 399–406.

[76] G.Z. Kyzas, N.K. Lazaridis, E.A. Deliyanni, Oxidation time effect of activated carbons for drug adsorption, Chem. Eng. J. 234 (December 2013) 491–499.

[77] G.Z. Kyzas, M. Kostoglou, N.K. Lazaridis, D.A. Lambropoulou, D.N. Bikiaris, Environmental friendly technology for the removal of pharmaceutical contaminants from wastewaters using modified chitosan adsorbents, Chem. Eng. J. 222 (April 2013) 248–258.

[78] G.Z. Kyzas, A. Koltsakidou, S.G. Nanaki, D.N. Bikiaris, D.A. Lambropoulou, Removal of beta-blockers from aqueous media by adsorption onto graphene oxide, Sci. Total Environ. 537 (December 2015) 411–420.

Chapter 2

Flotation: recent innovations in an interesting and effective separation process

George Z. Kyzas,[a] Nikolaos K. Lazaridis[b] and Kostas A. Matis[b,*]
[a]*Hephaestus Advanced Laboratory, Department of Chemistry, International Hellenic University, Kavala, Greece;* [b]*Division of Chemical Technology & Industrial Chemistry, Department of Chemistry, Aristotle University, Thessaloniki, Greece*
**Corresponding author*

1. Introduction

The importance of flotation, as a separation process, to the economy of the whole industrial world is considered to be enormous. Without this process, many familiar metals and inorganic raw materials would be exceedingly scarce and costly because the high-grade ores that could be processed by simple physical and mechanical methods have long since been used up. It is believed [1] that the historian Herodotus, in ancient Greece, is probably the first person who described a process similar to flotation in the 5th century B.C. for the separation of gold particles from sand using fatty substances. Later, in the 15th century A.D., the Arabs used resins for the selective separation of azurite from gangue. These are considered to be the historical roots of flotation.

Let us also give a reference to Bohemians, using a basket for wet sieving a pulverized ore, in perhaps the first published book in the area (*De Re Metallica*), in 1556 by Saxonian Georgius Agricola written in Latin; this is the beginning of the development of the mining and metallurgical sciences. Since then of course much time has passed with significant technical advances, which of course cannot be covered in a few lines. The increasingly complex mineralogy and lower grade of many current ore reserves have forced technology to evolve rapidly in the decades to treat these materials economically.

Various types of the flotation process exist, being tested at different scale and hence, the respective terminology, i.e., ion, precipitate, adsorbing colloid, froth, foam, sorptive, biosorptive flotation, and so on. While here, in the

Advanced Low-Cost Separation Techniques in Interface Science. https://doi.org/10.1016/B978-0-12-814178-6.00002-9

aforementioned list, due to the different (air or gas) bubble generation methods available, there should be included dispersed air, electrolytic, and dissolved air flotation (termed DAF). The latter usually requires previous flocculation, as it was examined in-depth in a recent review [2]. In the case of fine particles, their agglomeration may lead to greater capture efficiency. An overview of the effects of particles on bubble surfaces was published [3].

Fig. 2.1 illustrates graphically and in brief our group's history and contribution in the area. As "chemicals," one may include collectors, surfactants, dispersants, modifiers, and as "feed it could be oil and grease, plastics or inks (for paper recycling), etc. This scheme was inspired from a schematic representation of separation methods based on surface activity and wettability, by Fuerstenau [4]. The concept of unit operations, overtime, and subsequent concepts in chemical technology have evolved into a unified field of separation processes [5].

Certain significant activities in flotation research have been recently described, with main focus in the contribution of physical chemistry to flotation, such as zeta-potential measurements, contact angle, etc. [6]. There has been phenomenal progress in flotation chemistry and chemicals—both in design and manufacture; there are also many new concepts and tools on which we can capitalize [7]. Mineral, metal values, and by-products can be recovered by flotation separation. The economic aspect of the recovery of useful valuable minerals (such as gold), contributing to recycling meanwhile with environmental technology and wastewater treatment, was enlightened, for instance, of gold from arsenic-rich auriferous pyrite concentrates [8].

Furthermore, selective separation of pyrite from arsenopyrite in an auriferous bulk flotation concentrate, stockpiled in the mine area (in Stratoni Halkidiki, Greece), was found to be possible under carefully controlled flotation conditions; the latter could be beneficial to the gold recovering process. The removal of arsenic-bearing minerals from concentrates is becoming more important as environmental laws become ever stricter with regard to smelter emissions [9].

Nevertheless, to be fair, different extraordinary (even bizarre) flotation applications have been published: oil contaminated sand particles were noticed to self-assemble in order to form granular shells around air bubbles [10]. The

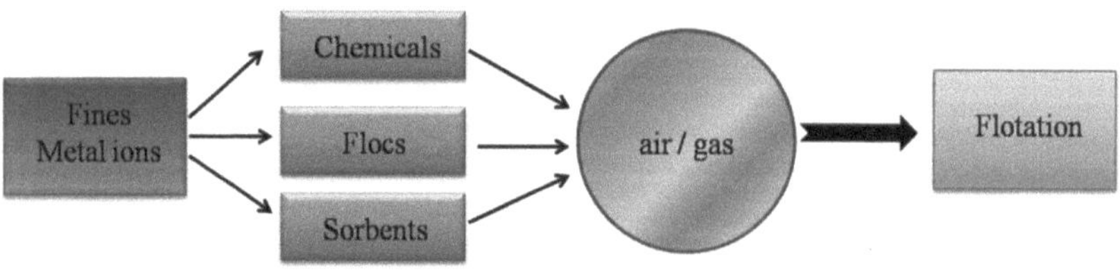

FIGURE 2.1 Scheme illustrating the experienced flotation techniques in the Laboratory.

most engineering application for the molecular self-assembly of sulfhydryl compounds on solid surfaces was elsewhere said to be froth flotation [11]. This idea was also forwarded by AFM measurements [12].

On the other hand, nowadays, flotation has been linked with sustainability; as for instance, in relation to deinking plants of the paper industry [13]; to the sulfidic mine tailings (among the largest mining wastes on Earth) that are prone to produce acid mine drainage [14]; and to the harvesting of algal biofuels [15], as for instance, harvesting imposes a major constraint in microalgal downstream processes (and production of various high value products). The possibility of reprocessing fine to ultrafine material from a tailing heap of an historic gravity separation circuit in Germany, by conventional flotation and by applying an oil-assisted agglomeration flotation approach, was investigated [16].

1.1 Background

Mineral particle attachment to air bubbles is the basis of the froth flotation process. A three-phase system is created, by various mechanisms, in which the following three interfacial tensions exert governing effects: (1) the gas–solid tension, (2) the solid–liquid tension, and (3) the gas–liquid tension, i.e., flotation being a quite unique ("peculiar") separation technique or the encyclopedia of colloid science, as it was called by Nguyen and Schulze [17].

The former tensions determine together the contact angle concept introduced by the Young equation at the point of conjunction of the three phases [18]. The larger the contact angle, the greater the tendency of the bubbles to adhere with the particles. The fractional flotation efficiency, i.e., the fraction of particles contained at the fluid volume that are scavenged by the bubble, which will be permanently attached to the bubble is the product of the collision efficiency and the attachment efficiency. The influence of different flotation variables on the speed of adhesion between particles and bubbles was presented [19].

Regarding the physicochemical aspect of the bubble–particle collision process, two models are usually considered important: the electrostatic model (according to it the energy for the coupling of the particle with the bubble is due to electrostatic attraction between the particle and an interfacial surface carrying opposite charge) and the hydrophobic model (according to it the particles become hydrophobic after the adsorption of the surfactant and this allows their adhesion with the bubbles) [20].

Furthermore, there are three mechanisms mainly by which bubbles can assist the suspended solids in raising them to the surface: (1) adhesion of a gas bubble to a suspended or solid phase, (2) trapping of a gas bubble in a floc structure as the gas bubble rises, and (3) adsorption of gas bubbles in a floc structure, as the floc structure is formed [21].

Whether or not a colliding particle sticks to a bubble is a question of which surface forces come into play, when the aqueous film between the two has thinned sufficiently. It was shown that if a solid is hydrophobic (defined as forming a finite contact angle), very thin water films spontaneously rupture and shrink into a series of droplets, leaving dry areas between them [22]. These films must be subject to a negative disjoining pressure, devoted possibly to electrostatic attraction where the solid–liquid and gas–liquid interfaces carry electrical double layer of opposite sign.

Nevertheless, extremely small particles will be carried along the streamlines and will not make a close encounter with a bubble, unless they cross the streamlines by executing Brownian motion (diffusion) [23]. Only when the process of particle detachment is well understood and modeled, minerals recovery using the flotation process can be modulated to achieve a high efficiency by suitably changing the operating parameters [24].

It was proposed—a NATO Advanced Study Institute in Cornwall, UK (with an ominous title)—that the fine particle recovery may be enhanced by using fine bubbles generated by the (then) unconventional techniques [25]. It was elsewhere observed that by decreasing the bubble diameters the possibility of collision is increased; a narrowly sized sample of very hydrophobic particles was used in these experiments [26].

The hydrodynamic performance of very fine gas required for dissolved air (pressure) flotation, which is produced by dissolving air in water under pressure and subsequent flushing to normal pressure through a nozzle, was studied [27]. The results clarified the effects of dissolved pressure, liquid flow rate, and nozzle geometry, and empirical equations were obtained for the number of generated bubbles.

In general, it is known that, after the formation of bubbles, they tend to coalesce [28]; coalescence is the phenomenon where bubbles grow in diameter by feeding on small bubbles. Hence, bubbles coalesce when two bubbles collide and form a single one. There are three main factors that favor this: (1) the presence of bubbles having nonuniform diameters, where the larger bubbles gobble up the smaller by collision; (2) violent agitation tends to coalesce bubbles; (3) the presence of a large surface in contact with the bubble–liquid surface. The large surfaces serve as deposition points for the microbubbles, where coalescence can readily take place. Also, as the bubbles move toward the surface of the solution, they collide with each other; therefore, it is quite possible for coalescence to occur.

These phenomena, however, can be inhibited with the addition of surfactants because they (1) are adsorbed on the bubble surfaces during the flotation process, having their polar groups oriented toward the liquid. The surfactants are hydrated and eventually the bubbles are surrounded with a hydrated layer. This layer stabilizes the surface of the bubbles by forming a wet hymenion and thus prevents adhesion and reduces the coalescence potential by protecting the bubbles from rupturing. (2) They allow better adhesion of the bubbles to the particles, ions, and solution constituents, thus permitting their effective transfer to the surface of the solution [29].

The importance of bubble charge in its effect on the rate of flotation of small particles was also investigated. In a single species pulp, i.e., an emulsion of oil in water, it could be possible to maximize the flotation rate by controlling the pH value of the emulsion, for example, to give essentially zero charge on the particle and perhaps on the bubbles [30]. In this case, the flotation rate would be maximum for the following two reasons: The electrostatic repulsion between bubbles and particles would be minimum. The conditions would also lead to coagulation of the oil droplets in order to form larger aggregates whose rate would be enhanced. Gas bubbles generally are reported to be negatively charged in pure water and in solutions of inorganic electrolytes [31].

The collision between a bubble and a particle is, therefore, the result of a combination of the interception phenomenon (finite particle size) and the tendency of particles to deviate from fluid streamlines (due to their velocity up and/or their inertia); the dependence of the attachment efficiency on the bubble diameter is not a general feature but depends on the particular conditions, for example, the particle density. A composite expression for the flotation frequency (frequency of successful bubble–particle collisions) in a combined gravitational and turbulent flow field was derived [32]. Flotation is, as mentioned, a complicated process combining many elementary physicochemical steps also with fundamental hydrodynamics.

Its modeling is certainly important to understand the process, as well as it is a necessary tool for equipment design and optimization. Flotation units, as the chemical reactors, are distinguished as cells of ideal and nonideal flow; the superiority of countercurrent contact flow was particularly highlighted [33]. Flotation efficiency is generally expected to be a function of the residence time distribution. The stimulus–response technique is often used, where the tracer (a dye, electrolyte, etc.) input is usually of a pulse or step function. In an investigation of electroflotation hydrodynamics, a theoretical multiparameter model (by Levenspiel) was applied, consisting of interconnected dead water, back mix and plug flow regions, with bypass, recycle, and cross flow through and around these regions.

2. ROLE of particles—separation applications

2.1 Minerals

Agitated flotation cells are widely used in the mineral processing industry for separating, recovering, and concentrating valuable particulate material from undesired gangue. Their performance is lowered, however, when part of the particulate system consists of fines, with particle diameters typically in the range from 30 to 100 μm. For example, it was observed difficult to float fine particles because of the reduction of middle particles (of wolframite) as carriers and the poor collision and attachment between fine particles and air bubbles; a new kinetic model was proposed [34].

As an alternative to agitated cells, bubble columns—used in chemical engineering practice as chemical reactors—were proposed for the treatment of fine particle systems. Flotation columns, as they came to be known, were "invented" back in the 1960s in Canada [35]. The main feature that differentiates the column from the mechanical flotation cell (of Denver type) is wash water, added at the top of the froth. It was thought to be beneficial to overall column performance since it helps clean the froth from any entrained gangue, while at the same time preventing water from the pulp flowing into the concentrate. In this way, it was hoped that certain cleaning flotation stages could be gained.

Let us note that the perhaps insistence here on mineral processing is only due to the fact that most of the available literature on flotation is from this area, where the process was originated and being widely practiced. The effect of particle size on flotation recovery is significant; it was shown that there exists a certain size range in which optimum results may be obtained in mineral processing. This range varies with the mineral properties such as density, liberation, and so on, but was said to be of the order of 10–100 μm [36].

Regulating the oxidation state of pyrite (FeS_2) and arsenopyrite (FeAsS), by the addition of an oxidation or reduction chemical agent and due to the application of a short-chain xanthate as collector (such as potassium ethyl xanthate, KEX), was the key to selective separation of the two sulfide minerals, pyrite and arsenopyrite [37]. Strong oxidizing agents can depress previously floated arsenopyrite. Various reagents were examined separately as modifiers and among them were sodium metabisulfite, hydrazinium sulfate, and magnesia mixture. The laboratory experiments were carried out in a modified Hallimond tube, assisted by zeta-potential measurements and, in certain cases, by contact angle measurements.

This conventional bench-scale flotation cell provides a fast, convenient, and low-cost method, based on small samples (around 2 g), usually of pure minerals and also artificial mixtures, for determining the general conditions under which minerals may be rendered floatable—often in the absence of a frother (to collect the concentrate in the side tube) [38]. This idea was later further modified in the lab replacing the diaphragm, in order to conduct dissolved air or electroflotation tests—see Section 3.

Pyrite concentrates sometimes contain considerable amounts of arsenic. Since they are usually used for the production of sulfuric acid, this is undesirable from the environmental point of view. However, gold is often associated with arsenopyrite, often exhibiting a direct relationship between Au content and As grade. There is, therefore, some scope for concentrating arsenopyrite since the ore itself is otherwise of little value (see Fig. 2.2). Note

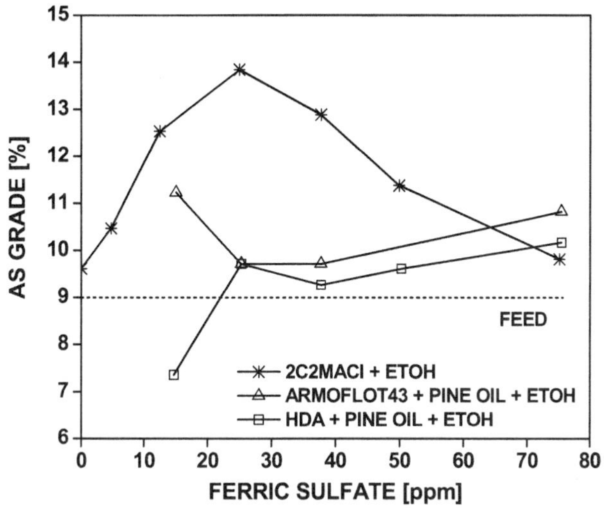

FIGURE 2.2 Effect of $Fe_2(SO_4)_3$, added as conditioner, on grade and recovery of arsenopyrite in column flotation. *Reprinted with permission; copyright Pergamon Press P. Mavros, K.A. Kydros, K.A. Matis, Arsenopyrite enrichment by column flotation, Miner. Eng. 6 (12) (1993) 1265–1277.*

that previous work on pyrites usually concentrated on the problem of floating pyrite [40].

In the aforementioned figure (shown as example), the following conditions were applied: (1) collector [2-coco 2-methyl ammonium chloride] 42 mg/L, frother (EtOH) 0.15% (v/v), superficial liquid velocity $u_L = 1.02$ cm/s, superficial gas velocity $u_G = 0.65$ cm/s, superficial wash water velocity $u_w = 0.53$ cm/s; (2) hexadecylamine, 45 mg/L; pine oil, 50 mg/L; EtOH, 0.025%; $u_L = 0.84$ cm/s; $u_G = 0.72$ cm/s; $u_w = 0.66$ cm/s; (3) Armoflot 43, 50 mg/L; pine oil, 50 mg/L; EtOH, 0.025%; $u_L = 0.84$ cm/s; $u_G = 0.71$ cm/s; $u_w = 0.66$ cm/s [39]. The pyrite (with a relatively important Au content of 21 g/ton) was a xanthate-floated concentrate. The presence of xanthates, however, might cause problems in the subsequent cyanidation of pyrites when recovering their Au value, which perhaps justified the need to find alternative collectors. In general, the amines exhibited a behavior similar to that of the xanthates (O-alkyl dithiocarbonates). The benefit of the amine was in its lower consumption, as compared with the xanthate systems.

The arsenic content of the pyrite was approximately 9% (from an initial 3.5% of the mixed sulfide ore). The material was sieved and the −75 μm fraction was used for the laboratory-scale cylindrical column experiments. The effect on metallurgical characteristics of the flotation concentrate of varying the amount of ferric sulfate added to the pulp was studied; three collectors were used and their performance was compared (in Fig. 2.2). Both hexadecylamine and Armoflot 43 (manufactured by Akzo) exhibited an increased recovery but a very low enrichment, whereas 2-coco 2-methyl ammonium chloride (Arquad-2C) showed a considerable enrichment; a compromise had to be made, therefore, between a high-grade and a low recovery.

Electroflotation (electrolytic flotation) is an unconventional separation process owing its name to the bubbles generation method it uses, i.e., electrolysis of the aqueous medium. In the bottom of the microcell, the two horizontal electrodes were made from stainless steel, the upper one being perforated. The current density applied was 300 A m^{-2}. It was observed that with lime used to control pH, different behavior was observed (see Fig. 2.3). Pyrite, with permanganate (a known depressant) also as modifier, remained activated from pH 5.0 to 8.0 at 80% recovery, while it was depressed at the pH range from 9.0 to 12.0. A conditioning of 30 min was applied in the presence of modifier alone and further 15 min after the addition of xanthate. The pure mineral sample, previously hand collected, crushed, and pulverized in the laboratory, was separated by wet sieving to the −45 to +25 μm particle size range.

Pyrite due to its very heterogeneous surface, consisting of a mosaic of anodic and cathodic areas, presents a strong electrocatalytic activity in the anodic oxidation of xanthate to dixanthogen. It is also possible that the presence of the electric field, during electroflotation, affected the reactions taking place. In order to explain this difference in flotation behavior thermodynamic calculations for the system Fe-EX-H_2O have been done [41]. It was concluded that electroflotation was capable of removing fine pyrite particles

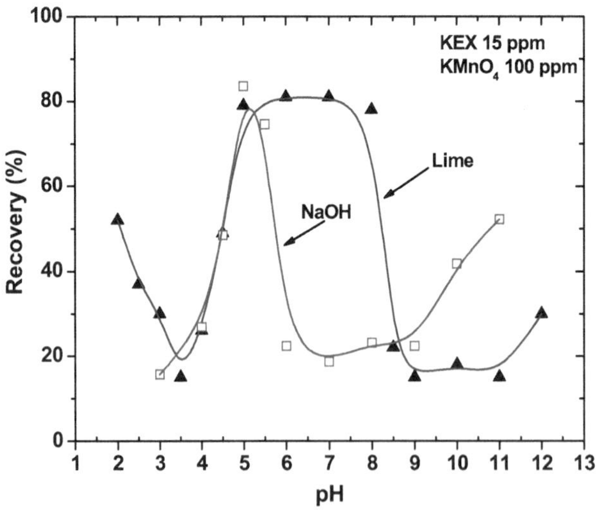

FIGURE 2.3 Electroflotation of pyrite: effect of the pH and regulators, lime, or sodium hydroxide. *Reprinted with permission; copyright Society of Chemical Industry K.A.Kydros, G.P. Gallios, K.A. Matis, Electrolytic flotation of pyrite, J. Appl. Chem. Biotechnol. 59 (1994) 223–232.*

from a dilute dispersion, under controlled conditions. Nevertheless, dispersed air and electroflotation presented apparent differences for the same application.

The size of the gas bubbles produced was of the order of 50 μm, in diameter [21]. Similar measurements were later carried out at Newcastle, Australia [42]; where it was also noted that a feature of electroflotation is the ability to create very fine bubbles, which are known to improve flotation performance of fine particles.

In fact, the two electrodes of a horizontal electrodes set, usually applied in electroflotation, could be separated by a cation exchange membrane, as only one of the produced gases is often necessary [43]. In the lower part/separated electrode, an electrolyte was circulated to remove the created gas, and in the meantime, increase the conductivity; hence having power savings (as the electric field is built up between the electrodes through the use of the suspension conductivity). Attention should be paid in this case to anode corrosion, mainly by the chloride ion (i.e., seawater).

Microorganisms have a tremendous influence on their environment through the transfer of energy, charge, and materials across a complex biotic mineral–solution interface; the biomodification of mineral surfaces involves the complex action of microorganism on the mineral surface [44]. Mixed cationic/anionic surfactants are also generating increasing attention as effective collectors during the flotation of valuable minerals (i.e., muscovite, feldspar, and spodumene ores); the depression mechanisms on gangue minerals, such as quartz, were focused [45].

2.2 Wastewaters

There is, in general, a "feeling" that a gap exists between froth (dispersed air) flotation and DAF; may be even for historical reasons, and this was the subtheme of the United Engineering Foundation Conference on Flotation (organized by Profs. R.W. Smith and J. Rubio, held in Tahoe City, California, 2001). The aforementioned feeling perhaps originates from the separate areas of flotation application or in other words, the completely separate traditions and of course, the seldom interaction among them. But, mainly we may say, it comes from the different aims in these two fields:

i) selective separation or not,
ii) difference in particles concentration and/or size,
iii) use or nonuse of collectors and other chemical reagents (modifiers, frothers) for environmental reasons, etc.; may be, the idea of collectorless flotation, if possible, is worthwhile.
iv) From an economic point of view, the things are completely different: as mineral processing is obviously economically attractive, while no one "really" wants to pay (at least too much) to treat the effluents.

Certainly, the process originated from mineral processing; hence, a lot has been learned from this scientific area, being almost the encyclopedia of flotation for the beginners.

According to the drinking water fact sheet of WHO (updated July 2017), 884 million people lack even a basic drinking water service, and at least 2 billion people use a drinking water source contaminated with feces, i.e., clean potable water is not self-evident. It has been well established nowadays that dissolved heavy metals escaping into the environment pose a serious health hazard, hence justifying perhaps the scope of our research.

DAF is known as an effective solid–liquid separation process used in water treatment as an alternative to sedimentation [46]; the experience in the area of the Water Research Center, UK, including relative costs, was described. Refineries commonly use the methods of DAF and/or dispersed air flotation to remove free and emulsified oil from refinery wastewater. Two waste streams from an oil refinery wastewater treatment system, float from a DAF unit (DAF sludge), and waste-activated sludge were investigated to determine toxicity and biogas production potential for anaerobic digestion through batch testing [47]. It is known for long time now that (dissolved air) flotation is effective for activated sludge plants, both for clarification and thickening [48].

A computational fluid dynamics (CFD) model incorporating flotation kinetic expressions was developed to simulate the performance of flotation tanks utilized in water and wastewater treatment plants [32]. The focus was on tanks that operate without any external means of flow mixing (impeller, etc.), such as the DAF tanks. The flotation kinetic concepts such as collision, attachment, and detachment efficiencies were employed first time in CFD studies by Koh and coworkers [49]. Elsewhere, it was argued that with the recent advances in nanotechnology, environmental and health consequences of nanomaterial disposal merit close attention [50].

The two main bubble generation techniques were examined: DAF is quite different from the dispersed air flotation [51]. In DAF processes, microbubbles of 10–80 μm in size were said to be generated [52]. The trend in the development of DAF technology for potable water was to move to very thick microbubble beds with high flow rates—even advertised as the third-generation technology; a flow rate of more than 60 m/h was given, where flotation operation was approaching turbulent flow conditions. A compact (downstream filtration being in the same basin) so-called reflotation plant with hydraulic flocculation was built, for instance, for Tampere Finland water and sewerage works in 1997 (by the company OY Rictor AB). One of the advantages of flotation was its speed, hence its ability to operate inside a factory, due to cold weather conditions.

The formation of macrobubbles, undesirable in DAF, was found closely to connect with the mechanisms operating in the depressurization zone [53]. The geometry of the nozzles had generally a strong influence on obtaining microbubbles: the greater the conic divergence angle, the smaller the bubbles and the more restricted the size distribution; as the distance from the expansion zone was increasing, bursting and coalescence occurred simultaneously. Water coming out of the nozzle has the aspect of a milky cloud of bubbles. The influence of bubble size on the fluid dynamic behavior in a DAF tank was examined in another study [54].

The process was applied, for instance, to cadmium ion (10^{-4} M) removal, as presented in Fig. 2.4A, following the mechanism of precipitate flotation as hydroxide at pH 10–11 by surfactant sodium dodecylsulfate, as collector 2×10^{-4} M (in presence of 0.1% v/v ethanol, as frother). Cadmium is a highly toxic and rather ubiquitous metal in its distribution throughout the world. These experiments were run in parallel and under the same conditions: Dispersed-1: airflow rate 600 cm^3 min^{-1} for a retention time of 10 min, in a column of 34-mm diameter and 0.4 L content; air was entering the porous diaphragm with an excess pressure of ~1 atm. Dispersed-2: airflow rate 200 cm^3 min^{-1} for a retention time of 4 min. While in DAF (unit purchased from Aztec Environmental Control Ltd.), a 10% recycle ratio was used, with 1 L content initially. A photograph was previously shown with detailed information of our laboratory arrangement, of a typical continuous flow flotation rig [57].

So, the removal of cadmium (initially up to 10 mg L^{-1}) was near 100%, while the remaining concentration in the solution after flotation was less than 0.10 mg L^{-1}. The parameters that determine the optimum flotation conditions, except the pH, appeared not to be very critical. Flotation was concluded as an effective method that deserves perhaps more attention as an alternative treatment to other conventional methods, especially concerning dilute metal ion solutions, even in hydrometallurgical processes. Later, the flotation of cadmium ion was reseen, using this time potassium ethyl xanthate as

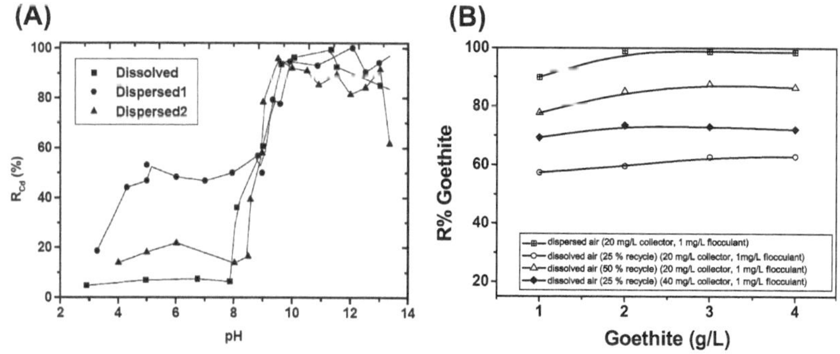

FIGURE 2.4 (A) Comparison of flotation techniques for cadmium ion removal: influence of solution pH. (B) Comparison of dispersed air with dissolved air flotation, for Cr(VI) removal on goethite. Presented in the 34th International Conference on Mining and Metallurgy, Bor, Serbia [55]. *(A) Reprinted with permission; copyright IAWQ A.I. Zoubouls, K.A. Matis, Removal of cadmium from dilute solutions by flotation, Water Sci. Technol. 31 (3–4) (1995) 315–326.*

precipitating agent [58]. The collectorless flotation of copper ion was earlier investigated by dissolved air, precipitated upstream as colloidal CuS [59].

The need for fresh sources of drinking water is becoming more and more urgent worldwide, including certainly many areas and/or countries across the Mediterranean Sea. Reverse osmosis membranes, on the other hand, are known to be very sensitive to foulants as colloids, inorganic scale, and biofouling; so, pretreatment (i.e., by flotation, among other) of their seawater feed is often a key step [60]. The economics is the usual worry in desalting water production [61].

2.2.1 Adsorbents

Carrier flotation (ultraflotation or "piggy-back" flotation) constitutes a technique to improve the flotability of fines; the known example of commercial application is that of anatase impurity being removed from kaolin clay by flotation using coarse calcite particles as an auxiliary mineral (the carrier) [62]. Adding an adsorbent (like powdered activated carbon) into the coagulation process is an emerging treatment solution for targeting hard-to-remove dissolved organic compounds from both drinking water and industrial wastewater [63]. The experimental tests were performed on synthetic oily wastewater.

Fig. 2.4B extends the aforementioned comparison between the two flotation techniques for the separation of metal-loaded goethite (1 g/L), at pH 4, following the mechanism of sorptive flotation. The advantage of dispersed air was obvious in this case; exactly similar results were obtained for the flotation of zinc-loaded goethite. The existing conditions applied were dodecylamine (20 mg/L) as collector, flocculant magnafloc 919 (1 mg/L); dissolved air: flocculation time = 5 min at 60 rpm− 5 min at 300 rpm; the recycle was the main parameter studied (at least 20% was necessary); dispersed air: flocculation time = 5 min 350 rpm−5 min 700 rpm, air = 425 cm^3/min. Meanwhile, the chromium(VI) removals were near 100% (from 50 mg/L initial concentration). The nanocrystalline sorbent, i.e., goethite, was previously synthesized [64] and had been tested as heavy metals (i.e., cadmium and pentavalent arsenic) adsorbent [65]. Other typical inorganic adsorbents satisfactory for heavy metals removal are certainly zeolites [66].

Perhaps, a special case of sorptive flotation constitutes the technique known as adsorbing colloid flotation (i.e., a type of carrier flotation), which involves the removal of the solute from the aqueous solution by adsorption on, coprecipitation with, or even occlusion in a carrier floc, produced by adding ferric or aluminum salts. A second stage involves the combination of flocs into larger aggregates, which are then floated with the help of an appropriate surfactant.

Bentonite and kaolinite clays were elsewhere published for the sorptive flotation removal of radioisotopes of europium [58] and pyrite for As(V) [67]. As well as, results were presented on the removal of Fe^{3+} ions, as precipitates and nanoparticles of $Fe(OH)_3$ by nanobubbles (generated at 2.5 bar, after their separation from microbubbles) and by DAF [68]. In parallel, the use of nanoparticles has attracted widespread interest for effectively removing pollutants (and in this treatment flotation was also included); so, carbon nanotubes were used for nickel ion [69] and silica nanoparticles for a dye [70].

2.2.2 Biosorbents

The ability of microorganisms to remove metal ions is a known phenomenon; biosorption was termed the process that makes use of dead biomass (contrary to bioaccumulation), coming usually from fermentation wastes or by-products. Metal recovery or removal from solution may involve various pathways [71]. A comparison of the various treatment processes for metal-laden wastewaters was published, among others, by Eccles [72] giving their characteristics and disadvantages when applied to dilute aqueous solution. May be the difficulty in biomass separation (harvesting) was advocated as one of the disadvantages of biosorption, and for this reason, flotation has been thoroughly examined in our lab.

A comparison of the floatability of various types of biomass has been carried out, over the whole pH range (presented as Fig. 2.5A). Four types of biomass were evaluated: *Streptomyces rimosus* (an actinomyces bacterium from tetracycline production), *Penicillium chrysogenum* (fungi used for antibiotics), *Saccharomyces carlsbergensis* (yeast for beer production), and grape stalks (a by-product of the winery industry). Improved biomass flotation recoveries were observed, some reaching almost 100%. These biomasses were tried as biosorbent for a synthetic aqueous mixture of metals in deionized water containing the following metals (in mmol dm^{-3}): Zn 0.765, Cu 0.157, Ni 0.034, Ca 2.49, and Na 4.35. The concentrations were selected as being representative of those commonly found in industrial wastewaters, whereas Ca and Na were also added to simulate the usual hardness of most metal-laden wastewaters.

Fig. 2.5B presents a conceptual flow diagram showing recycle routes of a total process (investigated under an EU environmental research program): the second cycle represents the applied flotation (or may be, sedimentation) separation. The solids are diverted to a filtration unit where they became immobilized as a filter cake [74]. The elution step regenerated the biomass for subsequent biosorption steps and also, greatly concentrated the cadmium in the eluate, with respect to the original wastewater. A rotating cathode cell was used downstream for metal recovery [75], resulting in cadmium powder and cadmium-depleted electrolyte, which is recycled as the eluant. The bacterial cell surfaces are known to be anionic due to the presence of ionized groups in the various cell wall polymers. This was also the reason that a cationic surfactant, such as cetyl-trimethyl ammonium bromide, was applied during

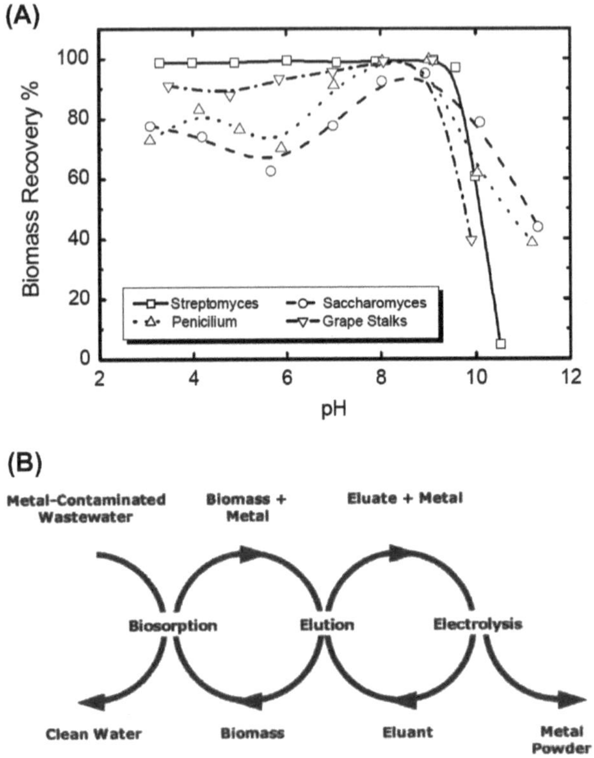

FIGURE 2.5 (A) A comparison of various biosorbent types, showing the obtained floatability, as a function of pH change. (B) An illustrative conceptual flow diagram showing recycle routes for a total process. *(A) Reprinted with permission; copyright Wiley D. Zamboulis, E.N. Peleka, N.K. Lazaridis, K.A. Matis, Metal ion separation and recovery from environmental sources using various flotation and sorption techniques, J. Appl. Chem. Biotechnol. 86 (3) (2011) 335–344; (B) Reprinted with permission; copyright Elsevier T.J. Butter, L.M. Evison, I.C. Hancock, F.S. Holland, K.A. Matis, A. Philipson, A.I. Sheikh, A.I. Zouboulis, The removal and recovery of cadmium from dilute aqueous solutions by biosorption and electrolysis at laboratory scale, Water Res. 32 (1998) 400–406.*

flotation. The presence of the heavy metals in the solution was also found to alter the electrokinetic properties of biomass.

The polluted aqueous system and the pretreated biomass were contacted countercurrently [76]. A two-stage countercurrent scheme was selected with one parallel of elution, based on extensive pilot plant studies; this scheme has similarities to a previously published activated carbon system for the separation of liquid contaminants. An advantage of the proposed process is that its basic unit processes are rather conventional and widely applied in the field. Multiple cycles of biosorption/flotation/elution were also examined with this system, applying appropriate elution between the treatment cycles of the flotation concentrate (i.e., the separated metal-loaded biomass). It was observed that supplementary addition of a surfactant during the third cycle was necessary, otherwise biomass flotation substantially dropped, as flotation is based on surface activity. The following mixture of common metals was examined in the laboratory as a simulated approach of industrial wastewaters: Cd 1.5, Zn 7.5, and Cu 30 mg/L. The used biosorbent was a large-scale fermentation waste from a pharmaceutical industry (with trade name Mycan).

Microbubbles, generated by a fluidic oscillator, were applied for the recovery of yeast cells from their growth medium using the bioflocculant, chitosan [77]; bubble size affects recovery efficiency in a different way. To decontaminate heavy metal–containing wastewater, a microbial biotechnology was developed by using the synergy between *sulfate-reducing* bacteria, *Bacillus cereus* and *Camellia oleifera* cake [78]; flotation has been led into this system for increasing the efficiency as well. Another work aimed to compare a biosorption method and a coagulation–DAF method (found preferable) as inexpensive and effective means of eliminating hexavalent chromium from industrial wastewater [79]; process optimization was carried out using an artificial neural network.

Researching on cadmium ion removal onto sewage sludge biomass, it was observed that electroflotation (in order to separate or concentrate the loaded biosorbent) was capable to operate even without the presence of surfactants (i.e., collectorless), giving promising results using 10 min as retention time, a small ethanol concentration as a convenient frothing agent (easily biodegradable), and applying at least 130 A/m^2 as current density [80]. In the absence of biomass, only ~25% of the metal was possible to be removed.

Today, there is a tendency for combined and more compact processes as, for example, membrane bioreactors (MBRs) [81]. The main limitation to the faster development of this process was attributed to membrane fouling [82]. Various hybrid systems that include membrane separation processes have been presented in literature in recent years, except the known MBR: so, a hybrid distillation-reverse osmosis system was published [83], a hybrid distillation-membrane pervaporation system [84] and also, a hybrid membrane-ozonation system [85]. A new hybrid process for cleaning wastewater combining membrane microfiltration and flotation was investigated in our lab [86].

3. Role of bubbles

3.1 The hybrid cell

The hybrid process combined the advantages of both membrane separation and flotation: the flotation cell removed a large proportion of suspended solid particles, while the membrane module produced a clean water permeate effluent. The proof of concept for the combined hybrid separation process was investigated using an aqueous suspension of ultrafine zeolite particles (from Ineos Silicas), and the main application was the treatment of heavy metal wastewaters. Air bubbling is one of the techniques used to limit membrane fouling, among the large number of techniques that have been used.

Certain limitations were imposed on the examined parameters, e.g., airflow, membranes type, and position, etc. The laboratory arrangement for the hybrid cell consisted of a mixing tank, peristaltic pump, thermometer, the flotation column with membrane microfiltration (shown in Fig. 2.6A),

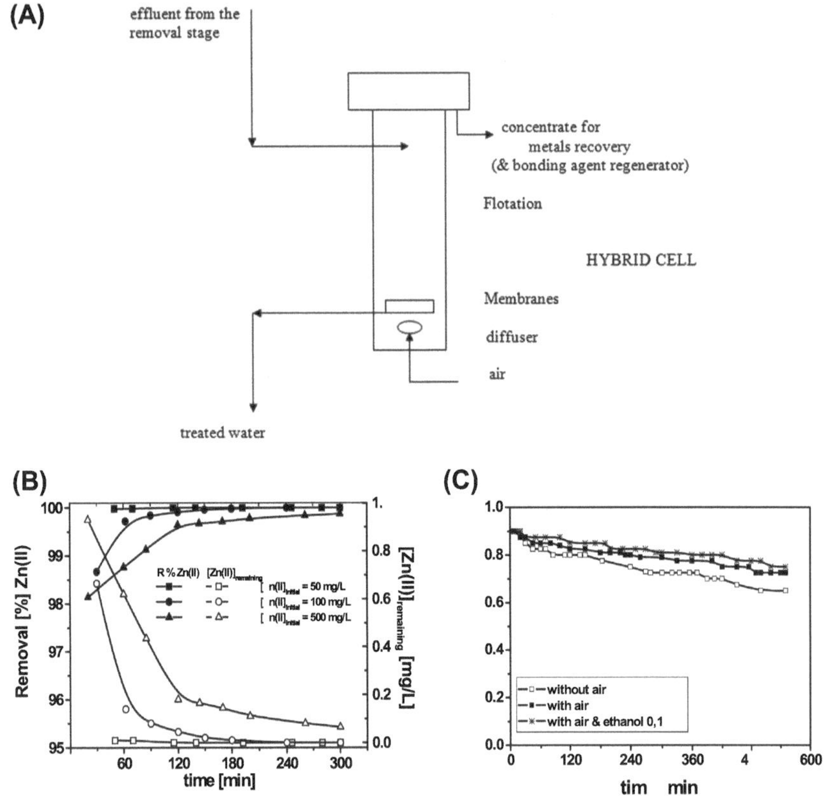

FIGURE 2.6 (A) Scheme of the innovative hybrid cell idea. (B) Obtained zinc removals in the retentate by the flotation-microfiltration cell; results expressed in two ways during precipitate flotation. (C) Effect of ethanol presence on membranes operation with metal-loaded zeolites: transmembrane pressure against time of operation. *(B) Reprinted with permission; copyright Taylor and Francis E.N. Peleka, N.K. Lazaridis, P. Mavros, K.A. Matis, A new hybrid flotation-microfiltration cell, Separ. Sci. Technol. 41 (2006a) 3229–3243; (C) Reprinted with permission; copyright Elsevier K.A.Matis, E.N. Peleka, D. Zamboulis, T. Erwe, V. Mavrov, Air sparging during the solid/liquid separation by microfiltration: application of flotation, Separ. Purif. Technol. 40 (2004) 1–7.*

manometer, air flow meter, nonreturn valve, slide valve, foam collection tank, and permeate collection tank. The microfiltration module, inserted inside the flotation column, consisted of a twin set of parallel, double-sided ceramic membranes, with a flat sheet multichannel geometry, having a mean pore size of 0.3 mm, a total surface area of 0.021 m^2, hydrophilic surface properties, and an expected pure water flow rate of ~2 L/h.

Fig. 2.6B presents an example of a similar work without a sorbent applied, highlighting also the successful contribution of precipitate flotation with the hybrid cell; metal precipitation of dissolved zinc as hydroxide was conducted by pH adjustment, at the pH value of 9. When varying the feed, i.e., metal, from 50 to 500 mg/L and hence, varying the zinc precipitate concentration, the membrane permeabilities were reduced increasing the concentration. When plotting as remaining metal in solution, with the higher feed concentration used, the obtained results are not as good (although the removals were over 99.5%). A comparison was also given, in the form of a table, of the above process with the one when zeolite was used (i.e., sorptive flotation) [86]. Advantage of the former was that less solids were present inside the hybrid cell and so the pressure drops were much lower.

With the latter system [87], it was found that the finer the air bubbles used, the lower the transmembrane pressure (TMP) through the membranes. Fig. 2.6C shows the influence of ethanol (as frother) on TMP, with 4 g/L loaded zeolites, the air sparger was porous diaphragm Schott D4; the corresponding figures for permeabilities were presented, too. It was obvious that the results were still improved when the frother was used, as higher TMP was noticed, as expected. The efficiency of flotation, as found macroscopically, meanwhile was shown to have as outcome more than 90% of solids to be floated, those escaping the membranes (see below the economics aspect).

During these experimental series of continuous flow separation, all the collected samples showed, due to the membranes filtration, 100% zeolites recovery; hence, efficient solid—liquid separation, while the metal removal depended just on the amount of used sorbent. At the end of this program, the idea was tested, with a hybrid rig constructed (by UPT Saarbrücken, Germany), on-site at the Assarel-Medet open pit copper mine effluent near Panagyurishte, Bulgaria. Dispersed air flotation was indeed capable for a preliminary solids recovery of the order of 90%, with the Cu content in the froth concentrate approaching 6%. It was found in this pilot-scale study that the residual heavy metal (Cu, Mn, Fe, and Pb) concentrations in the membranes permeate were below 0.05 mg L^{-1} [88,89].

3.2 Further discussion

Further to the information given above (due to overlapping), from an older work Fig. 2.7A is here presented; the attachment efficiency expresses the

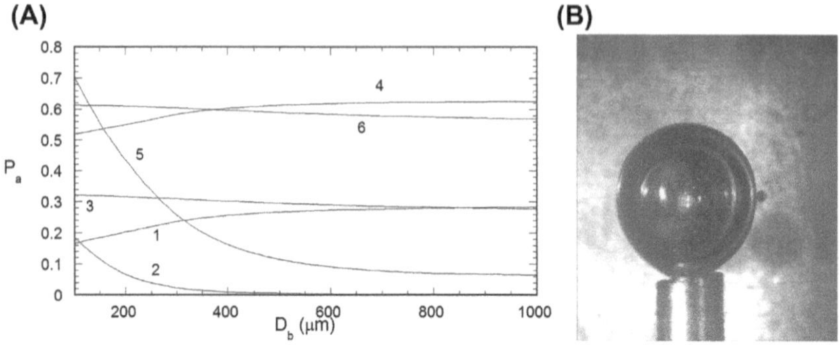

FIGURE 2.7 (A) Theoretical attachment efficiency, P_a, versus bubble diameter, D_b: (1) model (i), induction time $t_{ind} = 0.1$ s, particle size $D_p = 10$ μm; (2) model (iii), $t_{ind} = 0.1$ s, $D_p = 10$ μm; (3) model (ii), parameter $h_r = 1.98$, $D_p = 10$ μm; (4) model (i), $t_{ind} = 0.02$ s, $D_p = 30$ μm; (5) model (iii), $t_{ind} = 0.02$ s, $D_p = 30$ μm; (6) model (ii), $h_r = 1.45$, $D_p = 30$ μm. (B) Photograph of particles on a bubble; kindly offered from the flotation research of my good colleague Prof. Thodoris Karapantsios and coworkers, in the Lab. *(A) Reprinted with permission; copyright Elsevier M. Kostoglou, T.D. Karapantsios, K.A. Matis, Modeling local flotation frequency in a turbulent flow field, Adv. Colloid Interface Sci. 122 (2007) 79–91.*

probability of the collided particle to drain the thin liquid film and eventually aggregate with the bubble. Where, model (1) is a modification of the classical Yoon-Luttrell [26] approach by King [90]; model (2) was developed by Bloom and Heindel [91]; and model (3) was developed by Nguyen [92]. In the second model, the induction time t_{ind} (i.e., a parameter which collectively includes all physicochemical interactions between a bubble and a particle) has been replaced by another parameter h_r. This is the ratio of the initial thickness at the moment of collision of the bubble–particle liquid film to its thickness at the moment of disruption and formation of a particle–bubble aggregate. It is a purely empirical parameter and must be obtained (like t_{ind}) from experiments or empirical relations.

Obviously, $P_{a,iii}$ (for the third case) decreased as bubble size increases due to reasons mentioned. The efficiencies for the two other methods, however, showed a smaller variation (in opposite directions), but it was sure that the two methods cannot be assumed to be equivalent. Hence, it is apparent from the above figure that both the particle and the bubble size are of significance and affect the process and its separation results. On the other hand, the naive conclusion is perhaps that there should be some care as to what model is to be applied (and to the relative assumptions made). For bubble and particle diameters in the ranges 100–1000 μm and 2–40 μm, respectively, the new expression incorporated most appraised earlier existing theories which, however, have a limited extent of validity and to which the new expression can degenerate as particular asymptotic cases [32].

The single bubble–single particle interactions of interest to flotation applications were later investigated with a developed experimental device [93]. Two separate high-speed cameras were employed to monitor the bubble surface from two different Cartesian directions allowing thus a 3D perspective of particles trajectories and collisions with the bubble (see Fig. 2.7B). A single bubble was kept at a fixed position and the suspension of particles moved against it; the bubble was created at the tip of a vertical capillary tube of external diameter of 600 μm. The experiments in this work were performed using spherical hydrophobic particles (polystyrene cross-linked with divinylbenzene) (Micropore Technologies Ltd.) with an average diameter of 330 μm, $\rho = 1050\ kg/m^3$.

Understanding the limits of fine particle flotation is the key to their selective separation [94]. The low flotation rate and recovery of hydrophobic fine particles (<20 μm) is mainly due to their low collision efficiency with bubbles, which may be overcome with decreasing the bubble size; other influences investigated in this review were the particle aggregation, particle induction time, the action of surface and capillary forces on fine particle–bubble capture, and different flow conditions. Various bubble–particle collision and attachment efficiency models have been reviewed, and the factors influencing collision and attachment efficiencies were extensively discussed. Finally, Table 2.1 summarizes other recent literature publications on this area that are worth mentioning.

4. Reports on process costs

A techno-economic study was also conducted for the aforementioned hybrid cell (see, for example, Fig. 2.8), applying the software tool SuperPro Designer (from Intelligen). The study showed that the capital investment of the plant would be of the order of US$ 0.92M (2003 prices). The operating costs are about $ 1.76M per year. The processing rate is 68.83×10^6 kg/year of influent and the unit processing cost is 0.0256 $/MT of influent; detailed tables were published. The total revenues from water recycling and reuse are $ 68,800 per annum. It is noted that this only accounted for the environmental costs and did not consider any credit due to copper recovery. The cost of the combined process was not the mere addition of the individual processes costs: the required membrane surface, which is a major cost factor, is lower in the hybrid process since flotation removed the majority of solid particles, and therefore a smaller membrane area is required to yield the same permeate [106].

In Table 2.2, a brief account is given on costs from selected (from the very few available) papers in the recent literature; noting that the general trend, at least according to our opinion, is to avoid giving an idea on costs, but meanwhile speaking on low-cost effective processes. Finally, in an extensive review [112], a discussion on electroflotation costs can be traced, including some comparison with relative processes for effluent treatment.

TABLE 2.1 Selected papers on the role of particle/bubble size.

Brief flotation details/application	References
Fine particle capture and aggregation of the quartz ultrafines by nanobubbles were the main mechanisms responsible for higher recoveries; nanobubbles enhanced the contact angle of quartz.	[95]
Bubble diameter, bubble velocity, and turbulence suggested to be the key factors involved in particle–bubble interactions. The collision efficiency of ultrafine particles (1–10 μm) was generally improved with bubbles of finer sizes, e.g., $D_b = 0.08$ cm.	[96]
The effect of surfactants on bubble properties was examined in turbulent conditions, with the primary aim to identify the relationship between the size-scale surface properties leading to the decrease in bubble size.	[97]
The addition of frother to a synthetic sea salt reduced the measured bubble size, at certain mechanical conditions; in flotation, bubble size is a function of both coalescence and breakup phenomena.	[98]
Considering poorly modeled flotation systems and operation experience, a combined feed forward– and feedback-based control scheme of reagent dosage was presented to implement tracking for probability density function of the desired bubble size.	[99]
The effect of the addition of submicron silica particles of intermediate hydrophobicity was examined on the foaminess and gas dispersive properties of nonionic surfactants; the results indicated that this addition increased the gas holdup.	[100]
Nanomaterials were used that did not contribute to the flotation efficiency at shallow froth (i.e., 15 mm, in a Denver machine); contrary to this, nanosized Fe_2O_3 and Al_2O_3 provided a significant increase in barite recovery (7%–11%), at deep froth (35 mm).	[101]
A theory of combined microflotation was proposed to describe the kinetics of fine particles flotation, utilizing small quantities of fine bubbles in combination with coarse bubbles. Fine bubbles played the role of flotation carriers.	[102]
The effect of frothers on the degree of entrainment is important for selective flotation, and the mechanism of this was studied with gangue-only batch flotation tests, performed using fully liberated quartz.	[103]
A higher air fraction was shown to be needed at higher loading rates in DAF, to achieve the same removal efficiency. Model predictions, in the presence of stratified flow in the tank, showed that an optimum bubble size was present, which increased with an increase in particle size.	[104]

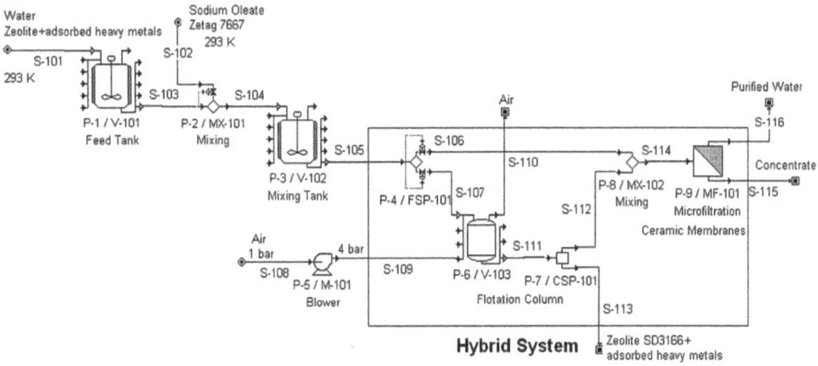

FIGURE 2.8 Technical economical flowsheet analysis of the hybrid process. *Reprinted with permission; copyright Elsevier Matis K.A., N.K. Lazaridis, A.I. Zouboulis, G.P. Gallios, V. Mavrov, A hybrid flotation-microfiltration process for metal ions recovery, J. Membr. Sci. 247 (2005) 29–35.*

5. Summary and conclusions

Flotation constitutes a scientifically interest (and perhaps, difficult) separation process, as we very often face a three-phase G/L/S system. The particles size to be floated should be rather small, so that they can be levitated by the bubbles produced, by different methods: dispersed air or froth, dissolved air, and electroflotation; a comparison was included. Fine particles usually present a (separation) problem in chemical industry and raw materials processing, often as by-products. Sustainability in this field is of significance, with emphasis given to recycling, because the need for fresh sources of drinking water is becoming urgent worldwide.

Some significant examples were presented in this work (see also Fig. 2.1), with satisfactory separation results, concerning heavy metal ions removal from aqueous solution, such as cadmium, zinc, and hexavalent chromium. Certainly, the process originated from mineral processing; i.e., the given reference to the further reprocessing of an arsenic-rich, gold-bearing pyrite concentrate by column flotation. This selective separation is of interest when, for instance, gold is associated selectively with one of the constituents; hence, contributing to recycling. The role and type also of bubbles, including their respective size, were discussed. The difference between wastewater treatment and mineral processing was highlighted.

Among the various available separation techniques, particularly the following were reviewed in some extent: the application upstream of sorbents (as zeolite, goethite, biosorbents, etc.) and of a combined flotation-microfiltration cell. Also, there were given hereby cost information for certain flotation techniques: DAF chemical dosage, solid waste remediation, flotation cell (as far as energy efficiency is concerned), posttreatment for DAF, electroflotation, and hybrid units.

TABLE 2.2 Some literature reports on relative costs of flotation applications.

Process and economics information	References
Work was carried out on a typical steel industry, in selection of the optimum amount of chemicals dosage used in a dissolved air flotation (DAF) unit to remove oil and grease; neural models were developed. The amount of alum required was ~104 kg/day-shift and of polyelectrolyte 1 kg/day-shift.	[107]
The distribution of turbulence was investigated in a mechanical (conventional) 3 m^3 flotation cell, considering the power number as a useful hydrodynamic indicator. Flotation was said not to be a particularly energy-intensive process.	[108]
Three soils polluted by municipal solid waste (MSW) incineration residues and containing various concentrations of Cu, Pb, Sb, Sn, and Zn were treated using magnetism, gravity separation (jig and shaking tables), and flotation/leaching. The estimated costs of the process, including direct and indirect costs, varied from $82 to $88 per ton of treated soil, depending on the proportion of MSW incineration residues mixed with the soil.	[109]
A sequential treatment of olive mill wastewater, comprising a DAF pretreatment and nanofiltration was published; involving operation costs of approximately 2.56–3.08 €/m^3 (depending on the working plan schedule and volume reduction factor) and requiring a footprint of approximately 52 m^2 to treat 1000 m^3 of olive mill wastewater.	[110]
Posttreatment of industrial wastewater polluted by metalworking fluids was performed using the photo-Fenton process, in following of the chemical addition-dissolved air flotation unit. The cost analysis of this process for the consumed energy (6 kWh) and chemicals (0.01818 kg $FeSO_4$ and 17.15 kg H_2O_2) was estimated at approximately 26 $ per 1 m^3 of DAF effluent.	[111]

Acknowledgments

Many and sincere thanks are due to all the colleagues in the (then) Laboratory of General and Inorganic Chemical Technology of Aristotle University of Thessaloniki (Greece).

References

[1] V.A. Glembotskii, V.I. Klassen, I.N. Plaksin, Flotation, Primary Sources, 1972. New York.
[2] E.A. Deliyanni, G.Z. Kyzas, K.A. Matis, Various flotation techniques for metal ions removal, J. Mol. Liq. 225 (2017) 260–264.
[3] G. Bournival, S. Ata, E.J. Wanless, The role of particles in multiphase processes: particles on bubble surfaces, Adv. Colloid Interface Sci. 225 (2015) 114–133.

[4] D.W. Fuerstenau, Fine particle flotation, in: P. Somasundaran (Ed.), Fine Particles Processing, SME-AIME, New York, 1980, pp. 669–705.

[5] E.N. Peleka, K.A. Matis, Water separation processes and sustainability, Ind. Eng. Chem. Res. 50 (2) (2011) 421–430.

[6] G.Z. Kyzas, E.A. Deliyanni, K.A. Matis, Research activities related to flotation process, Trends Green Chem. 1 (2015) 1–8.

[7] D.R. Nagaraj, R.S. Farinato, Evolution of flotation chemistry and chemicals: a century of innovations and the lingering challenges, Miner. Eng. 96–97 (2016) 2–14.

[8] G.Z. Kyzas, K.A. Matis, Methods of arsenic wastes recycling: focus on flotation, J. Mol. Liq. 214 (2016a) 37–45.

[9] G. Long, Y. Peng, D. Bradshaw, Flotation separation of copper sulphides from arsenic minerals at Rosebery copper concentrator, Miner. Eng. 66–68 (2014) 207–214.

[10] B. Tansel, D. Boglaienko, Self assembly, mobilization, and flotation of crude oil contaminated sand particles as granular shells on gas bubbles in water, Sci. Total Environ. 574 (2017) 437–442.

[11] G. Liu, J. Xiao, J. Liu, X. Qu, Q. Liu, H. Zeng, X. Yang, L. Xie, H. Zhong, Q. Liu, Z. Xu, In situ probing the self-assembly of 3-hexyl-4-amino-1,2,4-triazole-5-thione on chalcopyrite surfaces, Colloids Surf. A 511 (2016) 285–293.

[12] J. Wang, R.-H. Yoon, J. Morris, AFM surface force measurements conducted between gold surfaces treated in xanthate solutions, Int. J. Miner. Process. 122 (2013) 13–21.

[13] M. Ahola, K. Villforth, S. Schabel, Process modelling in deinking plants: forecasting recycled pulp quality and resource efficiency, Intl. Paperworld 2016 (2016) 26–35.

[14] B. Dold, Evolution of acid mine drainage formation in sulphidic mine tailings, Minerals 4 (2014) 621–641.

[15] C.A. Laamanen, G.M. Ross, J.A. Scott, Flotation harvesting of microalgae, Renew. Sustain. Energy Rev. 58 (2016) 75–86.

[16] T. Leistner, M. Embrechts, T. Leißner, S.C. Chelgani, I. Osbahr, R. Möckel, U.A. Peuker, M. Rundolph, A study of the reprocessing of fine and ultrafine cassiterite from gravity tailing residues by using various flotation techniques, Miner. Eng. 96–97 (2016) 94–98.

[17] A.V. Nguyen, H.J. Schulze, Colloidal Science of Flotation, Marcel Dekker, New York, 2004.

[18] J. Leja, Surface Chemistry of Froth Flotation, Plenum Press, New York, 1982.

[19] V.I. Klassen, V.A. Mokrushov, An Introduction to the Theory of Flotation, Butterworths, London, 1963, p. 43.

[20] A.N. Clark, D.J. Wilson, Foam Flotation – Theory and Applications, Marcel Dekker, New York, 1983.

[21] [a] K.A. Matis, P. Mavros, (i) Recovery of metals by ion flotation from dilute aqueous solutions, Separ. Purif. Methods 20 (1) (1991) 1–48;
[b] K.A. Matis, P. Mavros, Foam/froth flotation: Part II. Removal of particulate matter, Separ. Purif. Methods (1991) 163–198.

[22] H.J. Schulze, Elements of physically-based modeling of the flotation process, in: P. Mavros, K.A. Matis (Eds.), Innovations in Flotation Technology, Kluwer Academic, Dordrecht, 1992, pp. 171–180.

[23] B.V. Derjaguin, S.S. Dukhin, N.N. Rulyov, Kinetic theory of flotation of small particles, Surf. Colloid Sci. 13 (1984) 71–113.

[24] G. Wang, A.V. Nguyen, S. Mitra, J.B. Joshi, G.J. Jameson, G.M. Evans, A review of the mechanisms of bubble-particle detachment in froth flotation, Separ. Purif. Technol. 170 (2016a) 155–172.

[25] K.A. Matis, G.P. Gallios, Dissolved-air and electrolytic flotation, in: B.A. Wills, R.W. Barley (Eds.), Mineral Processing at a Crossroads, Martinus Nijhoff, Dordrecht, 1986, pp. 37–70.

[26] R.H. Yoon, G.H. Luttrell, The effect of bubble size on fine particle flotation, Miner. Process. Extr. Metall. Rev. 5 (1989) 101–122.

[27] T. Takahashi, I. Miyahara, H. Mochizuki, Fundamental studies of bubble formation in dissolved-air pressure flotation, J. Chem. Eng. Jpn. 12 (4) (1979) 275–280.

[28] K.A. Matis, A.I. Zouboulis, The role of bubble/particle size, in: K.A. Matis (Ed.), Flotation Science and Engineering, Marcel Dekker, New York, 1995, pp. 63–87.

[29] E. Valdes-Krieg, C.J. King, H.H. Sephton, Removal of surfactants and particulate matter from sea water desalination blowdown brines by foam fractionation, Desalination 16 (1975) 39–53.

[30] Y. Sato, Y. Murakami, T. Hirose, H. Yamamoto, Y. Urgun, Removal of emulsified oil particles by dissolved-air flotation, J. Chem. Eng. Jpn. 12 (1979) 454–459.

[31] S. Usui, H. Sasaki, Zeta-potential measurements of bubbles in aqueous solutions, J. Colloid Interface Sci. 65 (1978) 36–45.

[32] M. Kostoglou, T.D. Karapantsios, K.A. Matis, Modeling local flotation frequency in a turbulent flow field, Adv. Colloid Interface Sci. 122 (2007) 79–91.

[33] E.N. Peleka, K.A. Matis, Hydrodynamic aspects of flotation separation, Open Chem. 14 (2016) 132–139.

[34] G. Ai, X. Yang, X. Li, Flotation characteristics and flotation kinetics of fine wolframite, Powder Technol. 305 (2017) 377–381.

[35] D.A. Wheeler, Column flotation: the original column, in: S.H. Castro, J. Alvarez (Eds.), Froth Flotation, Proc. 2nd Latin-American Congr., Concepción, Chile, Elsevier, Amsterdam, 1988, pp. 17–40.

[36] W.J. Trahar, L.J. Warren, The floatability of very fine particles – a review, Int. J. Miner. Process. 3 (1976) 103–131.

[37] G.P. Gallios, K.A. Kydros, K.A. Matis, Electrokinetic behaviour of sulphide minerals: a contribution to the chemistry of flotation, in: G.P. Gallios, K.A. Matis (Eds.), Mineral Processing and the Environment, Kluwer Academic Publishers, Dordrecht, 1998, pp. 25–42.

[38] K.A. Matis, G.P. Gallios, K.A. Kydros, Separation of fines by flotation techniques, Sep. Technol. 3 (1993) 76–90.

[39] P. Mavros, K.A. Kydros, K.A. Matis, Arsenopyrite enrichment by column flotation, Miner. Eng. 6 (12) (1993) 1265–1277.

[40] G.W. Poling, M.J.V. Beattie, Selective depression in complex sulphide flotation, in: M.H. Jones, J.T. Woodcock, I.M.M. Australasian (Eds.), Principles of Mineral Flotation – the Wark Symposium, 1984, p. 137.

[41] K.A. Kydros, G.P. Gallios, K.A. Matis, Electrolytic flotation of pyrite, J. Appl. Chem. Biotechnol. 59 (1994) 223–232.

[42] MdS.K.A. Sarkar, G.M. Evans, S.W. Donne, Bubble size measurement in electroflotation, Miner. Eng. 23 (2010) 1058–1065.

[43] G.Z. Kyzas, K.A. Matis, Electroflotation process: a review, J. Mol. Liq. 220 (2016b) 657–664.

[44] K.H. Rao, A. Vilinka, I.V. Chernyshova, Minerals bioprocessing: R&D needs in mineral beneficiation, Hydrometallurgy 104 (2010) 465–470.

[45] L. Wang, R. Liu, Y. Hu, J. Liu, W. Sun, Adsorption behavior of mixed cationic/anionic surfactants and their depression mechanism on the flotation of quartz, Powder Technol. 302 (2016b) 15–20.

[46] T.F. Zabel, Flotation in water treatment, in: P. Mavros, K.A. Matis (Eds.), Innovations in Flotation Technology, Kluwer Academic, Dordrecht, 1992, pp. 431–454.

[47] L. Haak, R. Roy, K. Pagilla, Toxicity and biogas production potential of refinery waste sludge for anaerobic digestion, Chemosphere 144 (2016) 1170–1176.

[48] J. Bratby, G.v.R. Marais, Flotation, in: D.B. Purchas (Ed.), Solid/Liquid Separation Equipment Scale-Up, Uplands Press, Croydon, 1977, p. 155.

[49] P.T.L. Koh, M.P. Schwarz, CFD modeling of bubble-particle attachments in a flotation cell, in: Proc. Centenary Flotation Symp.; Brisbane, Australia, June, 2005, pp. 201–207.

[50] M. Zhang, P. Guiraud, Elimination of TiO_2 nanoparticles with the assist of humic acid: influence of agglomeration in the dissolved air flotation process, J. Hazard Mater. 260 (2013) 122–130.

[51] K.A. Matis, N.K. Lazaridis, Flotation techniques in water technology for metals recovery: dispersed-air vs. dissolved-air flotation, J. Min. Metall. A 38 (2002) 1–27.

[52] IWA, in: Proc. 4th Intl. Conf. Flotation in Water and Waste Water Treatment, Finnish Water and Wastewater Works Assoc., Helsinki, 2000.

[53] M. Ponasse, V. Dupre, Y. Aurelle, A. Secq, Bubble formation by water release in nozzle – II. Influence of various parameters on bubble size, Water Res. 32 (1998) 2498–2506.

[54] A. Chen, Z. Wang, J. Yang, Influence of bubble size on the fluid dynamic behavior of a DAF tank: a 3D numerical investigation, Colloids Surf., A 495 (2016) 200–207.

[55] K.A. Matis, A.I. Zouboulis, N.K. Lazaridis, Dispersed-air vs. dissolved-air flotation: a comparison applied to metal-loaded goethite, in: Proc. 34th Int'l. Conf. Mining and Metallurgy, Organized by Univ. Belgrade, Techn. Fac. Bor Serbia, 30 Sept.-3 Oct., 2002, pp. 240–246.

[56] A.I. Zoubouls, K.A. Matis, Removal of cadmium from dilute solutions by flotation, Water Sci. Technol. 31 (3–4) (1995) 315–326.

[57] A.I. Zouboulis, K.A. Matis, Removal of metal ions from dilute solutions by sorptive flotation, Crit. Rev. Environ. Sci. Technol. 27 (3) (1997) 195–235.

[58] M.R. Mahmoud, N.K. Lazaridis, K.A. Matis, Study of flotation conditions for cadmium(II) removal from aqueous solutions, Process Saf. Environ. Protect. 94 (2015) 203–211.

[59] N.K. Lazaridis, K.A. Matis, G.A. Stalidis, P. Mavros, Dissolved-air flotation of metal ions, Separ. Sci. Technol. 27 (1992) 1743–1758.

[60] E.N. Peleka, K.A. Matis, Application of flotation as a pretreatment process during desalination, Desalination 222 (2008) 1–8.

[61] M.A. Darwish, H.K. Abdulrahim, A.S. Hassan, A.O. Sharif, Needed seawater reverse osmosis pilot plant in Qatar, Desalin. Wat. Treatment 57 (2016) 3793–3819.

[62] Y.H.C. Wang, P. Somasundaran, A study of carrier flotation of clay, in: P. Somasundaran (Ed.), Fine Particles Processing, SME/AIME, New York, 1980, pp. 1112–1128.

[63] J.M. Younker, M.E. Walsh, Effect of adsorbent addition on floc formation and clarification, Water Res. 98 (2016) 1–8.

[64] D.N. Bakoyannakis, E.A. Deliyanni, A.I. Zouboulis, K.A. Matis, L. Nalbandian, T. Kehagias, Akaganeite and goethite-type nanocrystals: synthesis and characterization, Microporous Mesoporous Mater. 59 (2003) 35–42.

[65] K.A. Matis, D. Zamboulis, A.I. Zouboulis, N.K. Lazaridis, Goethite mineral as a sorbent for heavy metals, in: P. Misaelides, F. Macašek, T.J. Pinnavaia, C. Colella (Eds.), Natural

Microporous Materials in Environmental Technology, Kluwer Academic Publishers, The Netherlands, 1999, pp. 425–433.

[66] A.I. Zouboulis, D. Zamboulis, K.A. Matis, Foam flotation of zeolites: application on zinc ion removal, Separ. Sci. Technol. 26 (3) (1991) 355–365.

[67] G. Bulut, Ü. Yenial, E. Emiroğlu, A.A. Sirkeci, Arsenic removal from aqueous solution using pyrite, J. Clean. Prod. 84 (2014) 526–532.

[68] R. Etchepare, A. Azevedo, S. Calgaroto, J. Rubio, Removal of ferric hydroxide by flotation with micro and nanobubbles, Separ. Purif. Technol. 184 (2017) 347–353.

[69] M.H. Dehgani, B. Karimi, A. Rafieepour, Continuous removal of toxic nickel from industrial wastewater by flotation-adsorption with multi-walled carbon nanotubes: kinetic and adsorption isotherms, Desalination Wat. Treatment 71 (2017) 107–115.

[70] N. Hu, W. Liu, L. Ding, Z. Wu, H. Yin, D. Huang, H. Li, L. Jin, H. Zheng, Removal of methylene blue from its aqueous solution by froth flotation: hydrophobic silica nanoparticle as a collector, J. Nanoparticle Res. 19 (2) (2017) article no. 46.

[71] M.N. Hughes, R.K. Poole, Metals and Microorganisms, Chapman and Hall, London, 1989, pp. 328–395.

[72] H. Eccles, Treatment of metal-contaminated wastes: why select a biological process? Trends Biotechnol. 17 (1999) 462–465.

[73] D. Zamboulis, E.N. Peleka, N.K. Lazaridis, K.A. Matis, Metal ion separation and recovery from environmental sources using various flotation and sorption techniques, J. Appl. Chem. Biotechnol. 86 (3) (2011) 335–344.

[74] T.J. Butter, L.M. Evison, I.C. Hancock, F.S. Holland, K.A. Matis, A. Philipson, A.I. Sheikh, A.I. Zouboulis, The removal and recovery of cadmium from dilute aqueous solutions by biosorption and electrolysis at laboratory scale, Water Res. 32 (1998) 400–406.

[75] F.S. Holland, The development of the Eco-Cell process, Chem. Ind. 1 (1978) 453–456.

[76] A.I. Zouboulis, N.K. Lazaridis, K.A. Matis, Removal of toxic metal ions from aqueous systems by biosorptive flotation, J. Appl. Chem. Biotechnol. 77 (2002) 958–964.

[77] H. Hanotu, E. Karunakaran, H. Bandulasena, C. Biggs, W.B. Zimmerman, Harvesting and dewatering yeast by microflotation, Biochem. Eng. J. 82 (2014) 174–182.

[78] M. Wu, J. Liang, J. Tang, G. Li, S. Shan, Z. Guo, L. Deng, Decontamination of multiple heavy metals-containing effluents through microbial biotechnology, J. Hazard Mater. 337 (2017) 189–197.

[79] A. Esmaeili, E. Hejazi, Y. Vasseghian, Comparison study of biosorption and coagulation/air flotation methods for chromium removal from wastewater: experiments and neural network modeling, RSC Adv. 5 (111) (2015) 91776–91784.

[80] A.I. Zouboulis, K.A. Matis, Cadmium ion removal by electroflotation onto sewage sludge biomass, Int. J. Environ. Waste Manag. 9 (3/4) (2012) 245–256.

[81] I. Chang, S. Judd, Air sparging of a submerged MBR for municipal wastewater treatment, Process Biochem. 37 (2002) 915–920.

[82] T. Ueda, K. Hata, Y. Kikuoka, O. Seino, Effects of aeration on suction pressure in a submerged membrane bioreactor, Water Res. 31 (1997) 489–494.

[83] E. Essam, A. Mahmoud, E. Sadeq, E. Ahmad, Performance evaluation of two RO membrane configurations in a MSF/RO hybrid system, Desalination 128 (2000) 231–245.

[84] A.M. Eliceche, M.C. Daviou, P.M. Hoch, I.O. Uribe, Optimisation of azeotropic distillation columns combined with pervaporation membranes, Comput. Chem. Eng. 26 (2002) 563–573.

[85] B. Schlichter, V. Mavrov, H. Chmiel, Study of a hybrid process combing ozonation and membrane filtration-filtration of model solutions, Desalination 156 (2003) 257–265.
[86] E.N. Peleka, N.K. Lazaridis, P. Mavros, K.A. Matis, A new hybrid flotation-microfiltration cell, Separ. Sci. Technol. 41 (2006a) 3229–3243.
[87] K.A. Matis, E.N. Peleka, D. Zamboulis, T. Erwe, V. Mavrov, Air sparging during the solid/liquid separation by microfiltration: application of flotation, Separ. Purif. Technol. 40 (2004) 1–7.
[88] N.K. Lazaridis, E.N. Peleka, T.D. Karapantsios, K.A. Matis, Copper recovery from effluents by various separation techniques, Hydrometallurgy 74 (2004) 149–156.
[89] V. Nenov, N.K. Lazaridis, C. Blöcher, B. Bonev, K.A. Matis, Metal recovery from a copper mine effluent by a hybrid process, Chem. Eng. Process 47 (2008) 596–602.
[90] R.P. King, Modeling and Simulation of Mineral Processing Systems, Butterworth-Heinemann, London, 2001.
[91] F. Bloom, T.J. Heindel, A theoretical model of flotation deinking efficiency, J. Colloid Interface Sci. 190 (1997) 182–197.
[92] A.V. Nguyen, Hydrodynamics of liquid flows around air bubbles in flotation: a review, Int. J. Miner. Process. 56 (1999) 165–205.
[93] Z. Brabcová, T. Karapantsios, M. Kostoglou, P. Basařová, K. Matis, Bubble–particle collision interaction in flotation systems, Colloids Surf. A 473 (2015) 95–103.
[94] T. Miettinen, J. Ralston, D. Fornasiero, The limits of fine particle flotation, Miner. Eng. 23 (2010) 420–437.
[95] S. Calgaroto, A. Azevedo, J. Rubio, Flotation of quartz particles assisted by nanobubbles, Int. J. Miner. Process. 137 (2015) 64–70.
[96] A. Hassanzadeh, B.V. Hassas, S. Kouachi, Z. Brabcova, M.S. Çelik, Colloids Surf. A 498 (2016) 258–267.
[97] Z. Jávor, N. Schreithofer, K. Heiskanen, Multi-scale analysis of the effect of surfactants on bubble properties, Miner. Eng. 99 (2016) 170–178.
[98] J.M. Sovechles, M.R. Lepage, B. Johnson, K.E. Waters, Effect of gas rate and impeller speed on bubble size in frother-electrolyte solutions, Miner. Eng. 99 (2016) 133–141.
[99] J. Zhu, W. Gui, J. Liu, H. Xu, C. Yang, Combined fuzzy based feed forward and bubble size distribution based feedback control for reagent dosage in copper roughing process, J. Process Control 39 (2016) 50–63.
[100] G. Bournival, Z. Du, S. Ata, G.J. Jameson, Foaming and gas dispersion properties of non-ionic frothers in the presence of hydrophobized submicron particles, Int. J. Miner. Process. 133 (2014) 123–1131.
[101] E.C. Cilek, S. Karaca, Effect of nanoparticles on froth stability and bubble size distribution in flotation, Int. J. Miner. Process. 138 (2015) 6–14.
[102] N.N. Rulyov, Combined microflotation of fine minerals: theory and experiment, Miner. Proc. Extractive Metal. C 125 (2016) 81–85.
[103] L. Wang, Y. Peng, K. Runge, The mechanism responsible for the effect of frothers on the degree of entrainment in laboratory batch flotation, Miner. Eng. 100 (2017) 124–131.
[104] B. Lakghomi, Y. Lawryshyn, R. Hofmann, A model of particle removal in a dissolved air flotation tank: importance of stratified flow and bubble size, Water Res. 63 (2015) 262–272.
[105] K.A. Matis, N.K. Lazaridis, A.I. Zouboulis, G.P. Gallios, V. Mavrov, A hybrid flotation-microfiltration process for metal ions recovery, J. Membr. Sci. 247 (2005) 29–35.
[106] E.N. Peleka, M.M. Fanidou, P.P. Mavros, K.A. Matis, A hybrid flotation-microfiltration cell for S/L separation: operational characteristics, Desalination 194 (2006b) 135–145.

[107] R. Wadhwani, B. Mohanty, S. Khanam, Optimization of the amount of chemical dosage to improve the performance of dissolved air flotation unit, Separ. Sci. Technol. 50 (2015) 1038–1049.

[108] E. Tabosa, K. Runge, P. Holtham, K. Duffy, Improving flotation energy efficiency by optimizing cell hydrodynamics, Miner. Eng. 96–97 (2016) 194–202.

[109] P. Jobin, L. Coudert, V. Taillard, J.-F. Blais, G. Mercier, Remediation of inorganic contaminants and polycyclic aromatic hydrocarbons from soils polluted by municipal solid waste incineration residues, Environ. Technol. 37 (2016) 1983–1995.

[110] S. Sanches, M.C. Fraga, N.A. Silva, P. Nunes, J.G. Crespo, V.J. Pereira, Pilot scale nanofiltration treatment of olive mill wastewater: a technical and economical evaluation, Environ. Sci. Pollut. Res. 24 (2017) 3506–3518.

[111] M.M. Amin, M.M.G. Mofrad, H. Pourzamani, S.M. Sebaradar, K. Ebrahim, Treatment of industrial wastewater contaminated with recalcitrant metal working fluids by the photo-Fenton process as post-treatment for DAF, J. Ind. Eng. Chem. 45 (2017) 412–420.

[112] K.A. Matis, E.N. Peleka, Alternative flotation techniques for wastewater treatment: focus on electroflotation, Separ. Sci. Technol. 45 (2010) 2465–2474.

Chapter 3

Flotation in the 2010s: focus on mineral processing

George P. Gallios,[a] George Z. Kyzas[b,*] and Kostas A. Matis[a]
[a]*Division of Chemical Technology & Industrial Chemistry, Department of Chemistry, Aristotle University, Thessaloniki, Greece;* [b]*Hephaestus Advanced Laboratory, Department of Chemistry, International Hellenic University, Kavala, Greece*
**Corresponding author*

1. Introduction

All over the years, there have been various flotation techniques developed, and some of them are worth mentioning here, such as foam flotation [1], ion flotation [2], adsorbing colloid flotation [3], dissolved air flotation [4], electroflotation [5], combined microflotation [6], oil agglomerate flotation [7], two-phase flotation [8], flotation-microfiltration [9], and certainly, the Jameson cell [10]. The process has been recently reviewed, based on almost 40 years research experience on it [11].

Nevertheless, among various modern applications of flotation (apart from water and wastewater treatment), the original approach in mineral processing is believed that it is the most interesting. Although there are certain similarities between the fields of conventional mineral beneficiation and effluent treatment, as far as flotation is concerned, several differences (see, for instance, Table 3.1) should be kept in mind. It is true that a large volume of experimental work on the role of particle/bubble size, among others, has its origin in the application of flotation to water and wastewater purification [12].

Scope of the present review paper is the flotation of salt-type minerals and particularly, magnesite. Different physicochemical parameters affecting flotation were discussed. An investigation of salt-type mineral flotation may be interesting both from the practical and theoretical points of view.

Magnesite is an important economic nonmetallic mineral, since it is the main source of magnesium oxide (magnesia, MgO), which is widely used mainly as a refractory material. It is one of the important products of the Greek mineral wealth. Magnesite appears in veins or stockworks in serpentine host

Advanced Low-Cost Separation Techniques in Interface Science. https://doi.org/10.1016/B978-0-12-814178-6.00003-0

TABLE 3.1 Differences of flotation from the point of view of fine processing.

	Conventional mineral processing	Effluent treatment
Concentration (pulp density)	15%–40%	ppm range
Particle size	25–500 μm	Colloidal
Mixing	Intense, continuous	Unnecessary
Separation	Selective	S/L
Economy	Attractive	Unattractive

rock [13]. Beneficiation products with commercial quality are directly calcined in rotary kilns. Raw materials are of course crucial to Europe's economy. They form a strong industrial base, producing a broad range of goods and applications used in everyday life and modern technologies. Reliable and unhindered access to certain raw materials is a growing concern within the EU and across the globe. Magnesium is belonging to the critical raw materials (http://ec.europa.eu/growth/sectors/raw-materials/specific-interest/critical_en).

Salt-type minerals, which is the general category examined in the following, are important raw materials for the metallurgical, chemical, construction, and agricultural industries. Many millions of tonnes of these minerals are consumed each year. Examples are phosphates or apatites, $Ca_2(PO_4)_6(OH,Cl,F)_2$; baryte, $BaSO_4$; fluorite, CaF_2; gypsum, $CaSO_4 \cdot 2H_2O$; magnesite, $MgCO_3$; calcite, $CaCO_3$; dolomite, $CaCO_3$ $MgCO_3$; scheelite, $CaWO_4$; etc. Among the existing phosphates, mention is given to hydroxyapatite, fluorapatite, francolite, dahlite, dehmite, lewistonite, wilkeite, and ellestadite.

A large proportion of these minerals are produced by flotation process, typically by fatty acid collector. The magnesite ore deposits usually contain a variety of gangue minerals, mostly other carbonates (calcite, dolomite, etc.), silicates, and oxides. A classical flow sheet of salt-type mineral processing plant was presented [14]. Nevertheless, cationic flotation with amine collectors was applied to succeed separation of magnesite from silicate gangue (i.e., reverse flotation), as, for instance, had been used in Mantoudi Evia, Greece [15].

Certain significant activities in flotation research were reviewed [16], with main focus in the contribution of physical chemistry to flotation. Also, the impact of chemical speciation during various flotation applications was stressed [17]. It is noted that this chapter is a continuation of the other one on flotation, coming from the same laboratory (formerly, of General and Inorganic Chemical Technology) at AUTh [18].

2. Mineral processing

Arguing on the "the death of mining (?)," during his presentation in Australia, threats to mining were said by Davies [19] to be posed by the declining materials intensity of economic growth, by recycling and by the long downward trend in real mineral prices. Although it is not certain whether the above have been much exaggerated, certainly we believe that justifies the present paper, as the new economy activities are in fact surprisingly dependent on traditional raw materials [20]; for instance, a home PC typically contains around 30 mineral ingredients. Despite the heavy dependence of our economies on minerals, the responsibility for mineral production and the development of mining technology is being rather handed over to predominantly developing countries.

The dramatic reduction in mining activities in Western Europe is perhaps illustrated by the hard coal production, which has dropped from 495 million tonnes in 1970 to 99 in 1999. The main reasons for this reduction in production were more difficult geological conditions, an increase in depth of mining, increased labor costs, the development of an international coal trade, and the reduced political will of the European countries to subsidize coal production; there apparently exists a marked imbalance between the production and consumption figures [21]. In mineral processing, with the increasing demand for minerals, the continuously diminishing grade of ores and various economic pressures (i.e., cost of fuel), the problem of fines processing became important [22]; may be the title of the aforementioned book (originated from the corresponding NATO Advanced Study Institute held at Falmouth, UK) was prophetic.

Elsewhere, a paper [23] reflected on how sustainability is being interpreted both theoretically and in practice for the primary extraction industries. Two contrasting perspectives can be adopted: one states that continued extraction of nonrenewable resources is a necessary part of sustainable development, while the other states that extraction of these resources must be greatly reduced or even eliminated. There is now a burgeoning literature that examines sustainable development in the context of minerals and mining, most of which are concerned with sustainability at global and national scales. Sustainable development in the corporate mining context was reported [24].

So, another possible aim of the present work was the reprocessing of once deposited tailings, usually existing at a fine particle size. Published examples are the case of cassiterite from a historic tin operation [25] and that of gold-bearing arsenopyrite at Northern Chalkidiki area, Greece [26]. The increasingly complex mineralogy and lower grade of many current ore reserves have forced technology to evolve rapidly in the decades to treat these materials economically. Flotation, which is a quite unique separation process, constitutes one of the most significant operations in mineral beneficiation, termed froth flotation.

Flotation of minerals is usually performed in agitated cells. This may prove inconvenient when treating fine particles, especially when entrainment of gangue or loss of valuable mineral is not desired. Also, the solubility problem is enhanced with the finer range. Modern (then) flotation techniques permitted the mining of low-grade and complex ores, overcoming the associated problems such as the existence of fine particles; the former was at first accomplished by the introduction of column flotation [27].

Perhaps, the best example is the Florida phosphate industry where approximately one-third of the phosphate was discarded as slime. It was argued that one-fifth of the world's tungsten and one-half of Bolivian tin were lost in slimes because of the lack of a suitable method of separation at fine particle sizes [28]. As the particle size is reduced, two characteristics begin to dominate: the specific surface becomes large and the mass of the particle becomes very small.

Ideally, one would like fine mineral particles to be produced only for the purpose of liberation of valuables from gangue minerals. However, large amounts of fines are generated at the mine site itself. Advances in mining techniques, in order to reduce the production of fines, are one area that needs to be explored. Also, to avoid unwanted fines, grinding and classification circuits have to be made more efficient [29]. So, the processing of ores containing fines can be classified into two broad categories, namely (1) those in which the desired mineral almost entirely occurs in the fine size range, and in this instance, the process (ideally) should be designed to recover virtually all the mineral particles at this size range; (2) those in which the desired mineral extends from coarse sizes into the fine size range. In this case, the fines often interfere with the concentration of coarser particles also, poor flotation behavior of fines is observed. These problems, in general, have been handled either by elimination, minimizing, or ignoring the fines in the flotation circuit feed—noting that we are dealing with tonnages of ores [30].

Research on the mechanisms of particle–bubble interaction provided valuable information on how to improve the flotation of fine (and coarse particles) with novel flotation machines, which provides higher collision and attachment efficiencies of fine particles with bubbles and lower detachment of the coarse particles [31]. It was pointed out that fines can be collected more easily by finer bubbles. Dissolved air flotation, a process established already in effluent treatment, was examined for salt-type minerals; the technique offers certain advantages, principally the production of fine air bubbles (<120 μm), so it may be suitable for the processing of particles at the subsieve size range [32]. Selective separation of fines by electroflotation was presented [33], discussing the changes happening during such processing due to the oxidation–reduction reactions occurring on the electrodes; noting that this technique indeed had its origin in the ex-Soviet Union. Electroflotation (or electrolytic flotation) was also compared with the conventional one as far as costs are concerned.

The flotation of minerals from their ores in flotation cells involves understanding many chemical and physical processes ranging from electronic events in the bulk mineral and in the mineral surface régime, molecular events at interfaces, the statics and dynamics of thinning and rupture of microscopic to molecular interfacial films, the fluid dynamics of bubble–particle events, the dynamics of air–water–solid colloid sates in a flotation machine, etc. [34].

3. Case study: flotation of magnesite

3.1 Theoretical aspects and relative experimental work

The minerals mainly used in the following experiments were pure natural magnesite coming from Yerakini Chalkidiki, where the second biggest deposit of magnesite exists, and dolomite from Neraida Kozani (both in northern Greece). Their chemical analysis is mentioned in Table 3.2, certified also by X-ray diffraction analysis. Dolomite theoretically contains a combination of $CaCO_3$ (54.35%) and $MgCO_3$ (45.65%). The specific surfaces for magnesite and dolomite were, respectively, 1.93 and 0.0442 m^2/g, despite nominally similar size ranges. This was attributed to the numerous ultrafine particles, with size less than 10 μm, which existed on the larger magnesite particles, as shown under a microscope [14].

Magnesite and dolomite belong to the calcite minerals group; they crystallize in the rhombohedral system. The crystal structure of calcite is considered to be derived from that of NaCl, while in magnesite all the Ca^{2+} cations of the former are replaced by Mg^{2+}. In dolomite, only about half of them are replaced. The bonds between the cations and the anion radical are ionic, while the bonds within each complex CO_3^{2-} are covalent. Because of the

TABLE 3.2 Chemical analysis of dolomite and magnesite, wt%.

	Dolomite	Magnesite
Loss on ignition	46.74	5.158
Silicates and insolubles	0.28	0.26
R_2O_3	0.12	0.20
CaO	30.50	0.40
MgO	21.86	47.52
Total	99.50	99.96

difference in the ionic radius of magnesium and calcium cations, dolomite has an ordered structure and lower symmetry than magnesite.

Since any fraction (due to grinding, for example) breaks bonds that have at least a partial ionic character, the minerals have strongly hydrophilic surfaces and collectors must be used in order to float them. In the case of weakly ionizable collectors, which are widely used, it is important to consider their ionomolecular composition and its effect on the formation of insoluble salts or complexes. Fatty acids and amines are known to undergo dissociation [35]:

$$RCOOH \rightleftharpoons RCOO^- + H^+ \qquad pK = 4.7(at\ 298K)$$

$$RNH_2 + H_2O \rightleftharpoons RNH_3{}^+ + OH^- \qquad pK = 10.6$$

and the resulting ions may form insoluble salts with multivalent metal ions. The principal mechanism for oleate attachment to magnesium- and calcium-bearing salt-type minerals was considered to be chemisorptions at both neutral and basic pH values, shown by IR studies [36]. The major reactions involved are the following:

$$Mg^{2+} + 2RCOO^- \rightleftharpoons Mg(RCOO)_{2_{(s)}} \quad K_{sp} = 10^{-13.8}\frac{n!}{r!(n-r)!}$$

$$Ca^{2+} + 2RCOO^- \rightleftharpoons Ca(RCOO)_{2_{(s)}} \quad K_{sp} = 10^{-15.4}$$

which may take place either in solution or on particle surfaces, where the solubility products are coming from the literature [37]. An ethanolamine-based surfactant was introduced into the flotation of magnesite ore, and the possibility of improving its purity was studied; this collector was adsorbed on the surfaces of magnesite and dolomite by weak electrostatic interactions [38].

Few flotation reagents can match the standards of purity required in other sectors of chemical industry; it was ascertained that impurities have frequently proved beneficial, while purer chemicals did not perform as well [39]. Oleic acid, of course, is a weakly acidic insoluble compound, which forms very soluble salts with monovalent alkali metal ions, such as sodium; the species considered in the hydrolysis reactions are the oleate ion, oleic acid, acid soap, and oleate dimer. Oleic acid is often applied due to its double bond reactivity. A possible chemical mechanism of its oxidation on mineral particle surfaces was reported [40] and given in Table 3.3. Some evidence also suggested [41] that cross-linking may occur between oleate chains via an epoxide linkage, leading to a polymerized layer of collector molecules with greater hydrophobic character. The adsorption of fatty acids on magnesium carbonates was extensively studied [42].

On the other hand, solubility properties of minerals are of importance in flotation because of their role in determining both the chemical composition of

TABLE 3.3 Mechanism of oleic acid oxidation.

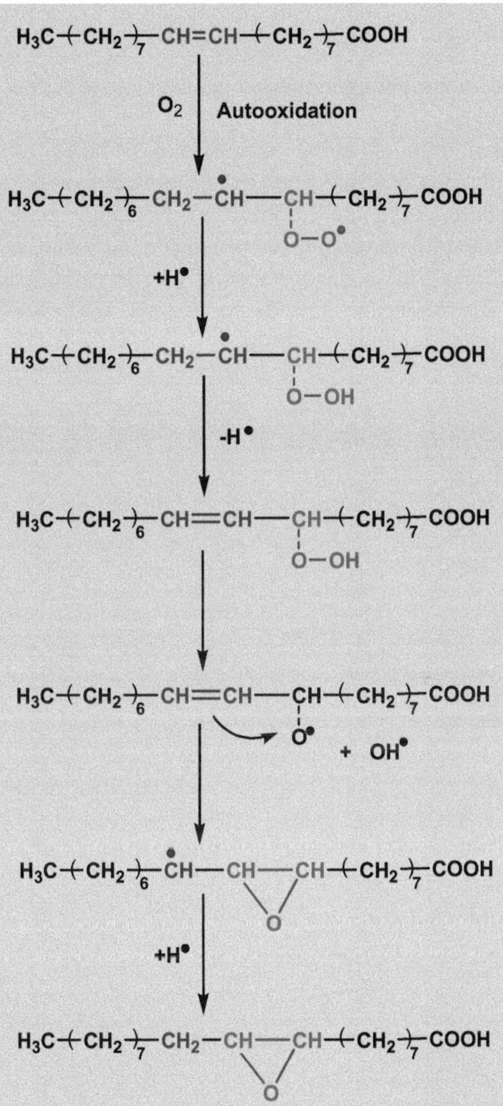

the aqueous phase and the charge characteristics of the interface. In general, the solubility can be expressed as follows:

$$M_xH_y \rightarrow xM^{y+} + yA^{x-}$$

where M stands for certain cations, such as Mg^{2+}, Ca^{2+}, and A for an anion such as CO_3^{2-}, PO_4^{3-}, WO_4^{2-}, etc. An idea for the salt-type minerals, under study, on the distribution of various species may be taken by the reactions of Table 3.4. Calculations of solubilities from thermodynamic data, in a system open to atmosphere, using the equilibrium constants were conducted [43]. Their solubilities are higher than those of silicate and oxide minerals. For pH values below 9.2, only the concentrations of calcium and magnesium cations were shown to be significant, and above this pH, the solubility controlling species were $(MgCO_3)_{aq}$ and $(CaCO_3)_{aq}$.

TABLE 3.4 Hydrolysis equations of magnesite[a] (Eqs. 3.1–3.11) and dolomite[a] (Eqs. 3.1–3.17).

Dolomite hydrolysis (Eqs. 3.1–3.17)	Magnesite hydrolysis (Eqs. 3.1–3.11)	$MgCO_3(s) \rightleftharpoons MgCO_3(aq)$ (3.1)
		$MgCO_3(aq) \rightleftharpoons Mg^{2+} + CO_3^{2-}$ (3.2)
		$CO_3^{2-} + H_2O \rightleftharpoons HCO_3^- + OH^-$ (3.3)
		$HCO_3^- + H_2O \rightleftharpoons H_2CO_3 + OH^-$ (3.4)
		$H_2CO_3 \rightleftharpoons CO_2(g) + H_2O$ (3.5)
		$Mg^{2+} + HCO_3^- \rightleftharpoons MgHCO_3^+$ (3.6)
		$MgHCO_3^+ \rightleftharpoons H^+ + MgCO_3(aq)$ (3.7)

TABLE 3.4 Hydrolysis equations of magnesite[a] (Eqs. 3.1–3.11) and dolomite[a] (Eqs. 3.1–3.17).—cont'd

		$Mg^{2+} + OH^- \rightleftharpoons MgOH^+$ (3.8)
		$MgOH^+ + OH^- \rightleftharpoons Mg(OH)_2(s)$ (3.9)
		$MgOH^+ + OH^- \rightleftharpoons Mg(OH)_2(aq)$ (3.10)
		$MgOH_2(aq) \rightleftharpoons Mg(OH)_2(s)$ (3.11)
		$CaCO_3 \cdot MgCO_3(s) \rightleftharpoons Ca^{2+} + Mg^{2+} + 2CO_3^{2-}$ (3.12)
		$Ca^{2+} + HCO_3^- \rightleftharpoons CaHCO_3^+$ (3.13)
		$CaHCO_3^+ \rightleftharpoons H^+ + CaCO_3(aq)$ (3.14)
		$Ca^{2+} + OH^- \rightleftharpoons CaOH^+$ (3.15)
		$CaOH^+ + OH^- \rightleftharpoons Ca(OH)_2(aq)$ (3.16)
		$Ca(OH)_2(aq) \rightleftharpoons Ca(OH)_2(s)$ (3.17)

[a] $pK_1 = -4.51$.

However, it was argued that these minerals do not dissolve very fast [44]. It was observed that the rate of change of the pH, during conditioning, was greater for dolomite than that for magnesite. Because of the minerals' dissolution, the fracture of the bonds, and different species adsorption from solution, the surface of minerals is charged; the latter is usually expressed as zeta-potential. It was surprisingly reported that the zero point of charge of magnesite may range from 2 up to even 11.5, depending on the method of measurement and origin of the samples [45].

The effect of modifying reagents, such as carboxymethyl cellulose (as the sodium salt), in the flotation of magnesium carbonates by fatty acid collectors was examined, combined with electrokinetic measurements. As presented, for instance, in Fig. 3.1, it was shown that this modifier (also sodium hexametaphosphate, $Na_6P_6O_{18}$) increased the negative charge of dolomite, although it did not much affect that of magnesite, i.e., showing a preferential adsorption. Possibly, the action of the modifier was due to adsorption on sites where calcium existed, as the surface of the minerals was negatively charged in the pH range investigated [35].

The solution pH did not seem to have any substantial influence on the adsorption of collector on magnesite, as presented in Fig. 3.2; however, pH affected significantly its flotation. Similar results were obtained with dolomite. The removal of the oleic acid from the solutions examined was very good, over

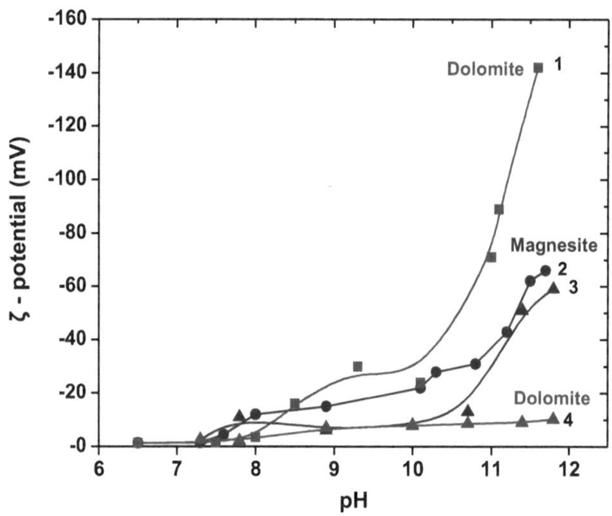

FIGURE 3.1 Zeta-potential measurements, by the method of electrophoretic mass transfer, for magnesite (designated as M) and dolomite (D): with (in lines 1 and 2) carboxymethyl cellulose modifier and without (in lines 3 and 4). *Reprinted, after modification, with permission; copyright Elsevier K.A. Matis, T.N. Balabanidis, G.P. Gallios, Processing of magnesium carbonate fines by dissolved-air flotation, Colloids Surface. 29 (1988) 191–203.*

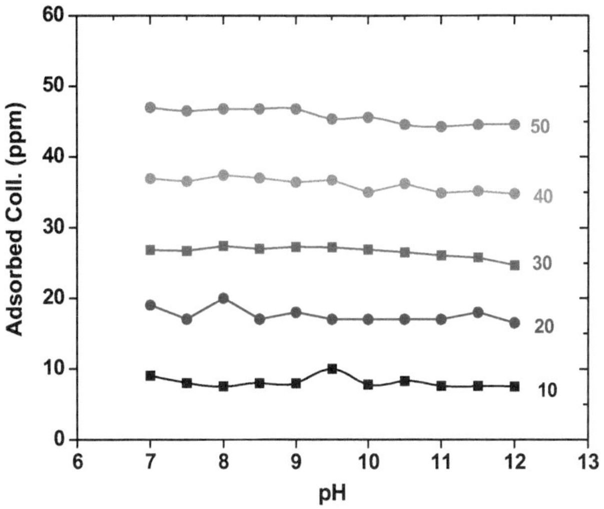

FIGURE 3.2 Adsorption of sodium oleate collector on magnesite at various pH values: effect of different concentrations (different color lines). *Reprinted with permission; copyright Gordon and Breach Matis K.A., G.P. Gallios, A.I. Zouboulis, Adsorption of fatty acids on fine mineral particles as a purification method of industrial wastewaters, Toxicol. Environ. Chem. 20–21 (1989a) 425–435.*

90% when the concentration initially was over 100 mg/L, but only 40%–60% in the more dilute cases [46].

In a study of oleate on calcite, fluorite, and baryte, it was found that the removal of oleate from solution could be attributed to the precipitation of calcium or barium oleate on the surface [47]. It was elsewhere reported that adsorption equilibrium is established quickly in alkaline environment in which the collector appears in the form of ions and the flotation under such conditions depends very little on the conditioning time [48].

The effect of solution pH on the floatability of magnesium carbonates was investigated and compared for two extreme collector concentrations, i.e., 10 and 40 mg/L sodium oleate [49]; in the latter case, both minerals had recoveries over 90% in the pH range examined of 5–12, while with the former a low rate was noticed (of about 20%) at pH 8–9. However, the reagents used for pH adjustment could also influence the minerals flotation. An example of the aforementioned was given during the flotation of calcite, fluorite, and apatite by oleate and use of either sodium hydroxide or carbonate for pH variation [50].

Service water certainly contains significant quantities of calcium and magnesium ions, which may be detrimental for fatty acid flotation, because they produce insoluble salts that decrease the available collector, as said above. This is presented, for instance, in Fig. 3.3, while studying calcium

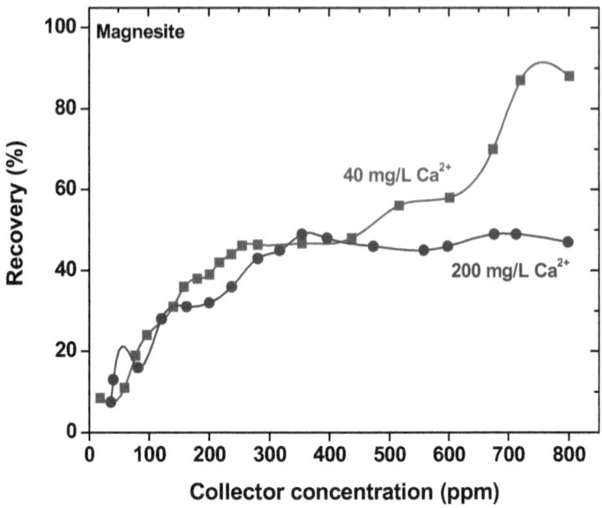

FIGURE 3.3 Use of hard water (red line: 40 mg/L, blue line: 200 mg/L Ca^{2+}) in magnesite flotation, at pH 10: influence of collector concentration. After crushing and grinding, the −180 to +125 μm particle size fraction was taken by wet sieving. *Reprinted with permission; copyright Elsevier Matis K.A., G.P. Gallios, Anionic flotation of magnesium carbonates by modifiers, Int. J. Miner. Process. 25 (1989) 261–274.*

sulfate addition. This showed that the precipitation of oleate by calcium ions does not mean that it will be ineffective as collector; however, it may well mean that its efficiency will be lowered, noting that the chemical reaction of oleate with calcium is quantitative, i.e., for 40 mg/L Ca^{2+} approximately 600 mg/L sodium oleate are required, noting that the use of 40 mg/L Ca^{2+} corresponds to the French hardness grade 10. The effect of Ca^{2+} on the separation of magnesite and dolomite was examined, and sodium hexametaphosphate was used as regulator for complexation [51]. The fundamentals of mineral surface complexation and their meaning in flotation were discussed [52].

In the same meeting, the mechanisms of adsorption–abstraction of surfactants on heterogeneous surfaces were investigated in depth [53]. It was also reported that it is difficult to achieve effective separation of magnesite from dolomite in a single-stage flotation, which may be enhanced by controlling the solution chemistry; as due to their same crystal structure, similar chemical composition, sparingly soluble nature showed similar flotation behavior [54]. A study was conducted aiming to understand why the fines respond well to flotation with oleate, but do not have the same performance with usual cationic collectors [55].

Knowledge of the wetting characteristics of the minerals helps to understand the efficiency of separation valuable mineral from gangues [56]. Fig. 3.4 illustrates magnesite contact angle measurements and flotation recovery versus

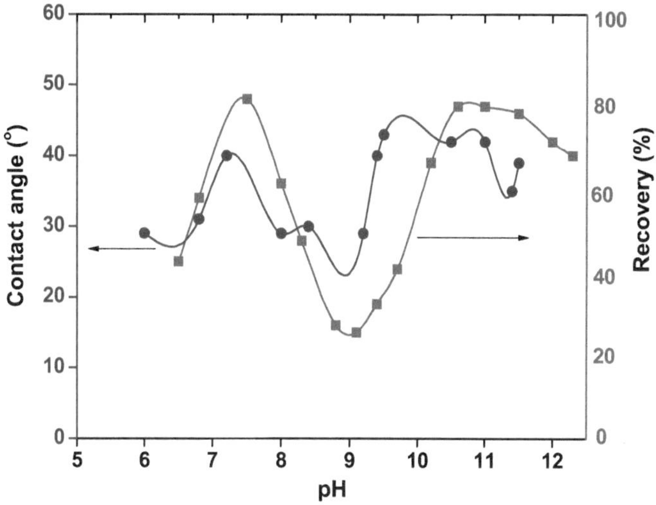

FIGURE 3.4 Correlation of contact angle and flotation recovery of magnesite against the solution pH, at sodium oleate concentration of 30 mg/L. *Reprinted with permission; copyright IMM Matis K.A., G.P. Gallios, G.A. Stalidis, C.T. Hollick, Flotation of magnesite and dolomite by fatty acids, Trans. IMM C 98, 99–105 (1989b).*

solution pH. The bubbles used in these experiments were characterized as being of medium size (diameter 1 mm). It was observed that even with no collector in solution, magnesite exhibited a small contact angle, of approximately 10 degrees [49]. With 30 mg/L sodium oleate, the curve obtained had two maxima, one about 40 degrees at pH 7–7.5 and the other of 45 degrees at a pH of about 10. Nevertheless, at higher collector concentrations, of 100 mg/L, the obtained curve was entirely different, probably reflecting the formation of a polymolecular adsorbate structure.

Contact angle was a concept introduced by the Young equation, at the point of conjunction of the three phases [16]. Contact angle measurements often serve in flotation as a measure of the hydrophobicity of mineral surfaces, although this was disputed by some researchers [57]. Both theoretical and experimental studies in the literature showed that the bubble–particle attachment efficiencies in potential and Stokes flow conditions increase with decreasing particle and bubble size and increasing particle contact angle and electrolyte concentration [58].

Many small-scale tests (around 2 g of solids sample) are available in flotation testwork, and among them, in mineral processing the Hallimond tube (with a G4 porous frit) is widely used [59]. It provides a fast, convenient, and low-cost method based on small samples usually of pure minerals and often with artificial mixtures to determine during testing (usually with distilled water) the general conditions under which minerals may be rendered floatable.

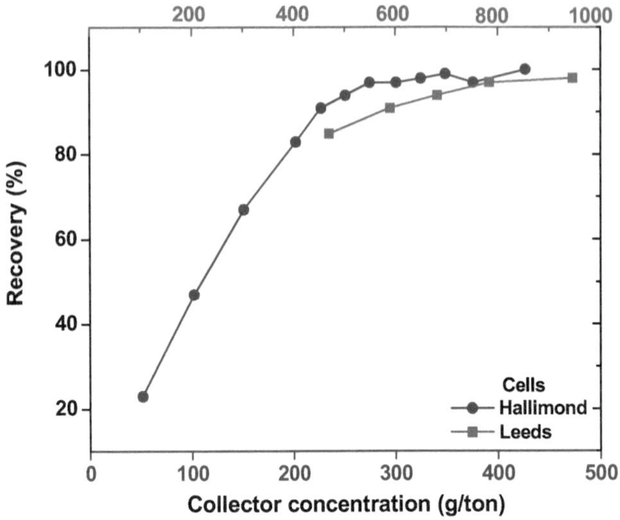

FIGURE 3.5 Influence of (commercial) fatty acid collector concentration on calcite flotation in two bench-scale cells of different size: top the laboratory, bottom the microcell-modified Hallimond tube. *Reprinted with permission; copyright Kluwer Gallios G.P., K.A. Matis, Flotation of salt-type minerals, in: P. Mavros and K.A. Matis (Eds.), Innovations in Flotation Technology, Kluwer Academic, Dordrecht, 1992, pp. 357–382.*

The use of the Hallimond microcell had been investigated in anionic pure calcite for floatability tests in the laboratory and compared with results obtained with a Leeds flotation cell; the −300 μm particle size was used in the larger cell, while in this case a tall oil fatty acid was the collector (Acintol FA-1 by Arizona Chemical Co.) (Fig. 3.5). Although different collector concentrations were of course needed, in order to obtain the same recovery in the two cells, it is observed that the results are quite similar. Hence, indicating that the Hallimond tube can be effectively used to study the flotation behavior of minerals and reach preliminary conclusions, before scaling up the work.

3.2 Selective separation by flotation

Modifying agents are added in flotation circuits of salt-type minerals to achieve selective separation; sodium hexametaphosphate belongs to the group of cyclic polyphosphates and has a very high negative charge (more than the pyrophosphate) [60]. It was found that the former reagent depressed dolomite anionic flotation (by sodium oleate) in the alkaline region of pH [43]. Further, to what was discussed above on the influence of sodium hexametaphosphate on zeta-potential of magnesite, it is also possible that its action is related to the solution rather (than the solids) because it is known that phosphate and polyphosphate ions can bond calcium from solution, giving complexes or insoluble salts and even redissolve insoluble salts of calcium. These thoughts can be described by the following reactions:

$$Ca^{2+} + phospate\ ion \rightleftharpoons calcium\ phospate\ complex$$
$$Ca^{2+} + precipitate\ ion \rightleftharpoons insoluble\ salt\ (precipitate)$$

It was reported that phosphates did not adsorb on salt-type minerals while the calcium concentration in solution increased [61].

Sodium fluorosilicate was also applied as modifier, acting as a depressant, although its flotation behavior is basically different from that of sodium silicate; the presence of fluorides catalyzes its polymerization to colloidal silicate. A comparison for the two magnesium carbonates is shown in Fig. 3.6. Fluorosilicate seemed to present the combined action of fluorides and silicates [62]. This modifier in aqueous solution is hydrolyzed according to the following reaction:

$$NaSiF_6 + 4H_2O \rightleftharpoons Na + +Si(OH)_4 + 4H^+ + 6F^-$$

while the so formed orthosilicate is then further hydrolyzed, as mentioned above.

The abovementioned modifiers are more or less common; another reagent, Calcon used in analytical chemistry as an indicator (in the volumetric determination of calcium by EDTA) was also tested [62]. Although this did not affect, as expected, the flotation of magnesite (which has no calcium in the

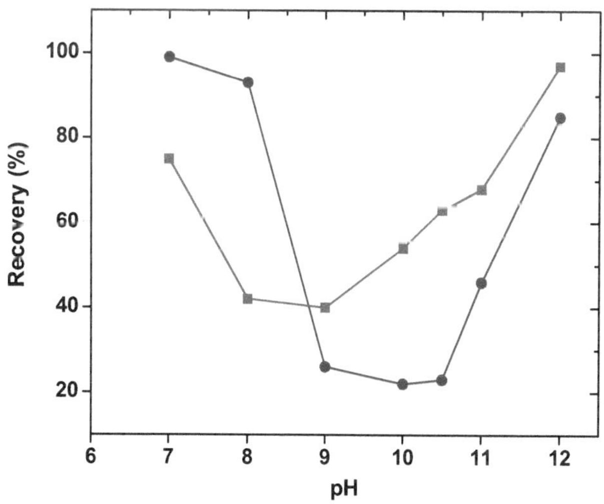

FIGURE 3.6 Influence of sodium fluorosilicate (Na_2SiF_6, concentration 10 mg/L) on flotation of carbonates by 40 mg/L oleate. *Reprinted with permission; copyright Kluwer Gallios G.P., K.A. Matis, Flotation of salt-type minerals, in: P. Mavros and K.A. Matis (Eds.), Innovations in Flotation Technology, Kluwer Academic, Dordrecht, 1992, pp. 357–382.*

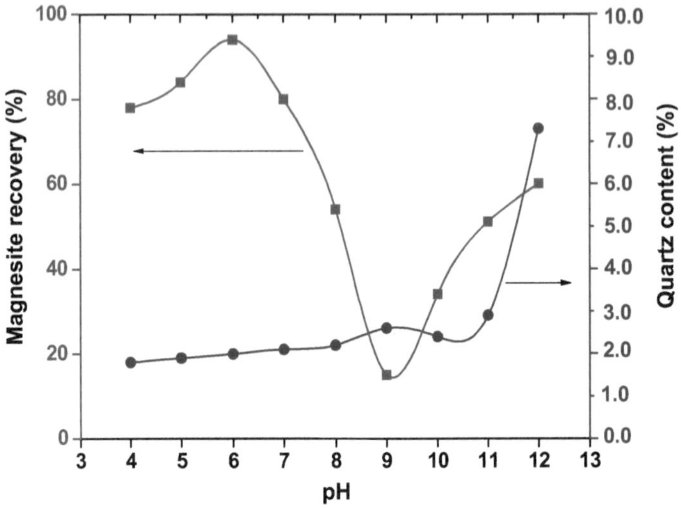

FIGURE 3.7 Flotation of a binary mixture of magnesite/quartz by sodium oleate; the particles size range was −250 + 100 μm. *Reprinted with permission; copyright Kluwer Gallios G.P., K.A. Matis, Flotation of salt-type minerals, in: P. Mavros and K.A. Matis (Eds.), Innovations in Flotation Technology, Kluwer Academic, Dordrecht, 1992, pp. 357–382.*

structure), it strongly depressed dolomite flotation. For instance, at a pH of about 10 the recovery dropped from 95% to 25% approximately.

The selective separation of salt-type minerals both from silicates and from other salt-type minerals is of interest; the former is usually much easier [63]. Fig. 3.7 presents some obtained results for anionic flotation of a magnesite/quartz artificial mixture; in this case, the laboratory work was conducted at a 250 g Denver flotation cell. High recoveries over 90% were noticed in acidic pH of 4–6, as well as a quite good grade of the concentrate—of less than 2% quartz content.

Amines were also applied in magnesite separation problems with certain silicates, as, for instance, with sepiolite, a hydrophobic magnesium silicate [64]. A process to reduce the content of gangue minerals was proposed, which at first used laurylamine (dodecylamine) as the collector to remove most talc and quartz (and part of the dolomite and serpentine), and then used sodium oleate to separate magnesite from dolomite and serpentine [65]. Molecular dynamics simulation and the results of zeta-potential and contact angle measurements indicated that adsorption of dodecylamine on magnesite surface could be strengthened by monohydric alcohols [66]. Reverse flotation combined with silicate-bacteria pretreatment was also utilized to achieve desiliconization of magnesite ore [67].

Fine mineral particles have interactive effects with coarse particles of other minerals; the extended DLVO theory was applied to calculate the interaction energy between the particles [68]. Further, in a comparative work of low-grade dolomite fines with the aim to remove the siliceous gangue content, direct flotation (by oleic acid) showed better control over the quality of the product, while reverse flotation (by dodecylamine) had an edge on the product yield [69]. The upgrading of magnesite deposits is often achieved through cationic silica flotation; however, when the ore contains substantial amounts of dolomite, the bulk of it remains with magnesite. It is noted that the ratio CaO/SiO_2 affects significantly the quality of the product; the best refractory properties are obtained with a ratio of 2:1.

The biomodification of mineral surfaces involves the complex action of microorganism on the mineral surface [70]. So, another possibility is perhaps the application of bioflotation [71]; *Rhodococcus opacus* is a unicellular gram-positive bacterium with different types of compounds on its surface (polysaccharides, carboxylic acids, lipid groups, and micolic acids in the cell wall). The ability of microorganisms to remove metal ions from solution is a well-known phenomenon; applications of biosorption make use of dead biomass, which does not require nutrients and can be exposed to environments of high toxicity [72].

Sodium silicate also is one of the common modifiers, acting generally as a depression agent, e.g., of silicates in scheelite flotation [73]. Its hydrolysis reactions may be found in following:

$$SiO_2 + 2H_2O \rightleftharpoons Si(OH)_4$$
$$Si(OH)_4 + OH^- \rightleftharpoons SiO(OH)_3^- + H_2O$$
$$SiO(OH)_3^- + OH^- \rightleftharpoons SiO_2(OH)_2^{2-} + H_2O$$
$$Si(OH)_4 + 2OH^- \rightleftharpoons SiO_2(OH)_2^{2-} + 2H_2O$$
$$4Si(OH)_4 + 2OH^- \rightleftharpoons Si_4O_6(OH)_6^{2-} + 6H_2O$$
$$4Si(OH)_4 + 4OH^- \rightleftharpoons Si_4O_8(OH)_4^{4-} + 8H_2O$$
$$2Si(OH)_4 + 2OH^- \rightleftharpoons Si_2O_3(OH)_4^{2-} + 3H_2O$$

In large concentrations sodium silicate is polymerized. In the case of dolomite, sodium silicate was a strong depressant for pH values under approximately 9. As found from zeta-potential measurements, sodium silicate adsorbs on calcite and dolomite [74]. However, it is possible that a salt-like calcium silicate may be formed, in salt-type minerals, either on the particle surface by chemisorption or in solution and adsorbed onto the mineral surface [75].

Depending on the existing conditions (i.e., pH, concentration) in the solution different species dominate. The action of this inorganic modifier is possible to be due to one or more chemical species and also be different from a mineral to another one [40]. Hence, the application of sodium silicate on salt-type minerals for their (selective) flotation was widely investigated. Typical chemical reactions explaining the process mechanism and its influence are:

$$Ca^{2+}{}_{surf} + SiO(OH)_3{}^- \rightleftharpoons CaSiO(OH)_3{}^+{}_{surf}$$

$$CaOH^+{}_{surf} + Si(OH)_4 \rightleftharpoons CaSiO(OH)_3{}^+{}_{surf} + H_2O$$

which describe the formation of calcium silicate on minerals particle surface.

Fig. 3.8 shows the application of sodium silicate in dolomite flotation by oleate. The modifier was indeed a strong depressant at pH less than 9. This could be attributed to silicic acid, as mentioned, which dominates in aqueous solutions in the pH range 4–10. Sodium silicate appeared to have a small activating action on magnesite.

Results for binary mixtures of magnesite/dolomite are presented in Fig. 3.9. An interaction of the minerals was observed when the flotation results were compared with values calculated theoretically from experiments with a single mineral. Thus, the concentrate was beneficiated in magnesite, although this was not expected. With higher concentrations of the modifier, the concentrate was enriched in magnesite for initial contents of magnesite in the mixture of 50%–80%. A similar interaction was also reported for the calcite–apatite system [76].

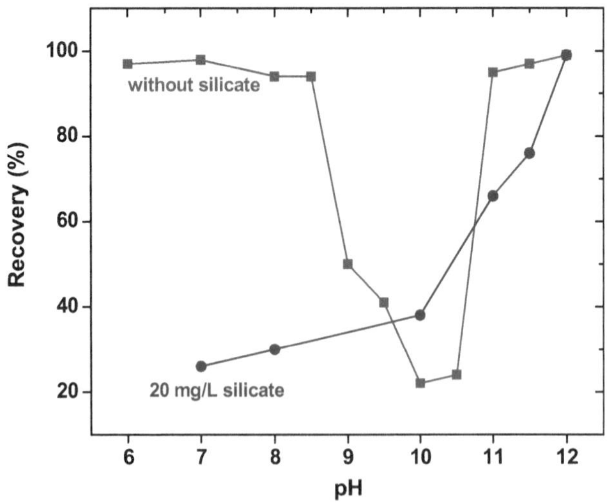

FIGURE 3.8 Use of sodium silicate ($Na_2O.xSiO_2$, with x = 3–5) modifier in the dolomite flotation, by 20 mg/L collector: red line, without silicate; blue line, with 120 mg/L silicate. *Reprinted with permission; copyright Elsevier K.A. Matis, G.P. Gallios, Anionic flotation of magnesium carbonates by modifiers, Int. J. Miner. Process. 25 (1989) 261–274.*

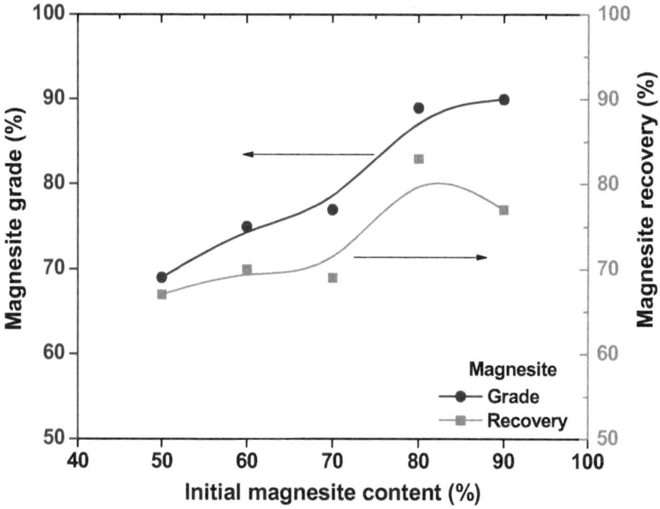

FIGURE 3.9 Effect of the initial content in magnesite of a binary mixture with dolomite: lon the magnesite grade of the concentrate (left y-axis) and magnesite recovery (right y-axis: with 20 mg/L sodium oleate, 240 mg/L sodium silicate, pH 11). *Reprinted with permission; copyright Elsevier K.A. Matis, G.P. Gallios, Anionic flotation of magnesium carbonates by modifiers, Int. J. Miner. Process. 25 (1989) 261–274.*

The interaction in the magnesite–dolomite system may be due to the higher concentration of calcium and magnesium ions in solution and the possible partial consumption of oleate by these ions [62]. When more reagents were added in solution, as for example silicate, then the interactions of their ions and the hydrolysis products with Ca^{2+} and Mg^{2+} should be taken into account. It is finally possible that the two salt-type minerals act antagonistically toward the collector.

Further results from single-mineral flotation with sodium carboxymethyl cellulose are given in Fig. 3.10. The modifier (as also shown in Fig. 3.1) acted as a strong depressant for the basic pHs; with the same conditions, only a slight activation was observed for magnesite. The mentioned zeta-potential measurements indicated that carboxymethyl cellulose increases the negative charge of dolomite in the basic region of pH. The adsorption study showed a significant decrease of oleate adsorption on dolomite. So, it probably acts antagonistically to oleate for calcium sites on dolomitic mineral surface. This organic modifier is known to be ionized completely and possesses a high negative charge [60]. In the acidic region, where carboxymethyl cellulose is not charged, its adsorption is rather small and does not influence the carbonates flotation.

Elsewhere, more explanation was presented [43]: at pH 7–11, magnesite and dolomite had the same charge and similar crystal structure; if carboxymethyl cellulose adsorption was due only to electrostatic forces, it should

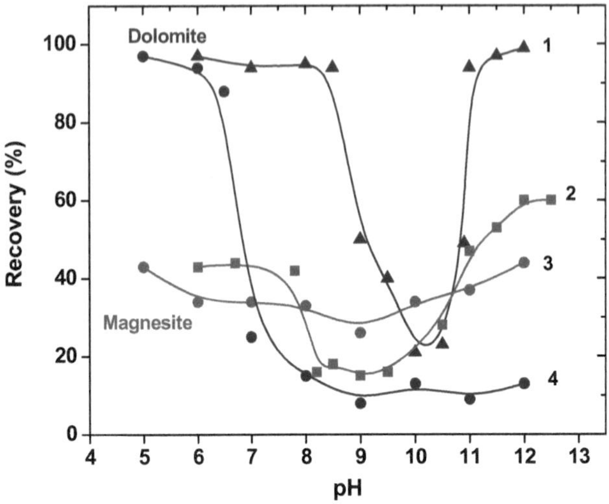

FIGURE 3.10 Influence of carboxymethyl cellulose on carbonates flotation, by 20 mg/L sodium oleate: 1 and 2, without modifier; and 3 and 4, with 100 mg/L cellulose. *Reprinted after modification, with permission; copyright Dekker Gallios G.P., K.A. Matis, Floatability of magnesium carbonates by sodium oleate in presence of modifiers, Separ. Sci. Technol. 24 129–143 (1989).*

adsorb on both minerals. However, its nonadsorption on magnesite is an indication of chemisorption in relation to the existence of calcium ions on the particles surface. Adsorption of the modifier on a dolomite surface, in preference to sites where calcium was present, reduced the available sites for collector adsorption, with result that there was in this way a reduction of flotation recovery in the alkaline pH region.

Fig. 3.11 shows some work on simple binary mixtures without any modifier added other than sodium hydroxide, for pH adjustment. A quite promising separation of magnesite from dolomite may be achieved by reverse flotation at pH 11, using only small quantities of sodium oleate as collector, with magnesite remaining in the underflow (i.e., the nonfloated fraction). The experimental determinations indicated that in the absence of collector there was a significant surface potential gap between the dolomite and the magnesite, the dolomite being much closer to electrical neutrality. Translated to a mineral mixture, the presence of such a gap, promoting decrease in magnesite surface area, had reached the point at which calcium activation of the magnesite began to affect the result [49].

Pyrophosphate addition as a modifier was also studied (see, for instance, Fig. 3.12). The depressive action was more noticeable with dolomite than with magnesite at pH greater than 9. Pyrophosphate, in contrast to hexametaphosphate, belongs to the group of polyphosphates having a linear chain [60]. It generally acts as a strong dispersing agent in pulps of fine mineral particles.

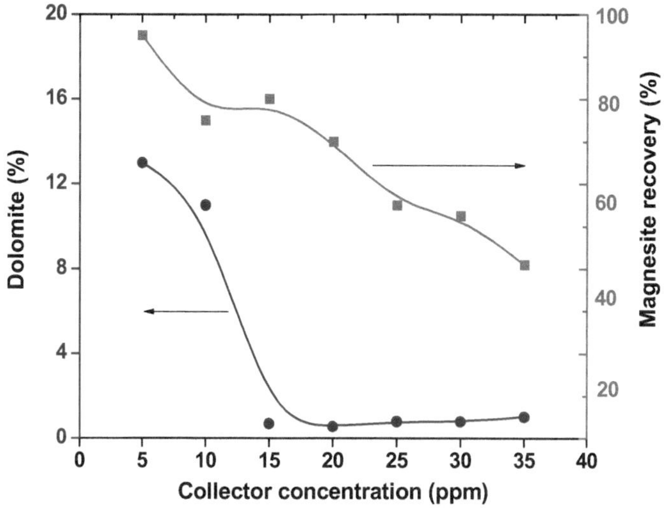

FIGURE 3.11 Influence of collector concentration on the dolomite grade and recovery in the tails of magnesite, with binary mixture of initially 20 wt% dolomite at pH 11. *Reprinted with permission; copyright IMM Matis K.A., G.P. Gallios, G.A. Stalidis, C.T. Hollick, Flotation of magnesite and dolomite by fatty acids, Trans. IMM C 98 (1989b) 99–105.*

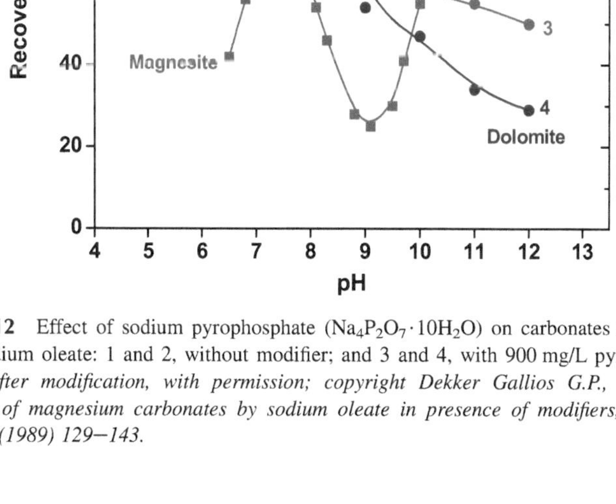

FIGURE 3.12 Effect of sodium pyrophosphate ($Na_4P_2O_7 \cdot 10H_2O$) on carbonates flotation, by 30 mg/L sodium oleate: 1 and 2, without modifier; and 3 and 4, with 900 mg/L pyrophosphate. *Reprinted after modification, with permission; copyright Dekker Gallios G.P., K.A. Matis, Floatability of magnesium carbonates by sodium oleate in presence of modifiers, Separ. Sci. Technol. 24 (1989) 129–143.*

It is also a polyelectrolyte with a high negative charge and forms characteristically chelates with metal cations.

From zeta-potential measurements [43] was observed that phosphates significantly increased the negative charge of dolomite, as was also found for scheelite [61]. The solubility of these complexes is high, and calcium is rapidly dissolved from dolomite surface, resulting in a negative surface charge. The solubility rate of calcium is a function of the crystal structure of the mineral. In relatively high concentrations, it was found that precipitates between the phosphate and calcium ions were formed in solution and then adsorbed on the mineral surface [77]. Either of both of the aforementioned two mechanisms could happen in this case.

4. Concluding remarks

The two minerals mainly investigated in the present, magnesite and dolomite, although of the same type (i.e., salt-type minerals) did not behave similarly as far as flotation is concerned. Dolomite, being here mineralogically crystalline, was more amenable to processing, giving high flotation recoveries at the size range used. The presence of modifying reagents could significantly depress its floatability, as expected.

The existence of very fine particles presented difficulties in flotation. The former can be created even during the conditioning (or mixing), particularly in the case of cryptocrystalline minerals, such as magnesite in the present, being friable. The salt-type minerals solubility constitutes a serious problem for flotation, if selectivity is required. It was also said that fine particles float poorly and less selectively under normal (reverse) flotation conditions, having detrimental effects on recovery of other minerals, to leave magnesite in suspension [78].

It was observed generally that the flotation response of pure magnesite with 30 mg/L oleate collector was rather similar to that of dolomite with 20 mg/L collector; peak magnesite recoveries being at pH 7.5 and 11 [49]. Readsorption of magnesium dioleate molecules by hydrophobic aggregation is likely to be important in the enhanced recovery at near neutral pH, but the lower solubility of magnesite caused both the intermediate trough (at pH 9) and subsequent recovery in flotation performance to occur at a lower pH than with dolomite.

A review of the available evidence suggests that the responses of the two carbonate minerals investigated at the various separating conditions were dominated by a rapid decrease in the solubility of the $CaCO_3$ component of the dolomite with rising pH and a much stronger affinity between oleate ions than between oleate ions and magnesium ions. The ultrafine particles (mostly of magnesite) detached during conditioning and then affected flotation in different ways.

In conclusion, the knowledge of chemistry and its respective use may help not only in minerals flotation research but also in the relevant application of this process. Flotation, with its various techniques, is a flexible and effective separation process, as indicated in the present review with the application of binary mineral mixtures and the investigation of modifier's influence.

References

[1] C. Micheau, A. Schneider, L. Girard, P. Bauduin, Evaluation of ion separation coefficients by foam flotation using a carboxylate surfactant, Colloids Surf. A 470 (2015) 52–59.

[2] F.S. Hoseinian, M. Irannajad, A.J. Nooshabadi, Ion flotation for removal of Ni(II) and Zn(II) ions from wastewaters, Int. J. Miner. Process. 143 (2015) 131–137.

[3] M. Santander, L. Valderrama, M. Guevara, J. Rubio, Adsorbing colloid flotation removing metals ions in a modified jet cell, Miner. Eng. 24 (2011) 1010–1015.

[4] B. Lakghomi, Y. Lawryshyn, R. Hofmann, A model of particle removal in a dissolved air flotation tank: importance of stratified flow and bubble size, Water Res. 63 (2015) 262–272.

[5] G.Z. Kyzas, K.A. Matis, Electroflotation process: a review, J. Mol. Liq. 220 (2016a) 657–664.

[6] N.N. Rulyov, Combined microflotation of fine minerals: theory and experiment, Miner. Process. Extr. Metall. 125 (2016) 81–85.

[7] J. Choi, E. Lee, S.Q. Choi, S. Lee, Y. Han, H. Kim, Arsenic removal from contaminated soils for recycling via oil agglomerate flotation, Chem. Eng. J. 285 (2016) 207–217.

[8] S.Y. Lee, I. Khoiroh, T.C. Ling, P.L. Show, Aqueous two-phase flotation for the recovery of biomolecules, Separ. Purif. Rev. 45 (2016) 81–92.

[9] E.N. Peleka, K.A. Matis, Bioremoval of metal ion and water treatment in a hybrid unit, Separ. Sci. Technol. 44 (2009) 3597–3614.

[10] R. Clayton, G.J. Jameson, E.V. Manlapig, The development and application of the Jameson cell, Miner. Eng. 4 (1991) 925–933.

[11] E.N. Peleka, G.P. Gallios, K.A. Matis, A perspective on flotation: a review, J. Chem. Tech. Biotech. 24 (October 2017). https://doi.org/10.1002/jctb.5486.

[12] K.A. Matis, G.P. Gallios, Dissolved-air and electrolytic flotation, in: B.A. Wills, R.W. Barley (Eds.), Mineral Processing at a Crossroads, Martinus Nijhoff, Dordrecht, 1986, pp. 37–70.

[13] E.T. Stamboliadis, Impact of preconcentration on the efficiency of magnesite reverse flotation, Can. Metall. Q. 47 (2008) 1–8.

[14] G.P. Gallios, K.A. Matis, Flotation of salt-type minerals, in: P. Mavros, K.A. Matis (Eds.), Innovations in Flotation Technology, Kluwer Academic, Dordrecht, 1992, pp. 357–382.

[15] T. Gambopoulos, A. Nestoridis, Separation of Magnesite From its Contaminants by Reverse Flotation, 1976. US Pat. 3976251.

[16] G.Z. Kyzas, E.A. Deliyanni, K.A. Matis, Research activities related to flotation process, Trends in Green Chemistry 1 (2015) 1–4.

[17] K.A. Matis, A.I. Zouboulis, Flotation techniques in water technology for metals recovery: the impact of speciation, Separ. Sci. Technol. 36 (2001) 3777–3800.

[18] G.Z. Kyzas, N.K. Lazaridis, K.A. Matis, Flotation: Recent Innovations in an Interesting and Effective Separation Process, 2018 chapter in present issue.

[19] D. Davies, Minerals: Hi-Tech and sustainable, in: AMIRA Presentation, February 2002.

[20] D. Humphreys, New horizons for an old industry, in: MassMin2000 Conf. Austral, IMM, Brisbane, October 2000.

[21] K.A. Matis, Mining and metallurgy: considerations of the past activity, present state and future, in: Perspectives in 34th Intl. October Conf. Min. Metall., Techn. Fac. Bor, 30 September–3 October 2002.

[22] P. Somasundaran, An overview of the ultrafine problem, in: B.A. Wills, R.W. Barley (Eds.), Mineral Processing at a Crossroads, Martinus Nijhoff, Dordrecht, 1986, pp. 1–36.

[23] S.J. Cowell, W. Wehrmeyer, P.W. Argust, J.G.S. Robertson, Sustainability and the primary extraction industries: theories and practice, Resour. Pol. 25 (1999) 277–286.

[24] G. Hilson, B. Murck, Sustainable development in the mining industry: clarifying the corporate perspective, Resour. Pol. 26 (2000) 227–238.

[25] T. Leistner, M. Embrechts, T. Leißner, S.C. Chelgani, I. Osbahr, R. Möckel, U.A. Peuker, M. Rudolph, A study of the reprocessing of fine and ultrafine cassiterite from gravity tailing residues by using various flotation techniques, Miner. Eng. 96–97 (2016) 94–98.

[26] G.Z. Kyzas, K.A. Matis, Methods of arsenic wastes recycling: focus on flotation, J. Mol. Liq. 214 (2016b) 37–45.

[27] R. Espinosa-Gomez, J.A. Finch, N.W. Johnson, Column flotation of very fine particles, Miner. Eng. 1 (1988) 3–18.

[28] D.W. Fuerstenau, S. Chander, A. Abouzeid, Recovery of fine particles by physical separation methods, in: P. Somasundaran, N. Arbiter (Eds.), Beneficiation of Mineral Fines – Problems and Research Needs, AIME, Ann Arbor, 1979, p. 3.

[29] D.W. Fuerstenau, Fine particle flotation, in: P. Somasundaran (Ed.), Fine Particles Processing, SME/AIME, New York, 1980, p. 669.

[30] K.A. Matis, G.P. Gallios, K.A. Kydros, Separation of fines by flotation techniques, Sep. Technol. 3 (1993) 76–90.

[31] D. Fornasiero, L.O. Filippov, Innovations in the flotation of fine and coarse particles, IOP Conf. Ser. J. Phys. 879 (2017) article ID 012002.

[32] K.A. Matis, T.N. Balabanidis, G.P. Gallios, Processing of magnesium carbonate fines by dissolved-air flotation, Colloids Surface. 29 (1988) 191–203.

[33] V.A. Glembotsky, A.A. Mamakov, A.M. Romanov, V.E. Nenov, Selective separation of fine mineral slimes using the method of electric flotation, in: Proc. 11th Int. Miner. Process. Cong., Univ. Cagliari, 1975, pp. 562–581.

[34] T.W. Healy, W.J. Trahar, Challenges in sulphide flotation fundamentals, in: K.V.S. Sastry, M.C. Fuerstenau (Eds.), Challenges in Mineral Flotation, SME/AIME, Littleton, 1989, pp. 3–13.

[35] H.S. Hanna, P. Somasundaran, Flotation of salt-type minerals, in: M.C. Fuerstenau (Ed.), Flotation – A.M. Gaudin Memorial Volume, SME/AIME, New York, 1976, pp. 197–271.

[36] A.S. Peck, M.E. Wadsworth, Infrared Studies of Oleic Acid and Sodium Oleate Adsorption on Fluorite, Barite and Calcite, US Bureau of Mines, 1963, p. RI 6202.

[37] C. Du Rietz, Chemisorption of collectors in flotation, in: Proc. 11th Int. Miner. Process. Cong., Univ. Cagliari, 1975 paper 13.

[38] W. Liu, W. Liu, S. Dai, T. Yang, Z. Li, P. Fang, Enhancing the purity of magnesite ore powder using an ethanolamine-based collector: insights from experiment and theory, J. Mol. Liq. 268 (2018) 215–222.

[39] D.R. Nagaraj, R.S. Farinato, Evolution of flotation chemistry and chemicals: a century of innovations and the lingering challenges, Miner. Eng. 96/97 (2016) 2–14.

[40] K.A. Matis, G.A. Stalidis, G.P. Gallios, Beneficiation of fine mineral particles by electrolytic flotation, Tech. Chron. 5 (4) (1985) C 35–56 (in Greek with English extended abstract).

[41] J.D. Miller, M.E. Wadsworth, M. Misra, J.S. Hu, Flotation chemistry of the fluorite/oleate system, in: M.H. Jones, J.T. Woodcock (Eds.), Principles of Mineral Flotation – the Wark Symposium, Australasian IMM, Parkville, 1984, pp. 31–42.

[42] P.R.G. Brandao, Adsorption of Oleate on Magnesite and its Influence on Flotation, Univ. Brit. Columbia, 1982, p. 175. Ph.D. Thesis.

[43] G.P. Gallios, K.A. Matis, Floatability of magnesium carbonates by sodium oleate in presence of modifiers, Separ. Sci. Technol. 24 (1989) 129–143.

[44] J.J. Prédali, Adsorption des sels d' acides gras a l' interface solide liquid et flotation des carbonates, Sci. Terre. Memoire 27 (1973). Univ. Nancy.

[45] P.R.G. Brandao, G.W. Poling, Anionic flotation of magnesite, Can. Metall. Q. 21 (1982) 211–220.

[46] K.A. Matis, G.P. Gallios, A.I. Zouboulis, Adsorption of fatty acids on fine mineral particles as a purification method of industrial wastewaters, Toxicol. Environ. Chem. 20–21 (1989a) 425–435.

[47] K.I. Marinakis, The Action of Sodium Silicate on the Flotation of Salt-type Minerals With Oleic Acid, Imperial College London, 1980, p. 227. Ph.D. Thesis.

[48] J.S. Laskowski, G.A. Nyamekye, Colloid chemistry of weak electrolyte collectors: the effect of conditioning on flotation with fatty acids, Int. J. Miner. Process. 40 (1994) 245–256.

[49] K.A. Matis, G.P. Gallios, G.A. Stalidis, C.T. Hollick, Flotation of magnesite and dolomite by fatty acids, Trans. IMM C 98 (1989b) 99–105.

[50] V.I. Klassen, V.A. Mokrousov, An Introduction to the Theory of Flotation, Butterworths, London, 1963, p. 303.

[51] N. Luo, D. Wei, Y. Shen, C. Han, C. Zhang, Elimination of the adverse effect of calcium ion on the flotation separation of magnesite from dolomite, Minerals 7 (2017) 150. https://doi.org/10.3390/min7080150.

[52] Z.-X. Sun, W. Forsling, Mineral surface complexation in flotation, in: P. Mavros, K.A. Matis (Eds.), Innovations in Flotation Technology, Kluwer Academic, Dordrecht, 1992, pp. 263–282.

[53] J.M. Cases, F. Villieras, The mechanisms of collector adsorption-abstraction (ionic and non-ionic surfactants) on heterogeneous surfaces, in: P. Mavros, K.A. Matis (Eds.), Innovations in Flotation Technology, Kluwer Academic, Dordrecht, 1992, pp. 25–56.

[54] G. Chen, D. Tao, Effect of solution chemistry on floatability of magnesite and dolomite, Int. J. Miner. Process. 74 (2004) 343–357.

[55] A.R.S. de Medeiros, C.A.M. Baltar, Importance of collector chain length in flotation of fine particles, Miner. Eng. 122 (2018) 179–184.

[56] N. Gence, Wetting behaviour of magnesite and dolomite surfaces, Appl. Surf. Sci. 252 (2006) 3744–3750.

[57] I. Wark, Contact angles and flotation research, in: M.H. Jones, J.T. Woodcock (Eds.), Principles of Mineral Flotation – the Wark Symposium, Australasian IMM, Parkville, 1984, pp. 1–6.

[58] T. Miettinen, J. Ralston, D. Fornasiero, The limits of fine particle flotation, Miner. Eng. 23 (2010) 420–437.

[59] D.W. Fuerstenau, P.H. Metzger, G.D. Seele, How to use this modified Hallimond tube …for better flotation testing, Eng. Min. J. March 93 (1957).

[60] J. Leja, Surface Chemistry of Froth Flotation, Plenum Press, New York, 1982 pages 618, 326 and 758.

[61] L. Changeen, L. Yongxin, Selective flotation of scheelite from calcium minerals with sodium oleate as a collector and phosphates as modifiers. II. The mechanism of the interaction between phosphate modifiers and minerals, Int. J. Miner. Process. 10 (1983) 219–235.

[62] K.A. Matis, G.P. Gallios, Anionic flotation of magnesium carbonates by modifiers, Int. J. Miner. Process. 25 (1989) 261–274.

[63] G.P. Gallios, A Study of the Behaviour of Carbonate Minerals of Magnesium during Their Flotation, Dept. Chem., Aristotle Univ. Thessaloniki, 1987, p. 245. Ph.D. Thesis.

[64] F.O. Emrullahoglou, Concentration by Flotation of Magnesite Ore From Cayirbaf (Konya) District in the Middle Anatolia of Turkey, Anadolu Univ. Eskisehir, 1984. Ph.D. Thesis.

[65] J. Yao, Y. Hou, Y-l. Wang, W-x. Zhong, W-z. Yin, Research on the floatability of magnesite and its gangue minerals with sodium oleate and lauryl amine as collectors, Adv. Mater. Res. 798–799 (2013) 328–332.

[66] H. Zhang, W. Liu, C. Han, D. Wei, Intensify dodecylamine adsorption on magnesite and dolomite surfaces by monohydric alcohols, Appl. Surf. Sci. 444 (2018) 729–738.

[67] Q. Teng, Y. Feng, H. Li, Effects of silicate-bacteria pretreatment on desiliconization of magnesite by reverse flotation, Colloids Surf. A 544 (2018) 60–67.

[68] J. Yao, J. Xue, D. Li, Y. Fu, E. Gong, W. Yin, Effects of fine-coarse particles interaction on flotation separation and interaction energy calculation, Part. Sci. Technol. 36 (2018) 11–19.

[69] S.K. Tripathy, S.I. Angadi, N.K. Patra, D.S. Rao, Comparative separation analysis of direct and reverse flotation of dolomite fines, Miner. Process. Extr. Metall. Rev. (2018). https://doi.org/10.1080/08827508.2018.1435534.

[70] K.H. Rao, A. Vilinka, I.V. Chernyshova, Minerals bioprocessing: R&D needs in mineral beneficiation, Hydrometallurgy 104 (2010) 465–470.

[71] A.E.C. Botero, M.L. Torem, L.M.S. de Mesquita, Surface chemistry of biosorption of *Rhodococcus opacus* and its effect in calcite and magnesite flotation, Miner. Eng. 21 (2008) 83–92.

[72] M.X. Loukidou, E.N. Peleka, T.D. Karapantsios, K.A. Matis, Biosorption of metal ions, Trends Chem. Eng. 13 (2011) 53–64.

[73] H.L. Shergold, Flotation in mineral processing, in: K.J. Ives (Ed.), The Scientific Basis of Flotation, Proc. NATO ASI, Cambridge, 1982 preprint.

[74] P. Parsonage, D. Melven, A.F. Healey, D. Watson, Depressant function in flotation of calcite, apatite and dolomite, in: Proc. Conf. Reagents in the Mineral Industry, IMM, Rome, 1984, pp. 33–40.

[75] K.I. Marinakis, H.L. Shergold, Influence of sodium silicate addition on the adsorption of oleic acid by fluorite, calcite and barite, Int. J. Miner. Process. 14 (1985) 177–193.

[76] P. Somasundaran, J.O. Amankonah, K.P. Ananthapadmanabhan, Calcite-apatite interactions and their effects in selective flotation using oleate, in: Proc. XV Int. Miner. Process. Cong., Cannes, 1985, pp. 244–254.

[77] A.I. Arol, I. Iwasaki, Control of montmorillonite via complexation and ultrasonics in the selective flocculation of iron ores, in: SME Annual Meeting, Louisiana, 1986.

[78] J. Yao, W. Yin, E. Gong, Depressing effect of fine hydrophilic particles on magnesite reverse flotation, Int. J. Miner. Process. 149 (2016) 84–93.

Chapter 4

Fundamentals and applications of nanobubbles

Elisavet D. Michailidi,[a,b,*] George Bomis,[c] Athanasios Varoutoglou,[c] Eleni K. Efthimiadou,[a,b] Athanasios C. Mitropoulos[c] and Evangelos P. Favvas[a,c,*]

[a]*Institute of Nanoscience and Nanotechnology, NCSR "Demokritos", Terma Patriarchou Grigoriou and Neapoleos, Attica, Greece;* [b]*Inorganic Chemistry Laboratory, Chemistry Department, National and Kapodistrian University of Athens, Zografou, Greece;* [c]*Hephaestus Advanced Laboratory, Department of Chemistry, International Hellenic University, Kavala, Greece*

**Corresponding author*

1. Introduction

At first glance, it is completely incontrovertible that a surface, for example, the internal side of a glass of water, is completely wetted. This is a fact regarding what could be visible by using the human eye for this observation. Zooming into the "wetted" part of the surface with an optical microscope confirms the macroscopic observation: the surface is indeed wet. However, by a further investigation about what in fact happens into the surface−water interface the previous conclusion has been changed. Zoom further, now by using more accurate techniques, which can access to the nanoscale, nanoscopic gaseous domains trapped at the interface are found [1]. These new nanoscopic gaseous domains come in two basic forms: (1) surface nanobubbles (NBs) [2−4] and (2) micropancakes [5−7]. Their main difference is that the observed surface NBs are spherical cap-shaped with dimensions of about 10−50 nm in diameter, whereas the micropancakes are spread like spots of several microns in diameter but restricted to only 1−2 nm in height [1].

An accurate definition for NBs: *NBs are defined as nanoscopic gaseous cavities, which diameter is less than* 1 μm. An equivalent term that can also be found in the literature is "ultrafine bubbles." The international standards organization (ISO) has evaluated standards for submicrometer bubbles (ISO/TC281). As it has already mentioned, NBs can exist both on surfaces (surface

Advanced Low-Cost Separation Techniques in Interface Science. https://doi.org/10.1016/B978-0-12-814178-6.00004-2

or interfacial NBs) [8] and as dispersed phase in a liquid bulk phase (bulk NBs) [9,10]. Nowadays, NBs have drawn the attention of the scientific community due to their extraordinary longevity which ranges up to several weeks and even months. The paradox is that the prolonged stability of such bubbles is not in consonance with classical thermodynamics, as according to the Young−Laplace law, the internal pressure is inversely proportional to the diameter and, thus, submicron bubbles should dissolve immediately. In addition to their longevity, it has been observed that NBs have an effect to the physicochemical properties of the medium in which they are dispersed. The growing significance of nanotechnology and the special properties of NBs have drawn huge attention in many sectors due to their wide range of potential applications.

Some of these fields, where NBs' potential applications are spread, are the surfactant-free cleaning, the mineral processing, the intracellular drug delivery, the enhanced drug susceptibility of cancer cells, etc. [11−18]. The NBs' conception mechanism plays the most important role on the properties of the NBs. That is why emphasis must be given to the design and the construction of the NBs generator. To this end the focus has been turned to the design and production of the NB generators, which are able to produce uniform, in size and shape, and stable NBs. In this section, the NB generator which has been designed, produced, and protected by European patent, by two coauthors of this chapter, is described and its sketches are also presented. In the current chapter, the authors provide a condense overview regarding the history, the production methods, important properties, and potential/candidate uses of NBs.

2. The nanobubbles history

The research concerning the properties and formation of NBs is clearly defined between surface and bulk NBs. References of bulk NBs predated reports of surface NBs.

Obviously, in order to discuss about NBs, it could not be omitted to mention the fundamental principles of bubble kinetics theory. According to Laplace pressure bubble catastrophe (LPBC), even in the case initial equilibrium is achieved, a small perturbation causes the bubble either to grow or shrink and disappear. LPBC theory still has a major impact on NB research as it implies that very small bubbles cannot be thermodynamically stable. This has been dealt with in a landmark paper from 1950 by Epstein and Plesset who stated "a gas bubble in a liquid-gas solution will grow or shrink by diffusion according as the solution is oversaturated or undersaturated" [19].

Sette and Wanderlingh in 1962 demonstrated that high-energy neutrons present in cosmic rays, or introduced artificially, reduced the sound energy required to initiate the cavitation of bulk water [20]. Conceivably, the first direct evidence of the existence of stable bubbles, sized less than 1 μm, within bulk phase was reported by Johnson and Cooke in 1981 [21]. The authors observed that bubbles generated by shear in saline water remain stable for a period of more than 22 h. They claimed that, in presence of naturally existent surfactants, a film is formed on the bubble–water interface, thus explaining the bubbles' stability.

Concerning bulk NBs, only a few studies were conducted for the following decade, when Bunkin et al. [22,23] confirmed that stable microbubbles (MBs) can exist in dilute electrolyte solutions. After 2000, a significant number of publications were accented to the generation and properties of NBs and emphasis was given to explain the mechanisms behind these properties. Kim et al. [24] reported the generation of bulk nanoparticles, attributed to NBs, by sonication. Kikuchi and his team [25–29] experimented upon the generation of NBs through electrolysis while Oeffinger and Wheatley [30] employed surfactant-stabilized NBs as ultrasound (US) contrast agents produced by the sonication of a perfluorocarbon gas. Najafi et al. [31] generated NBs in a closed cuvette by increasing the temperature, thus reducing the solubility of dissolved gases and precipitating NBs. Ohgaki et al. [32] achieved the formation of NBs by the injection of gas (N_2, CH_4, and Ar) in water. They referred that the concentration of NBs was ~1013 bubbles per mL and persisted for up to 2 weeks. Later, Uchida et al. [33] produced NB solutions by injection of ultrapure oxygen. The most unequivocal evidence, that bulk NBs are actually gaseous, was given by Kobayashi et al. [34], using a particle measuring instrument which uses the resonant mass measurement method to detect density relative to the solvent of individual nanoparticles as they passed one by one through a microresonator.

With regard to surface NBs, the first clue of their existence derived from surface force measurements between hydrophobic surfaces immersed in aqueous solutions, by Israelachvili and Pashley in 1982 [35]. During the next decade, scientists were trying to give an explanation for the unexpected attractive forces between hydrophobic surfaces until Parker et al. [3] suggested that the attraction was a result of the presence of NBs which are bridging the hydrophobic surfaces. Ljunggren and Eriksson [36] claimed that the NBs, required to explain the observed attraction, were expected to exist for less than a second. However, experiments proved that the attraction persisted for a long time, up to a few hours, after the immersion.

Surface NBs were depicted, for the first time, at the early 2000s. In Japan, Ishida et al. [37] used tapping mode AFM and proved the correlation between surface NBs and the long-range attractive force. Furthermore, they showed that at hydrophobic surfaces NBs have an anomalous contact angle. At around the same time, Lou et al. [2] produced surface NBs, by the solvent exchange method, on highly ordered pyrolytic graphite. They also used atomic force microscopy (AFM) to image the produced bubbles, and their technique became a standard on surface NB production since then. These groups established the foundations for the future research as they highlighted the

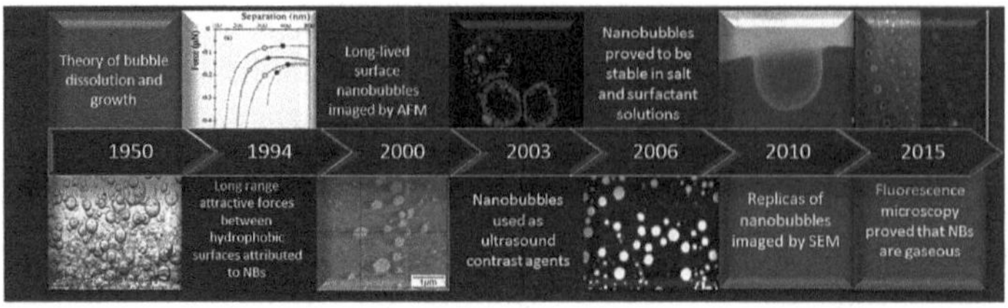

FIGURE 4.1 Timeline of a brief selection of significant publications in the field of surface and bulk nanobubbles (NBs) [20].

stability of surface NBs and their anomalous contact angle. Following that, one of the most significant works, concerning the stability of surface NBs, is from Zhang et al. [7]. They examined a hydrophobic surface by infrared spectroscopy to show the existence of CO_2 NBs on it.

As mentioned, research is clearly districted between surface and bulk NBs. However, a selection of landmark papers is shown on a short timeline, as created by Alheshibri et al. [20] (Fig. 4.1).

To conclude, although many issues, such as the origin of the extreme stability, are still not fully resolved, research upon both bulk and surface NBs exhibits a rapid increase during the last decade (Fig. 4.2). While only six articles were published in 2000, the number raised up to 98 in 2016.

3. Nanobubbles production

3.1 Bulk nanobubbles

The formation of NBs is often achieved when the homogenous liquid phase undergoes a phase change caused by sudden pressure reduction below a critical value, which is known as cavitation [13]. The cavitation is commonly induced by the passage of ultrasonic wave (acoustic cavitation) or by high

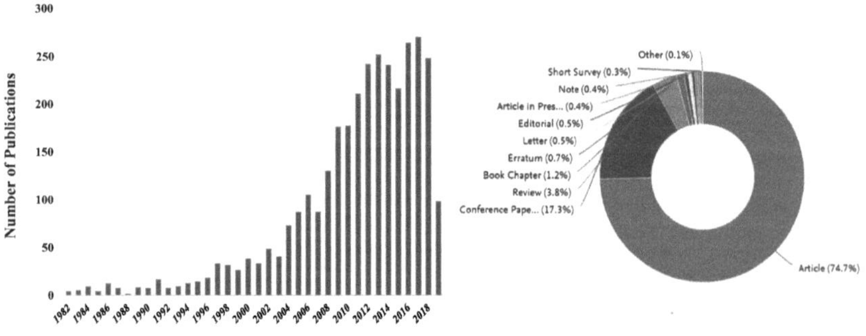

FIGURE 4.2 Documents by year and type with the keywords "nanobubbles," "ultrafine bubbles," and "microbubbles." *Retrieved from Scopus, April, 2019.*

pressure variations in the flowing liquid (hydrodynamic cavitation) [38,39]. Sonication is one of the methods for the generation of bulk NBs. Kim et al. produced NBs by sonication (42 kHz, 70 W) in the presence of an electrode coated by palladium, directly in a cuvette which was placed in a dynamic light scattering (DLS) apparatus [24]. In a later work, using sonication at 20 kHz frequency and a different generator, Cho et al. found that the NBs size increases with US power and production time. In this manner, the NB size was >700 nm [40]. Oeffinger and Wheatley generated NBs with a mean diameter of 450–700 nm via ultrasonication of a mixed surfactant solution with regular purging using octafluoropropane gas [30].

Based on hydrodynamic cavitation mechanism, Venturi tubes are widely used for MB and NB generation. In the Venturi-type generator system, both gas and liquid are passed simultaneously via the Venturi tube to generate the bubble. When pressurized fluid is introduced in the tubular part, the liquid flow velocity in the cylindrical throat becomes higher, whereas pressure becomes lower compared with the inlet section, thus resulting in cavitation. Ahmadi et al. produced NBs with a mean diameter of 130–545 nm via a Venturi tube apparatus [41]. Fan et al. were able to generate NBs of a mean diameter with less than 50 nm using Venturi tube [39]. Moreover, Wu et al. succeeded to produce NBs ($d < 500$ nm) with a generator based on batch high-intensity cavitation [42].

From 2001 to 2009, Kikuchi and his team experimented upon the generation of NBs through electrolysis [25–29]. By DLS, they detected stable nanoparticles, attributed to NBs, both in the cathodic and the anodic solutions. According to this study, particles in the anodic solution are O_2 NBs and particles in the cathodic solution are H_2 NBs. Najafi et al. generated NBs in a closed cuvette by increasing the temperature, thus reducing the solubility of dissolved gases and precipitating NBs [31]. Ohgaki et al. [32] achieved the formation of NBs by the injection of gas (N_2, CH_4, and Ar) in water. They referred that the concentration of NBs was ~1013 bubbles per mL and persisted for up to 2 weeks. Later, Uchida et al. produced NB solutions by injection of ultrapure oxygen [33].

For the bulk NB investigation, numerous techniques are available. The most accurate is the DLS, a certified technique for the determination of the well-dispersed and well-defined phase confirmation into liquid phases. The DLS method is based on photon correlation spectroscopy [43]. The intensity I of light, scattered at a specific angle θ in a liquid sample, is measured. Since the particles obey the diffusion law, the frequencies of the interfering waves are shifted due to the Doppler effect. In addition, confocal microscopy can provide a high spatial resolution bottom view of dyed bulk NBs. In Fig. 4.3, two images from confocal microscopy are presented. The images are referred to two dyed, one with the red rhodamine B and one with FITC, pigment samples of deionized water samples enhanced with air NBs. The average size of the NBs of these two samples was of about 600 nm, as this was verified by

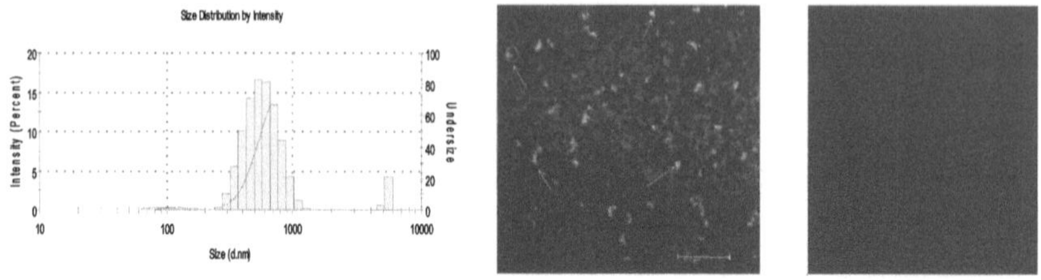

FIGURE 4.3 Dynamic light scattering plot and air nanobubble images from confocal microscopy.

DLS measurements (Fig. 4.3). Both samples were produced by our team by using the NB generators which are presented in the fourth section.

3.2 Surface nanobubbles

When a solid is fully immersed in a liquid tank, one expects that the liquid is in direct contact with the surface of the solid. However, this is often not the case. Contrarily, nanoscopic spherical cap-shaped gaseous domains are present at the interface. These existences are called surface NBs. Their existence seems to be paradoxical, as a simple classical estimate suggests that they should dissolve in microseconds, due to the large Laplace pressure inside these nanoscopic spherical-cap-shaped objects [44]. In one of the earliest studies concerning the AFM imaging of surface NBs, Lou et al. also proposed a production method which became known as the solvent exchange technique. NBs are generated when a hydrophobic surface is immersed in ethanol, subsequently displaced by water; this results in the supersaturation of dissolved gas and the nucleation of NBs [2]. This method has since become a standard. However, its effectiveness is variable; the saturation levels that can be obtained are up to three times the equilibrium concentration according to An et al. [45]. Irradiation with electron beam induces radiolysis, resulting to the nucleation and growth of NBs on the surface [46].

As Loshe and Zhang report [44], if the height H is fluctuated between 1 and 10 nm and the diameter L has a size between some hundreds of nanometers up to several microns, then the volume V (see Fig. 4.4) is in the order of 10^{-18} L (attoliters). By taking into account that an ambient pressure is occurred into the NB (which is an underestimated pressure due to Laplace pressure) and by apply the ideal gas law the number of gas molecules is calculated at about only $N \sim 10^3$. According to the Brunauer–Emmett–Teller theory [47] for the adsorption of nitrogen, whose cross section is 16 nm^2, within the footprint area $\pi L^2/4$, the number is about 500 molecules of nitrogen.

For the surface NB investigation a wide range of techniques are available. The most accurate are the AFM and the scanning electron microscopy which provides a high spatial resolution. By these two techniques a real picture of the formatted surface NBs is observed and both shape and size are observable and

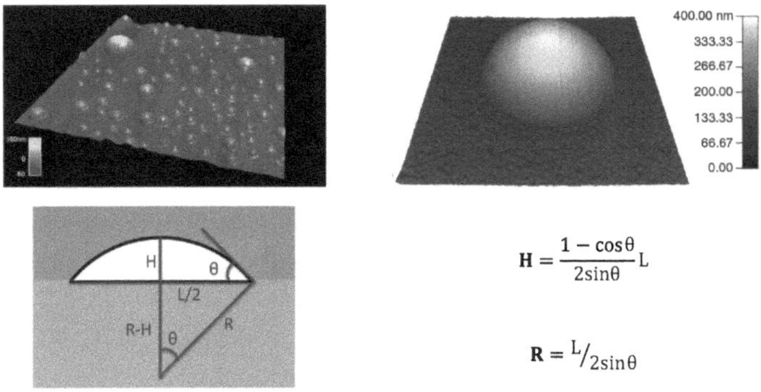

$$H = \frac{1 - \cos\theta}{2\sin\theta} L$$

$$R = {}^{L}/_{2\sin\theta}$$

$$V = \frac{\pi}{24} H(3L^2 + 4H^2)$$

FIGURE 4.4 Surface nanobubbles (NBs). Up left: AFM image of nanobubbles produced by the solvent exchange method on highly oriented pyrolytic graphite (HOPG). The imaged area is $4 \times 4\ \mu m^2$. Up right: AFM image of a single large surface nanobubble on a hydrophobized gold surface. The mapped area is $3 \times 3\ \mu m^2$. Down: Sketch of a surface nanobubble or nanodroplet. *Adapted from D. Lohse, X. Zhang, Surface nanobubbles and nanodroplets, Rev. Mod. Phys. 87 (3) (2015) 981.*

measurable. A typical optical microscopy can resolve only the bubbles with a large lateral diameter wider of about 1.5 μm.

By using other techniques, such as X-ray scattering (SAXS), neutron reflectivity, attenuated total reflection infrared spectroscopy (ATR-FTIR), and surface plasmon resonance spectroscopy, the investigation of the surface NBs takes place indirectly through the change of the physicochemical properties of the studied system. Confocal microscopy can also provide a high spatial resolution bottom view of dyed surface nanodroplets [48]. In situ transmission electronic microscopy has recently been used for the characterization of the NB nucleation process, possessing superb spatial and temporal resolution [49,50]. The absorption on the crystal surface could be quantitatively monitored with quartz crystal microbalance (QCM). In 1989, Carr et al. [51] demonstrated the feasibility of electrochemical QCM for detection of bubble nucleation.

4. Nanobubbles generators

In general, three ways of generating MBs and NBs are reported. The most common way uses compression of the gas stream to dissolve gas into liquid, which is subsequently released through a specially designed nozzle system, to nucleate small bubbles as potentially NBs, based on the cavitation principle. By the second way, power US is used to induce cavitation locally at points of extreme rarefaction in the standing ultrasonic waves. Finally, at the third method a gas stream is delivered under low offset pressure, and airs to break off the bubbles due to an additional feature, whether it is mechanical vibration, or flow focusing, or fluidic oscillation. The first two methods of MBs/NBs generation request high power demand than the third described method which should have the lowest power consumption [52,53].

The compression of a gas through an immersed, into a liquid, membrane is a method for NB production which is referred to the first described case. As is described by Kukizaki and Goto monodispersed NBs/MBs could be generated from shirasu-porous-glass (SPG) membranes with uniform pores in a system composed of dispersed gaseous and continuous water phases containing a surfactant [54]. Monodispersed NBs with a mean bubble diameter of 360–720 nm were stably produced from membranes with mean pore diameters of 43–85 nm. The mean bubble diameter was shown to be 8.6 times larger than the mean pore diameter; therefore, the NB diameter could be controlled by the membrane pore size. The experimental apparatus used for the bubble formation experiments is shown in Fig. 4.5. The gas–liquid dispersion system consisted of air as the dispersed gaseous phase and distilled water containing sodium dodecyl sulfate (SDS) as the continuous water phase.

Air was pressurized into a SDS solution ranging from 0.05 to 0.5 wt.% flowing through an SPG membrane with a transmembrane/bubble point pressure ratio of 1.1–2.0. The SDS concentration of the water phase ranged from 0.01 to 0.5 wt.%. A continuous water phase (SDS aqueous solution) of 3.0 dm^3 was flowed inside the membrane using a pump, while air was introduced on the outside and forced through the membrane pores.

Nowadays, there are numerous micro-NBs generators available in the global market. Among these, only few are referred as NBs generators, and the majority of them are limited to the production of MBs. Ultra-NBs is a completely new idea for radically new future technologies that have not been addressed so far by academia or industry. Regarding the protected generators only few patents are referred to this kind of instruments and the more relevant patents to the NB generators are the following five:

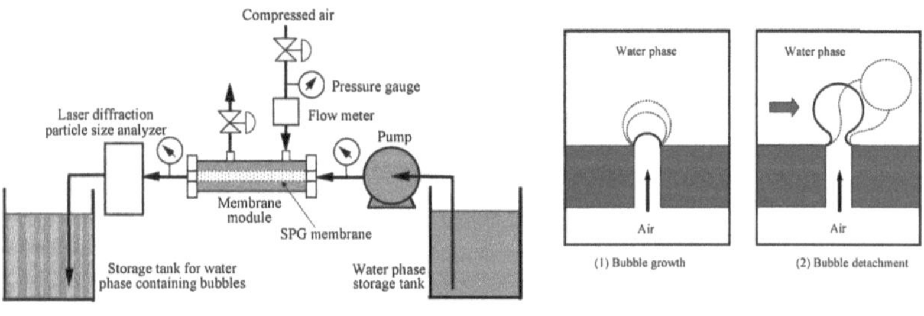

FIGURE 4.5 Schematic diagram of the experimental apparatus for nanobubbles production by using shirasu-porous-glass (SPG) membranes. *Adapted from M. Kukizaki, M. Goto. Size control of nanobubbles generated from Shirasu-porous-glass (SPG) membranes, J. Membr. Sci. 281 (1) (2006) 386–396.*

1. In 2007, the US Patent numbered 2007/0189972A1 [55]. This invention relates to a method of forming NBs that have potential utility in every industrial application and that impart special functions, especially to water. The present invention is a method of forming NBs by applying physical irritation to MBs contained in a liquid so that the MBs are abruptly contracted to form NBs.
2. In 2010, the US Patent numbered US 2010/0080759 A1 [56]. The present invention discloses a method for forming a nanobubble. The proposed method combines the using of inorganic nanoparticles, selected polymers, and solvents, and according to a freeze-drying process the NB is formed.
3. In 2012, the US Patent numbered US 2012/0086137A1 [57]. By this document a method for generating and dissolving NBs in a liquid was established. Bamboo fiber materials were used for passing the gas through their structure thereby forming NBs and at the same time dissolving the gas in the liquid.
4. In 2012, the US Patent numbered US8317165B2 [58]. By this invention an apparatus for producing a NB-containing liquid with use of general purpose products and which can be manufactured at low cost and in a short period of time was described and patented.
5. In 2014, the US Patent numbered US 2014/8678354B2 [59]. In this invention a technique for generating submicron bubbles was described and protected.

In recent experiments [60], it has been discovered that under certain circumstances drainage of a liquid from a porous medium is forced by ultra-NBs (<10 nm) that are formed by utilizing the energy difference between Baxter–Cassie and Wenzel wetting states (Fig. 4.6).

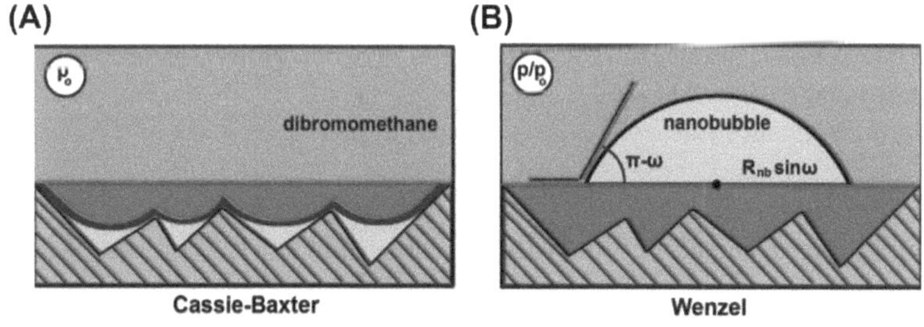

FIGURE 4.6 Sketch showing the formation of an interfacial nanobubble. Left: During adsorption, the surface roughness accommodates an amount adsorbed locally in a Cassie–Baxter wetting state. Right: During desorption, the tensile force overcomes the free energy barrier in passing from Cassie–Baxter to Wenzel states, resulting in the heterogeneous formation of an interfacial nanobubbles. *Adapted from A.C. Mitropoulos, K.L.Stefanopoulos, E.P.Favvas, E.F.Vansant, N. Hankins. On the formation of nanobubbles in vycor porous glass during desorption of halogenated hydrocarbons, Sci. Rep. (2015) 10943.*

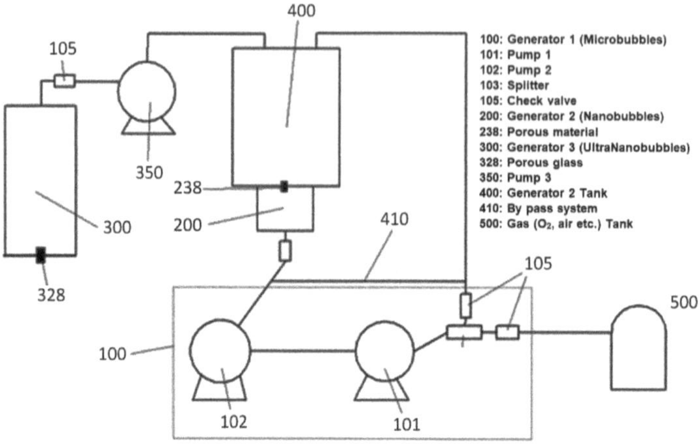

FIGURE 4.7 A schematic diagram illustrating the device. *Adapted from A. Mitropoulos and G. Bomis. Device for Generating and Handling Nanobubbles, 2016. EP2995369 A1.*

Based and on this evidence, Mitropoulos and Bomis designed and constructed a new apparatus for production of MBs, NBs, and ultra-NBs. They have protected their invention, entitled "A new device for generating and handling nanobubbles" by the European Patent EP20140386022 [61]. This invention relates to a device for generating and handling NBs and, more particularly, to a device for generating MBs, NBs, and ultra-NBs of desired sizes and a methodology of handling them in bulk solution (Fig. 4.7).

By this invention is provided a device for producing NBs. The NBs are generated in three stages at three different dimensions, said invention having three different bubble generators. In the first stage, in generator-1, a liquid is mixed with pressurized oxygen or other gas to generate bubbles in the liquid. The bubbles produced at this stage progressively decay in generator-1 to the MB size range. Said MB liquid is then passed through generator-2 where NBs are generated by passing the liquid through a metal (brass) porous material. In the third stage, ultra-NBs are generated either statically or through rotational and/or vibrational forces; the NBs produced in generator-2 are introduced into generator-3's fractal pump. Said pump consists of a porous plug with a fractal roughness on the pore walls and an oscillating piston.

This pump works by utilizing the energy differences between Baxter–Cassie and Wenzel wetting states to produce ultra-NBs under high pressure. The porous plugs can rotate for more effective production of NBs or ultra-NBs in the liquid, and the material from which they are made can be of various composition and pore size (e.g., porous glasses, porous gold, etc).

The current apparatus (Fig. 4.8) comprises the following:

1. Three bubble generators for generating bubbles by design and different sizes: MBs, NBs, and ultra-NBs, respectively.

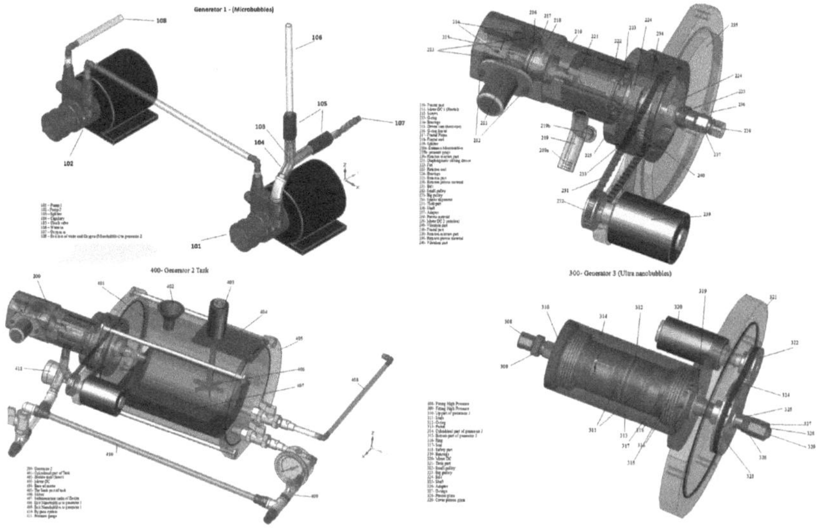

FIGURE 4.8 Sketches of the generators and the setup including the liquid main tank. *Adapted from A. Mitropoulos and G. Bomis. Device for Generating and Handling Nanobubbles, 2016. EP2995369 A1.*

2. A pressurized gas input.
3. A NB liquid-containing tank.
4. A sedimentation tank inside the previously stated NB liquid-containing tank, said sedimentation tank used to collect zeolite of other porous material which settles in the sedimentation tank.
5. An ultra-NB liquid-containing output.
6. A fractal pump for generating ultra-NBs.
   a. A pump having a porous plug with a fractal roughness on the pore walls;
   b. The porous plug also being combined with a piston to apply pressure to the liquid in order to produce ultra-NBs under high pressure.

5. Nanobubbles stability

The fact that NBs have such an extended lifetime has triggered the interest of the researchers, and a large number of studies discuss the reasons and the phenomena behind this property. As mentioned to the introduction, taking into consideration the classical thermodynamics, a paradox seems to appear in systems containing NBs due to the fact that the longevity of NBs is not in accordance with the Young–Laplace law [62]:

$$\Delta P = \frac{2\gamma}{R}$$

where $\Delta P = P_{vap} - P_{liq}$, the pressure inside (vapor phase) and outside (liquid phase) of the bubble, respectively, and γ is the surface tension.

Clearly, since the pressure inside a gas bubble is inversely proportional to its diameter, NBs should have extremely large internal pressure. As follows, the gas inside the NB cannot be in equilibrium with its surroundings. Hence, NBs should dissolve immediately, within a few microseconds, in favor of larger bubbles due to the "Ostwald ripening" phenomenon [63]. On the contrary, it is undisputable that NBs persist for several weeks and in some cases even for months [64].

One of the most prevailing theories claims that the reason of NB stability is their charge. Each NB is surrounded by an electrical double layer [65,66], which introduces an opposing force to the surface tension thus slowing or preventing their dissipation. It is believed that this double layer has an important role in the formation and stability of NBs in aqueous solutions by providing a fairly high repulsive force, which prevents interbubble aggregation and coalescence of the stable bubbles [67]. More specifically, it is believed that NBs are negatively charged on the surface due to the preferential adsorption of cations on the surface [24,68–70]. Cations (protons) are more likely hydrated and therefore have a tendency to stay in the bulk aqueous medium, whereas the smaller, less hydrated, and more polarized anions tend to adsorb on the bubbles' surfaces. In this way, zeta potential possesses a key role on NB stability.

Zeta potential assists in predicting long-term stability in colloidal system [71]. If all the particles in suspension have a high zeta potential (negative or positive), then they confer stability (the suspension or dispersion will resist agglomeration). In the case of NBs, zeta potential takes negative values and is varied depending the kind of gas [67] which is introduced and the pH of the liquid phase [40,72]. This is explained by the excess of OH- ions at the gas–water interface, in pure water. The charging mechanism of NBs was also believed to be due to preferential adsorption of OH^- in electrolytic solution [31]. According to Ushikubo et al. [67], bubbles with a high absolute value of zeta potential confers stability. Furthermore, they suggest that the stability of NBs is due to the magnification of electrostatic repulsive forces caused by overlapping electrical double layers of the neighboring bubbles.

In addition, Ohgaki et al. [32] stated that the stability of NBs is a result of the differentiated hydrogen bonding at the gas–water interface. It is reported that, in NBs, the length of hydrogen bonds is 0.273 nm instead of 0.295 nm, thus reducing the diffusivity of gases through the interfacial film [73]. In 2013, Wang, Liu, and Dong [74] report that the NB surface in kinetically stable and impermeable status. Tolman [75] calculated theoretically that the surface tension should decrease significantly at such small sizes. A reduction in surface tension of the gas–liquid interface reduces the Laplace pressure and could explain the NB stability [76,77].

In agreement with the low internal pressure hypothesis and based on molecular simulations, Nagayama et al. [78] concluded that the number of vapor atoms inside NBs is very low, hence the interior gas pressure would not be high enough to support the force balance of an NB. The authors attributed the disagreement with the Y–L equation to the fact that the liquid–vapor interface region plays an important role on the force balance at the curved liquid–vapor interface of an NB.

A very popular theory by Lohse and Weijs [79] supports that NBs remain stable due to the slow rate of dissolution of gas into the surrounding phase which is already supersaturated with that gas. Also, the experiments of Ushikubo et al. [67] indicated that the gas supersaturation condition of the water can reduce the gas transfer rate from the bubble to the liquid. However, this factor should be not the only one that can explain the NB stability, since NBs should exist even near or after reaching the saturation equilibrium. V. Craig disagrees claiming that "this theory wouldn't explain the stability of nanobubbles produced by electrolysis, where it is unlikely that significant supersaturation occurs" [80].

Seddon et al. [10] and Seddon (2011) provided a model for this remarkable NBs stability to bulk dissolution. Their argument is that the gas in an NB is of Knudsen type. This leads to the generation of a bulk liquid flow which effectively forces the diffusive gas to remain local.

6. Applications of nanobubbles

In the past few years, more and more attention has been given to the potential applications of the MBs and NBs. The applications are speeded to many sectors including mine industry, medical applications, wastewater treatment, surface cleaning, and agriculture. Some of the most prominent applications are mentioned and discussed in this section.

6.1 Flotation

Flotation is a technique, widely used by the mining industry, for the separation of minerals, phosphates, or coal from gangue. Coal and minerals particle size and surface hydrophobicity are two main parameters that affect three key steps in froth flotation process: particle–bubble collision, adhesion, and detachment. Flotation is also one of the processes used for the recovery of recycled paper and wastewater treatment.

A high froth flotation efficiency is limited to the narrow particle size range, between 50 and 600 μm for coal as shown in Fig. 4.9 [81]. The coarser the coal, the harder it is to stay attached to the air bubble. A finer coal size will lead to a poor particle air bubble collision. NB-coated fine particles will lead to particle aggregation and are thus more easily recovered due to an increased collision probability. In addition, NBs can increase the particle hydrophobicity [39,82] and thus increase the bubble–particle attachment probability and decrease the detachment probability.

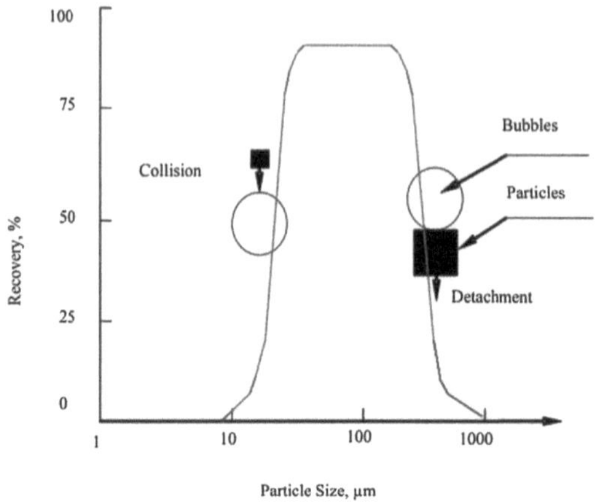

FIGURE 4.9 The effect of particle size on froth flotation recovery. *Adapted from A. Jowett, Formation and disruption of particle-bubble aggregates in flotation. Fine Particles Processing 1 (1980) 720–754.*

Fan et al. showed that by enhancing the flotation process with NBs, the coal and phosphate flotation particle size range is wider and the particle surface hydrophobicity is increased. Thus the coal and phosphate froth flotation recovery is improved [11].

The effect of NBs on the coarse particles was more significant than that on the finer, especially at lower collector dosages [12]. Tao stated that NBs could expand the applications of flotation [83].

6.2 Water treatment

Increasing attention has been given to the potential applications of MBs and NBs for the water treatment, given their ability to generate highly reactive free radicals. Microbubbles and NBs have been used for detoxification of water [84]. MNBs containing N_2 or CH_4 have been proven to enhance the activity of microorganisms in membrane bioreactors. It has been shown that N_2 MNBs can be as well used from fermentation, brewing, and human waste treatment. Furthermore, NBs are catalyzing chemical reaction thus enhancing the detoxification efficiency, improving the chemical treatment of water. In this context, air MBs/NBs as a pretreatment means has been shown to be highly beneficial for downsizing the water/wastewater treatment plants and improving the quality of product water by diminishing chemical, biological, and physical loads [85].

The collapse of MBs and NBs is associated with the generation of highly reactive free radicals providing a great potential for water disinfection. It has been found that ozone MNBs cause faster disinfection kinetics of *Escherichia coli*, leading to a reduced reactor size and smaller ozone dose compared with conventional ozone disinfection [86]. Also, high deactivation efficiency for *E. coli* has been achieved by producing NBs through hydrodynamic cavitation [87–90]. Moreover, ozone NBs have been proved to prevent pathogen growth in bathing pool assemblies [91].

Furthermore, the removal of organic pollutants with NBs has been widely studied. MNBs have been used for the removal of various organic compounds such as rhodamine B [92], p-nitrophenol [93], and alachlor [93a]. Sivakumar et al. investigated the decolorization of dye [94] and Takahashi et al. examined the decomposition of phenol using air-MBs in aqueous solutions, in the absence of dynamic stimulus [95].

6.3 Oil recovery

Crude oil is one of the most important sources of energy. From the first years of the crude oil production the same objective is still in place: to produce as more crude oil as it is possible. For this purpose, enhanced oil recovery (EOR) techniques have been developed, and they are used worldwide. Crude oil development and production in every oil reservoirs, worldwide, can include up to three distinct phases: the primary, the secondary, and the tertiary (or enhanced) recovery. At the first phase the natural pressure of the reservoir drives the oil into the wellbore. By the combination of pumps the produced oil is finally followed to the ground. Note that only about 10%–15% of the total reservoir oil could be produced by the primary recovery. The secondary recovery is taking place by the injection of water where the 20%–40% of the reservoir oil could be produced. In order to increase the oil recovery many techniques, mentioned as EOR techniques, have been attempted. Among these, three have been commercialized: (1) the thermal recovery, where the introduction of heat, by the injection of steam, for example, is one of these, (2) the gas injection, where gases such as carbon dioxide or natural gas or even nitrogen were injected into the reservoir for a further pushing of oil to the wellbore and finally to the ground, and (3) the chemical injection. By the chemical injection, special designed polymers are involved to the reservoir for increasing the effectiveness of waterfloods, otherwise the introduction of selected surfactants concluded to the decrease of the surface tension and the increase of the detachment of the oil droplets [96].

NB technology did not have introduced yet in the EOR phase but seems to be a very promising technique to be added in the list of OER techniques. Until today only few research works are referred to this promising idea. Instead of NBs, MBs are easily produced, they have different physicochemical properties and they did not provide long stability. Recently, Agarwala and Liu published their work entitled "Enhanced microbubbles assisted cleaning of diesel contaminated sand" [97]. In this work, they investigated the effect of low-

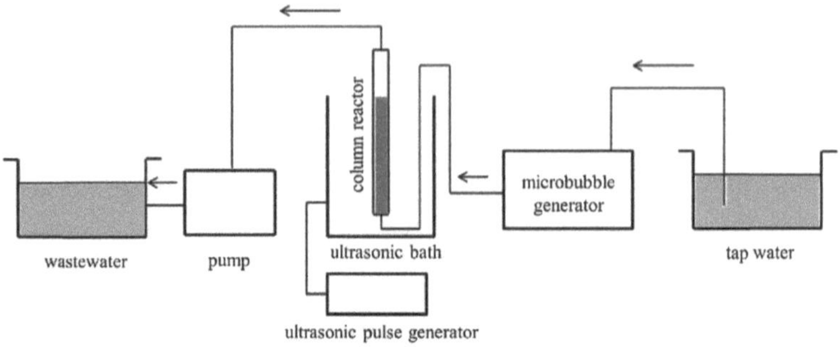

FIGURE 4.10 Experimental setup for cleaning of diesel-contaminated sand with ultrasound and microbubbles as was used by Agarwala and Liu. *Adapted from A. Agarwal, Y. Liu. Enhanced microbubbles assisted cleaning of diesel contaminated sand, Mar. Pollut. Bull. 124 (1) (2017) 331–335.*

intensity pulsed US, temperature, and salinity on cleaning efficacy of fine bubbles with diameter<50 μm for diesel-contaminated sands (Fig. 4.10).

The combination of treatment with ultrasound and MBs increased the diesel removal amount from 41% and 68% to 47% and 76%, respectively, for the cases of nonuse of US and by using respectively. The effect of high temperature was found to be prominent during the initial stages of cleaning. In addition, MBs generated in 599 mM saline water efficiently removed 85% diesel from fine sand within 30 min in contrast to only 41% diesel removal with MBs in freshwater. This study provides evidence. This is a technique which could be transferred to the EOR technology if for developing highly efficient MB-based chemical-free technology for diesel-contaminated sediments.

Overall, only few works are referred to the technology of oil recovery by using NB/MB technology. For example, Guet reported that oil recovery could benefit from gas-lift technology if sufficiently small MBs can be generated in situ downhole [98]. In addition to this, driving such innovations as new pump cavitation mechanisms could find an application for the oil recovery process [99]. Also, Zimmerman et al. suggest that CO_2 flooding for EOR with MBs will help avoid the channeling instability [100].

On the other hand, NBs technology has not been suggested yet as a solution in the EOR technology for oil production. It is an open field that scientists and engineers could focus on it at the next years. We believe that the results could be particularly attractive.

6.4 Surface cleaning

The electrochemical formation of NBs at conductive surfaces has recently been applied to the prevention of surface fouling and defouling surfaces. The studies by Wu et al. [17] show that proteins can be removed from a surface by

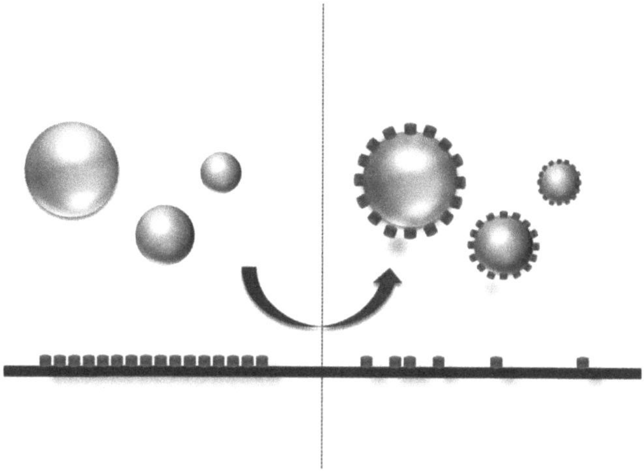

FIGURE 4.11 The effect of bulk nanobubbles on cleaning and preventing contamination on surfaces. *Adapted from J. Zhu, H. An, M. Alheshibri, L. Liu, P.M. Terpstra, G. Liu, V.S. Craig. Cleaning with bulk nanobubbles. Langmuir 32 (43) (2016) 11203–11211.*

electrochemically forming NBs at the surface. Preexisting NBs reduce the absorption of proteins at a surface. NBs have been applied for the prevention and removal of components adsorbed onto solid surfaces (Fig. 4.11). It has been shown that adsorption of various components onto surfaces could be inhibited by NBs, thus preventing the surfaces from fouling.

For example, NBs can block adsorption of bovine serum albumin on mica surface [38] while NBs also help remove organic contaminants from pyrolytic graphite [17,102] and gold surfaces [103]. Recently, a similar defouling effect of NBs was also observed on stainless steel surface [91]. It is demonstrated that NBs can be used as a novel means for the prevention of fouling and for removal of protein from a surface that is already fouled. Fouling prevention is achieved by covering a surface with NBs. Cleaning of conducting surfaces by NBs is promising in any system where fouling occurs in biomedia. On the basis of reports that NBs influence the adsorption of proteins and that NBs can be generated both in bulk and on surfaces, the defouling and cleaning effect of NBs is investigated by many researchers [102–104]. Most findings look notably important for flotation in the mineral and environmental area because NBs present a high surface area per volume and assist the capture (collision + adhesion) of particles by coarser bubbles due to their highly charged surface [11,12,39].

Wide use of oxygen NBs has been anticipated due to their extremely high bioactivity and mass transfer efficiency. Due to their ability to generate free radicals without the use of any toxic chemical, MBs have been proven to be a new environmentally friendly technique for oxidation of organic compounds, water disinfection, and fouling control [95].

Due to high mass transfer efficiency, they could potentially prevent fouling of ceramic membrane filtration processes. Ghadimkhani et al. examined the influence of air NBs on the defouling of this kind of membranes. To simulate fouling, humic acid, as an organic foulant, was applied to the membrane. Membrane defouling was achieved by feeding air NBs to the membrane. The surface of the ceramic membrane was successfully unclogged by NBs, as confirmed by atomic force microscope images before and after the treatment. It was assumed that $OH^{\bullet}$ radicals, which occur from the collapse of MNBs, bond with organic foulants such as humic acids and break down the organic matter [105].

Moreover, it has been shown that NBs have an effect on the surface tension of water. In a recent study, Ishizaki et al. investigated the cleaning potential of MNBs and confirmed that in the removal of oleic acid and salt, the removal effect was increased depending on the total surface area of the bubbles. The increased cleaning efficiency was attributed to the fact that water, containing fine bubbles, has a lower contact angle in comparison with normal water. More specifically, it was found that the contact angle of ion-exchanged water is 57.9 degrees, while 49.5 degrees is the contact angle (the average value of 20 measurements) for fine bubble water. Thus, it is deduced that MNB water has better wettability and lower surface tension, which is promising for cleaning applications [106]. Similar are the evidences based on our recently, unpublished yet, data. MBs and NBs may provide a promising path for a convenient, clean, cheap, and environmentally friendly technique suitable for cleaning conducting surfaces. Hence, it can be concluded that the use of MBs and NBs in developing new technology is still ahead to be explored.

7. Biological applications

Recently, the application of MNB technology in biological processes has been focused upon considerably. Due to their small diameter, bulk NBs have very high internal pressure compared with the bulk phase. This pressure difference (ΔP) contributes to the increase of gas solubility and was previously reported in supersaturation conditions. Yoshida et al. and Owerkowicz et al. demonstrated that hyperoxia promotes the growth of animals [107] and plants [108]. Based on that, Park et al. examined the effect of air NBs on the growth of leaf lettuce [109]. In this direction, Ebina et al. studied the effect of O_2 and air NBs on the growth of living organisms by improving the oxygen conditions through oral administration [110]. They used NB-enriched water for the cultivation of *Brassica campestris*; the plants were cultured hydroponically for a period of 4 weeks using air-NBs. They also studied the influence on mice and fish. DBA1/J mice were bred with normal free chaw and free drinking using O_2-NB water. Moreover, sweetfish and rainbow trout were kept in tanks with air-NB water. The results in all cases were significant. Compared with normal water, NBs promoted the height (19.1 cm for NBs, 16.7 for normal water), length of leaves (24.4 cm NBs, 22.4 cm normal water), and aerial fresh weight (27.3 g NBs, 20.3 g normal water) of *B. campestris*. Sweetfish's total weight was increased from 3.0 to 10.2 kg for NB-enriched water, while in normal water its weight only increased from 3.0 to 6.4 kg. Total weight of rainbow trout increased from 50.0 to 129.5 kg in normal water, whereas it increased from 50.0 to 148.0 kg in air-NB water. Finally, the weight and length of mice were also promoted. The weight of mice bred with NBs was 23.5 g, compared with 21.8 g for normal water, while the height was 17.0 and 16.1 cm, respectively. Liu et al. showed that water containing NBs promoted the size of barley seeds by 15%–25%, in comparison with distilled water with the same DO concentration. Liu et al. concluded that the germination rates of barley seeds dipped in water containing NBs were 15%–25% points greater than those of seeds dipped in distilled water, which verified the clear effect of NBs on the physiological activity [111]. Ushikubo et al. showed that when barley coleoptile cells were floated in water after the generation of oxygen MNBs, cytoplasmic streaming rates inside the cells were accelerated [112].

Moreover, NBs may provide a transport mechanism for gas delivery to a membrane or cell and thus affect transmembrane proteins or the membrane structure. Both effects considerably alter the cell function [10,113]. It is well known that hypoxia is a critical factor in the progression of several diseases, one of them cancer. Based on that, Owen et al. [114] assess the potential to reduce tumor hypoxia using an orally delivered suspension of submicrometer-sized oxygen NBs. By encapsulating oxygen into NBs it will achieve higher concentrations and stability in comparison with direct dissolution in water. In this study, SCID mice xenografts were developed by using human pancreatic cancer cells (BxPc-3 cells in male mice). Oxygen and argon NBs have been compared with oxygen-saturated water which administrated orally via gavage method. Animals were sacrificed and hypoxia-inducible factor-1α (HIF1α) protein measured indicating the ability of NBs' suspensions to influence tumor hypoxia. According to the published results, only in the case of oxygen NBs there is a statistical significant reduction of HIF1a and vascular endothelial growth factor (VEGF) expression.

In a study from Iijima et al. the effect of oxygen-NB water combined with radiation was studied in vitro. EBC-1 lung cancer and MDA-MB-231 breast cancer cell lines were cultured with and without the presence of NBs. Cell viability and HIF-1a levels were evaluated under normoxic and hypoxic conditions. The cells treated with oxygen NBs had shown a significant suppression of HIF-1a compared with the cells treated with normal culture media. Under hypoxic conditions, MDA-MB-231 and EBC-1 cells exhibited resistance to radiation, which was significantly suppressed with the use of oxygen NBs at various radiation doses [115].

Sharma et al. examined the ability of MBs to enhance the detachment of bacterial strains from glass surfaces, combined with fluid flow. Adhesion and detachment of *Actinomyces naeslundii* T14V-J1 and *Streptococcus oralis* J22 were studied, and it was proved that the presence of MBs in the flow increased the detachment of *A. naeslundii* from 40% to 98% [116]. Moreover, concerning the effect of MNBs on the removal of oral bacteria, it was shown that MBs in a fluid flow were more efficient at inducing single bacterial detachment, resulting in almost complete (97%) removal for *S. oralis* J22 and a 15-fold increase in *A. naeslundii* T14V-J1 removal as compared with the detachment caused by fluid flow alone [117].

The bactericidal activity of ozone NBs on periodontal pathogenic bacterial along with their cytotoxicity against human oral cells were presented in a study by Hayakumo et al. The bactericidal activity against *Porphyromonas gingivalis* and *Aggregatibacter actinomycetemcomitans* was evaluated by in vitro time-kill assays, and it was proved that the numbers of colony-forming units (CFUs)/mL of *P. gingivalis* and *A. actinomycetemcomitans* dropped below the limit of detection (<10 CFU mL^{-1}) after 0.5 min exposure to ozone NBs. This is attributed to free radical−mediated oxidation reactions. Conversely, the viability decrease of oral tissue cells, evaluated by the MTT assay, was minor after 24-h treatment with ozone NBs, indicating that ozone NBs could serve as a periodontal therapy by eliminating pathogen-containing biofilms [118].

MBs/NBs have wide applications in both disease diagnosis and therapy. By comparison with MBs, NBs have smaller carrying capacity. Where the fine bubble quite often leaves the system having transferred only a small fraction of its load, due to its small carrying capacity and high transfer rates (surface area per unit volume increases with decreasing diameter), frequently becomes capacity-limited [118a,119] and, inevitably, transfer with MBs is a transient process. With NBs, transfer rates are very high, yet in fact, the delivery of bulk materials with NBs is expensive due to the low volumetric flow rates and gas phase holdups, as well as the energy cost of their production. Yet, even NBs expensively produced with high energy density techniques have high transfer efficiencies and performance. Nevertheless, quite a lot of the biomedical uses of NBs stem from their ability to deliver materials in a controlled fashion.

7.1 Effect of NBs on plants' growth

The effective role of NBs has been investigated, from our group, in the case of tomato, oat, and tobacco plants' growth. In specific, marglobe tomato was cultivated both at the land and hydroponic farming methods. In the first case, four tomato seeds were placed into the plastic cells within a standard (pre-weighted) volume of soil of 125 mL. The plant nurseries were watered exactly the same time and by using the same water amount every day. For their watering, four different kinds of water were used. The first kind was raw tap; the second, water enhanced with oxygen NBs, the third, water enhanced with natural air NBs; and the fourth, water enhanced with nitrogen NBs. At the

TABLE 4.1 Weight of cultivated tomato plants.

Water enhanced with N_2 nanobubbles (NBs)	Water enhanced with air NBs	Water enhanced with O_2 NBs	Tap water	Type of watering water
870	730	690	550	Mass of tomato plant (mg)

point where the tomato plants reached 10 cm in height the plants were uprooted and their weight was recorded (Table 4.1).

As we can see in the above table, in the cases where enhanced with NBs water were used for the plant watering, the tomato weight increased by 25.5%, 32.7%, and 58% for air-, oxygen-, and nitrogen-enhanced NB water. It is noteworthy that the highest impact on the plant growth was happened in the case of the nitrogen NBs enhanced water, followed by the air NBs, and the lowest effect was observed in the case of oxygen-enhanced water.

In addition, the tomato growth was also investigated for semihydroponic cultivation. In two special arrays of styrofoam cells, three tomato seeds were planted, each cell was 5 mL of soil volume, and the containers were putted into two water reservoirs. One of them was full of raw tap water, and the second tank was full of tap water enhanced with air NBs. When the tomato plants achieve the height of 10 cm the plants were uprooted and their weight was recorded.

The weight increased by 31.25% in the case of cultivated tomatoes in the water enhanced with air NBs. Furthermore, as we can see in Fig. 4.12, the roots of the tomato plants were grown extremely only when the plants were into the tank of the tap water. On the other hand, the tomato plants which were

FIGURE 4.12 Tomato plants cultivation into water (left) and air NB–enhanced water (right).

into the tank with the NB-water did not grow more than few centimeters their roots (see Fig. 4.12), probably thanks to the overabundance concentration of the dissolved air, capable for determining the requested environment for the optimum cultivation conditions.

One more different experiment which took place was the following: eight tomato plants, 20 days grown, were transplanted at two hydroponic Wilma systems. At the one, tap water was used for watering, whereas at the second, again, water enhanced with air NBs was used. The watering took place automatically with a frequency of 2 h and an active time of 10 min with a dose of water used amount of 4 L.

The tomato plants were cultivated for 1 month, and after this period they were uprooted and specific characteristics were measured and the results are presented in Table 4.2. The average weight of the sprout, the height of the sprout, the average root weight, the root height, and the leaf surface were measured.

It is directly observable that the use of the air NB water gave an attractive push on the growth rate characteristics for the five cases of the different cultivation batches. As a next step, we plan to study systematically the effect of this kind of cultivation by going further, in collaboration with biologists and agronomists, up to the investigation of the characteristics of the produced fruits.

Finally, in another case, the use of enhanced water, with oxygen and nitrogen NBs, is presented in Fig. 4.13. Here three batches of oat seeds amounting 50 g were putted into three plates of 100-mL deionized water, deionized water enhanced with oxygen NBs, and deionized water enhanced with nitrogen NBs.

Also in this experiment it is remarkable that the use of the NB-enhanced water gave extremely big differences on the growth of the cultivated oat seeds. Here, it must be noted that the total mass, after 8 days, of the remained oat seeds together with the developed oat plants was 288 and 202 g for the cases of watering with O_2-NB/water and N_2-NB/water, respectively.

8. Conclusions

It is obvious that the field of NB research could be proved as one of the most important topics at many areas of the science and technology. Despite the fact that the field is being investigated for more than two decades, it is clear that it has still a long way to go to be fully understood. Additional work is needed for the standardization of the NBs production procedures, especially in the cases where extremely large volumes of NB-liquids are required. We now also need to invent new experimental procedures for an easy investigation not only of the NBs' size but also their population. Of course, large effort must be given for understanding better not only the creation mechanism but also the effective

TABLE 4.2 Typical growth rates for cultivated tomato plants.

Leaves' surface (cm^2)		Plant root height (cm)		Plant root weight (g)		Plant height (cm)		Weight of plant (g)		Series
Nanobubble (NB)-water	Tap water	NB-water	Tap water	NB-water	Tap water	NB-water	Tap water	NB-water	Tap water	1
21	13	12	12	0.8	0.7	80	60	27	13.4	2
17	11.1	11	10	0.8	0.6	44	65	21.1	10.4	3
19	10.2	14	12	0.7	0.6	60	55	11.6	8.2	4
18	8.8	15	11	0.9	0.5	90	57	26.1	9.2	5

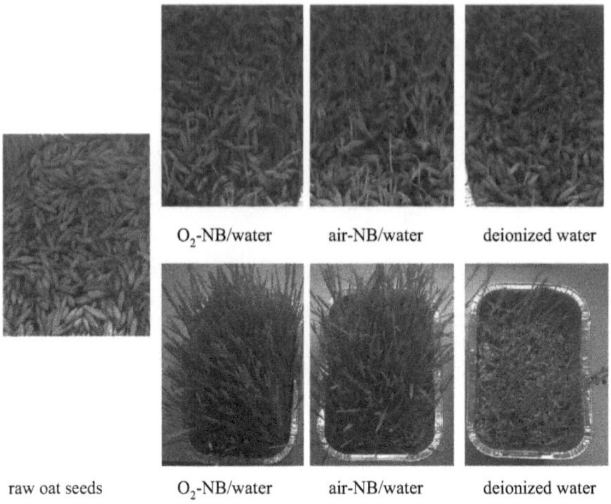

FIGURE 4.13 Oat cultivation after 4 days (up line) and 8 days (down line) from their immersion into different kinds of water.

activity of the NBs at all these processes, from the surface cleaning technology up to medical and agricultural applications.

Acknowledgments

E.D. Michailidi would like to thank the project "Study of the effect of nanobubbles enhanced water solutions on the cleaning of stainless steel and ceramic surfaces from oil-based contaminants," and E.P. Favvas would like to thank the project "Agro-NB." Both projects are supported by Stavros Niarchos Foundation. E.P. Favvas and E.K. Efthimiadou would also like to acknowledge the support of the project MIS 5002567, implemented under the "Action for the Strategic Development on the Research and Technological Sector," funded by the Operational Program "Competitiveness, Entrepreneurship and Innovation" (NSRF 2014–2020) and cofinanced by Greece and the European Union (European Regional Development Fund).

References

[1] J.R. Seddon, D. Lohse, Nanobubbles and micropancakes: gaseous domains on immersed substrates, J. Phys. Condens. Matter 23 (13) (2011) 133001.

[2] S.T. Lou, Z.Q. Ouyang, Y. Zhang, X.J. Li, J. Hu, M.Q. Li, F.J. Yang, Nanobubbles on solid surface imaged by atomic force microscopy, J. Vac. Sci. Technol. B 18 (5) (2000) 2573–2575.

[3] J.L. Parker, P.M. Claesson, P. Attard, Bubbles, cavities, and the long-ranged attraction between hydrophobic surfaces, J. Phys. Chem. 98 (34) (1994) 8468–8480.

[4] J.W. Tyrrell, P. Attard, Images of nanobubbles on hydrophobic surfaces and their interactions, Phys. Rev. Lett. 87 (17) (2001) 176104.

[5] J.R. Seddon, E.S. Kooij, B. Poelsema, H.J. Zandvliet, D. Lohse, Surface bubble nucleation stability, Phys. Rev. Lett. 106 (5) (2011a) 056101.

[6] J.R. Seddon, H.J. Zandvliet, D. Lohse, Knudsen gas provides nanobubble stability, Phys. Rev. Lett. 107 (11) (2011b) 116101.
[7] X.H. Zhang, A. Khan, W.A. Ducker, A nanoscale gas state, Phys. Rev. Lett. 98 (13) (2007) 136101.
[8] Z. Che, P.E. Theodorakis, Formation, dissolution and properties of surface nanobubbles, J. Colloid Interface Sci. 487 (2017) 123–129.
[9] S.H. Oh, J.M. Kim, Generation and stability of bulk nanobubbles, Langmuir 33 (15) (2017) 3818–3823.
[10] J.R. Seddon, D. Lohse, W.A. Ducker, V.S. Craig, A deliberation on nanobubbles at surfaces and in bulk, ChemPhysChem 13 (8) (2012) 2179–2187.
[11] M. Fan, T. Daniel, R. Honaker, L.U.O. Zhenfu, Nanobubble generation and its applications in froth flotation (part II): fundamental study and theoretical analysis, Min. Sci. Technol. (China) 20 (2) (2010a) 159–177.
[12] M. Fan, D. Tao, R. Honaker, Z. Luo, Nanobubble generation and its applications in froth flotation (part III): specially designed laboratory scale column flotation of phosphate, Min. Sci. Technol. (China) 20 (3) (2010b) 317–338.
[13] M. Fan, D. Tao, R. Honaker, Z. Luo, Nanobubble generation and its application in froth flotation (part I): nanobubble generation and its effects on properties of microbubble and millimeter scale bubble solutions, Min. Sci. Technol. (China) 20 (1) (2010c) 0001–0019.
[14] P. George, A.V. Nguyen, G.J. Jameson, Assessment of true flotation and entrainment in the flotation of submicron particles by fine bubbles, Miner. Eng. 17 (7) (2004) 847–853.
[15] A. Michalopoulou, E.P. Favvas, A.C. Mitropoulos, P. Maravelaki, V. Kilikoglou, I. Karatasios, A comparative evaluation of bottom-up and top-down methodologies for the synthesis of calcium hydroxide nanoparticles for the consolidation of architectural monuments, Mater. Today:. Proc. 5 (2018) 27425–27433.
[16] M. Suzuki, K. Koshiyama, F. Shinohara, S. Mori, M. Ono, Y. Tomita, T. Yano, S. Fujikawa, G. Vassaux, T. Kodama, Nanobubbles enhanced drug susceptibility of cancer cells using ultrasound, Int. Congr. Ser. 1284 (2005) 338–339.
[17] Z. Wu, H. Chen, Y. Dong, H. Mao, J. Sun, S. Chen, J. Hu, Cleaning using nanobubbles: defouling by electrochemical generation of bubbles, J. Colloid Interface Sci. 328 (1) (2008) 10–14.
[18] Z.A. Zhou, Z. Xu, J.A. Finch, J.H. Masliyah, R.S. Chow, On the role of cavitation in particle collection in flotation–A critical review. II, Miner. Eng. 22 (5) (2009) 419–433.
[19] C.E.P. Epstein, M.S. Plesset, On the stability of gas bubbles in liquid-gas solutions, J. Chem. Phys. 18 (11) (1950) 1505–1509.
[20] M. Alheshibri, J. Qian, M. Jehannin, V.S. Craig, A history of nanobubbles, Langmuir 32 (43) (2016) 11086–11100.
[21] B.D. Johnson, R.C. Cooke, Generation of stabilized microbubbles in seawater, Science 213 (4504) (1981) 209–211.
[22] N.F. Bunkin, F.V. Bunkin, Bubstons are stable gas microbubbles in highly diluted solutions of electrolytes, Zhurnal Eksperimentalnoi I Teoreticheskoi Fiziki 101 (2) (1992) 512–527.
[23] N.F. Bunkin, A.V. Kochergin, A.V. Lobeyev, B.W. Ninham, O.I. Vinogradova, Existence of charged submicrobubble clusters in polar liquids as revealed by correlation between optical cavitation and electrical conductivity, Colloids Surf., A 110 (2) (1996) 207–212.
[24] J.Y. Kim, M.G. Song, J.D. Kim, Zeta potential of nanobubbles generated by ultrasonication in aqueous alkyl polyglycoside solutions, J. Colloid Interface Sci. 223 (2) (2000) 285–291.

[25] K. Kikuchi, A. Ioka, T. Oku, Y. Tanaka, Y. Saihara, Z. Ogumi, Concentration determination of oxygen nanobubbles in electrolyzed water, J. Colloid Interface Sci. 329 (2) (2009) 306–309.

[26] K. Kikuchi, S. Nagata, Y. Tanaka, Y. Saihara, Z. Ogumi, Characteristics of hydrogen nanobubbles in solutions obtained with water electrolysis, J. Electroanal. Chem. 600 (2) (2007) 303–310.

[27] K. Kikuchi, H. Takeda, B. Rabolt, T. Okaya, Z. Ogumi, Y. Saihara, H. Noguchi, Hydrogen particles and supersaturation in alkaline water from an Alkali–Ion–Water electrolyzer, J. Chem. Soc. 506 (1) (2001) 22–27.

[28] K. Kikuchi, Y. Tanaka, Y. Saihara, Z. Ogumi, Study of hydrogen nanobubbles in solution in the vicinity of a platinum wire electrode using double-potential step chronoamperometry, Electrochim. Acta 52 (3) (2006b) 904–913.

[29] K. Kikuchi, Y. Tanaka, Y. Saihara, M. Maeda, M. Kawamura, Z. Ogumi, Concentration of hydrogen nanobubbles in electrolyzed water, J. Colloid Interface Sci. 298 (2) (2006a) 914–919.

[30] B.E. Oeffinger, M.A. Wheatley, Development and characterization of a nano-scale contrast agent, Ultrasonics 42 (1) (2004) 343–347.

[31] A.S. Najafi, J. Drelich, A. Yeung, Z. Xu, J. Masliyah, A novel method of measuring electrophoretic mobility of gas bubbles, J. Colloid Interface Sci. 308 (2) (2007) 344–350.

[32] K.E. Ohgaki, Physicochemical approach to nanobubble solutions, Chem. Eng. Sci. 65 (3) (2010) 1296–1300.

[33] T. Uchida, S. Oshita, M. Ohmori, T. Tsuno, K. Soejima, S. Shinozaki, K. Mitsuda, Transmission electron microscopic observations of nanobubbles and their capture of impurities in wastewater, Nanoscale Res. Lett. 6 (1) (2011) 295.

[34] H. Kobayashi, S. Maeda, M. Kashiwa, T. Fujita, Measurement and identification of ultrafine bubbles by resonant mass measurement method, in: International Conference on Optical Particle Characterization (OPC 2014), vol. 9232, 2014, p. 92320S.

[35] J. Israelachvili, R. Pashley, The hydrophobic interaction is long range, decaying exponentially with distance, Nature 300 (5890) (1982) 341–342.

[36] S. Ljunggren, J.C. Eriksson, The lifetime of a colloid-sized gas bubble in water and the cause of the hydrophobic attraction, Colloids Surf. A 129 (1997) 151–155.

[37] N. Ishida, T. Inoue, M. Miyahara, K. Higashitani, Nano bubbles on a hydrophobic surface in water observed by tapping-mode atomic force microscopy, Langmuir 16 (16) (2000) 6377–6380.

[38] A. Agarwal, W.J. Ng, Y. Liu, Principle and applications of microbubble and nanobubble technology for water treatment, Chemosphere 84 (9) (2011) 1175–1180.

[39] M. Fan, Y. Zhao, D. Tao, Fundamental studies of nanobubble generation and applications in flotation, Separ. Technol. Miner. Coal Earth Resour. (2012) 457–469.

[40] S.H. Cho, J.Y. Kim, J.H. Chun, J.D. Kim, Ultrasonic formation of nanobubbles and their zeta-potentials in aqueous electrolyte and surfactant solutions, Colloids Surf., A 269 (1) (2005) 28–34.

[41] R. Ahmadi, A. Khodadadi Darban, Modeling and optimization of nano-bubble generation process using response surface methodology, Int. J. Nanosci. Nanotechnol. 9 (3) (2013) 151–162.

[42] C. Wu, K. Nesset, J. Masliyah, Z. Xu, Generation and characterization of submicron size bubbles, Adv. Colloid Interface Sci. 179 (2012) 123–132.

[43] B.J. Berne, R. Pecora, Dynamic Light Scattering, Krieger, Malabar, FL, 1990.

[44] D. Lohse, X. Zhang, Surface nanobubbles and nanodroplets, Rev. Mod. Phys. 87 (3) (2015) 981.
[45] H. An, G. Liu, R. Atkin, V.S. Craig, Surface nanobubbles in nonaqueous media: looking for nanobubbles in DMSO, formamide, propylene carbonate, ethylammonium nitrate, and propylammonium nitrate, ACS Nano 9 (7) (2015) 7596–7607.
[46] J.M. Grogan, N.M. Schneider, F.M. Ross, H.H. Bau, Bubble and pattern formation in liquid induced by an electron beam, Nano Lett. 14 (1) (2013) 359–364.
[47] S. Brunauer, P.H. Emmett, E. Teller, Adsorption of gases in multimolecular layers, J. Am. Chem. Soc. 60 (2) (1938) 309–319.
[48] X. Zhang, J. Wang, L. Bao, E. Dietrich, R.C. van der Veen, S. Peng, D. Lohse, Mixed mode of dissolving immersed nanodroplets at a solid–water interface, Soft Matter 11 (10) (2015) 1889–1900.
[49] T.W. Huang, S.Y. Liu, Y.J. Chuang, H.Y. Hsieh, C.Y. Tsai, W.J. Wu, F.G. Tseng, Dynamics of hydrogen nanobubbles in KLH protein solution studied with in situ wet-TEM, Soft Matter 9 (37) (2013) 8856–8861.
[50] Y. Liu, S.J. Dillon, In situ observation of electrolytic H_2 evolution adjacent to gold cathodes, Chem. Commun. 50 (14) (2014) 1761–1763.
[51] M.W. Carr, A.R. Hillman, S.D. Lubetkin, M.J. Swann, Detection of electrolytically generated bubbles using an electrochemical quartz crystal microbalance, J. Electroanal. Chem. Interfacial Electrochem. 267 (1–2) (1989) 313–320.
[52] R. Parmar, S.K. Majumder, Microbubble generation and microbubble-aided transport process intensification—a state-of-the-art report, Chem. Eng. Process. Process Intensif. 64 (2013) 79–97.
[53] W.B. Zimmerman, V. Tesar, S. Butler, H.C. Bandulasena, Microbubble generation, Recent Patents on Engineering 2 (1) (2008) 1–8.
[54] M. Kukizaki, M. Goto, Size control of nanobubbles generated from Shirasu-porous-glass (SPG) membranes, J. Membr. Sci. 281 (1) (2006) 386–396.
[55] K. Chiba, M. Takahashi, Method of Forming Nanobubbles, 2007. US 20070189972 A1, Aug 16.
[56] W.H. Chang, Tao-Yuan, C.-A. Lin, Tao-Yuan, C.-H. Lee, Tao-Yuan, C.-K. Yeh, Tao-Yuan, Method for Forming Nanobubble, 2010. US 2010/0080759 A1, Apr. 1.
[57] R.Y.U. Sang-Ryul, Method and Apparatus for Generating Nanobubbles in Liquid, 2012. US 2012/0086137 A1, Apr. 12.
[58] K. Yamasaki, K. Chuhjoh, K. Iwata, U.S. Patent No. 8,317,165, U.S. Patent and Trademark Office, Washington, DC, 2012.
[59] W.B. Kerfoot, Nano-bubble Generator and Treatments, 2014. US 2014/8678354B2, Mar. 25.
[60] A.C. Mitropoulos, K.L. Stefanopoulos, E.P. Favvas, E.F. Vansant, N. Hankins, On the formation of nanobubbles in vycor porous glass during desorption of halogenated hydrocarbons, Sci. Rep. (2015) 10943.
[61] A. Mitropoulos, G. Bomis, Device for Generating and Handling Nanobubbles, 2016. EP2995369 A1, Mar 16.
[62] J.W. Schmelzer (Ed.), Nucleation Theory and Applications, John Wiley Sons, 2006.
[63] P.W. Voorhees, The theory of Ostwald ripening, J. Stat. Phys. 38 (1) (1985) 231–252.
[64] E. Duval, S. Adichtchev, S. Sirotkin, A. Mermet, Long-lived submicrometric bubbles in very diluted alkali halide water solutions, Phys. Chem. Chem. Phys. 14 (12) (2012) 4125–4132.

[65] P. Attard, M.P. Moody, J.W. Tyrrell, Nanobubbles: the big picture, Physica A 314 (1) (2002) 696–705.
[66] M.A. Hampton, A.V. Nguyen, Nanobubbles and the nanobubble bridging capillary force, Adv. Colloid Interface Sci. 154 (1) (2010) 30–55.
[67] F.Y. Ushikubo, M. Enari, T. Furukawa, R. Nakagawa, Y. Makino, Y. Kawagoe, S. Oshita, Zeta-potential of micro-and/or nano-bubbles in water produced by some kinds of gases, IFAC Proc. Vol. 43 (26) (2010) 283–288.
[68] A. Gray-Weale, J.K. Beattie, An explanation for the charge on water's surface, Phys. Chem. Chem. Phys. 11 (46) (2009) 10994–11005.
[69] S. Uddin, L. Jin, M. Mirnezami, J.A. Finch, An apparatus to measure electrical charge of bubble swarms, J. Colloid Interface Sci. 389 (1) (2013) 298–305.
[70] R.H. Yoon, J.L. Yordan, Zeta-potential measurements on microbubbles generated using various surfactants, J. Colloid Interface Sci. 113 (2) (1986) 430–438.
[71] W. Jia, S. Ren, B. Hu, Effect of water chemistry on zeta potential of air bubbles, Int. J. Electrochem. Sci 8 (2013) 5828–5837.
[72] G.H. Kelsall, S. Tang, S. Yurdakul, A.L. Smith, Electrophoretic behaviour of bubbles in aqueous electrolytes, J. Chem. Soc. Faraday Trans. 92 (20) (1996) 3887–3893.
[73] S. Nakashima, C.J. Spiers, L. Mercury, P.A. Fenter, M.F. Hochella Jr., Physicochemistry of Water in Geological and Biological Systems, Structures and Properties of Thin Aqueous Films, Universal Academy Press, 2004.
[74] S. Wang, M. Liu, Y. Dong, Understanding the stability of surface nanobubbles, J. Phys. Condens. Matter 25 (18) (2013) 184007.
[75] R.C. Tolman, The effect of droplet size on surface tension, J. Chem. Phys. 17 (3) (1949) 333–337.
[76] C. Fradin, A. Braslau, D. Luzet, D. Smilgies, M. Alba, N. Boudet, J. Daillant, Reduction in the surface energy of liquid interfaces at short length scales, Nature 403 (6772) (2000) 871–874.
[77] S.H. Park, J.G. Weng, C.L. Tien, A molecular dynamics study on surface tension of microbubbles, Int. J. Heat Mass Transf. 44 (10) (2001) 1849–1856.
[78] G. Nagayama, T. Tsuruta, P. Cheng, Molecular dynamics simulation on bubble formation in a nanochannel, Int. J. Heat Mass Transf. 49 (23) (2006) 4437–4443.
[79] J.H. Weijs, D. Lohse, Why surface nanobubbles live for hours, Phys. Rev. Lett. 110 (5) (2013) 054501.
[80] V.S.J. Craig, Very small bubbles at surfaces—the nanobubble puzzle, Soft Matter 7 (1) (2011) 40–48.
[81] A. Jowett, Formation and disruption of particle-bubble aggregates in flotation, Fine Part. Process. 1 (1980) 720–754.
[82] A.S.S. Ahmed, Cavitation Nanobubble Enhanced Flotation Process for More Efficient Coal Recovery, Mining Engineering, University of Kentucky, USA, 2013.
[83] D. Tao, Role of bubble size in flotation of coarse and fine particles-a review, Separ. Sci. Technol. 39 (4) (2005) 741–760.
[84] K.K. Yamasaki, Water Treatment Method and Water Treatment System, 2010. US patent, US7662288 B2, Feb 16, 2010.
[85] K.K. Yamasaki, Wastewater Treatment Equipment and Method of Wastewater Treatment, 2009. US Patent, US7485232 B2, Feb 3.
[86] M. Sumikura, M. Hidaka, H. Murakami, Y. Nobutomo, T. Murakami, Ozone micro-bubble disinfection method for wastewater reuse system, Water Sci. Technol. 56 (5) (2007) 53–61.

[87] S. Arrojo, Y. Benito, A.M. Tarifa, A parametrical study of disinfection with hydrodynamic cavitation, Ultrason. Sonochem. 15 (5) (2008) 903–908.

[88] K.K. Jyoti, A.B. Pandit, Ozone and cavitation for water disinfection, Biochem. Eng. J. 18 (1) (2004) 9–19.

[89] K.K. Jyoti, A.B. Pandit, Water disinfection by acoustic and hydrodynamic cavitation, Biochem. Eng. J. 7 (3) (2001) 201–212.

[90] L. Mezule, S. Tsyfansky, V. Yakushevich, T. Juhna, A simple technique for water disinfection with hydrodynamic cavitation: effect on survival of *Escherichia coli*, Desalination 248 (1–3) (2009) 152–159.

[91] K.K. Chen, U.S. Patent No. 7,488,416, U.S. Patent and Trademark Office, Washington, DC, 2009.

[92] X. Wang, J. Wang, P. Guo, W. Guo, G. Li, Chemical effect of swirling jet-induced cavitation: degradation of rhodamine B in aqueous solution, Ultrason. Sonochem. 15 (4) (2008) 357–363.

[93] K.M. Kalumuck, G.L. Chahine, The use of cavitating jets to oxidize organic compounds in water, J. Fluids Eng. 122 (3) (2000) 465–470.

[93a] X. Wang, Y. Zhang, Degradation of alachlor in aqueous solution by using hydrodynamic cavitation, J. Hazard Mater. 161 (1) (2009) 202–207.

[94] M. Sivakumar, A.B. Pandit, Wastewater treatment: a novel energy efficient hydrodynamic cavitational technique, Ultrason. Sonochem. 9 (3) (2002) 123–131.

[95] l. Takahashi, Method of Forming Nanobubbles, 2007. US 2007/0189972 A1, August 16.

[96] U.S. Department of Energy, Enhanced Oil Recovery, www.doe.gov.

[97] A. Agarwal, Y. Liu, Enhanced microbubbles assisted cleaning of diesel contaminated sand, Mar. Pollut. Bull. 124 (1) (2017) 331–335.

[98] S. Guet, G. Ooms, Fluid mechanical aspects of the gas-lift technique, Annu. Rev. Fluid Mech. 38 (2006) 225–249.

[99] G.R. Samuel, K.J. Saveth, Optimal design of progressing cavity pumps (PCP), J. Energy Resour. Technol. 128 (4) (2006) 275–279.

[100] W.B. Zimmerman, V. Tesař, H.H. Bandulasena, Towards energy efficient nanobubble generation with fluidic oscillation, Curr. Opin. Colloid Interface Sci. 16 (4) (2011) 350–356.

[101] J. Zhu, H. An, M. Alheshibri, L. Liu, P.M. Terpstra, G. Liu, V.S. Craig, Cleaning with bulk nanobubbles, Langmuir 32 (43) (2016) 11203–11211.

[102] Z. Wu, X. Zhang, X. Zhang, J. Sun, Y. Dong, J. Hu, In situ AFM observation of BSA adsorption on Hopg with nanobubble, Chin. Sci. Bull. 52 (14) (2007) 1913–1919.

[103] G. Liu, Z. Wu, V.S. Craig, Cleaning of protein-coated surfaces using nanobubbles: an investigation using a quartz crystal microbalance, J. Phys. Chem. C 112 (43) (2008) 16748–16753.

[104] M. Noble, A. Brown, P. Jauregi, A. Kaul, J. Varley, Protein recovery using gas–liquid dispersions, J. Chromatogr. B 711 (1) (1998) 31–43.

[105] A. Ghadimkhani, W. Zhang, T. Marhaba, Ceramic membrane defouling (cleaning) by air nano bubbles, Chemosphere 146 (2016) 379–384.

[106] T. Ishizaki, Y. Matsuda, T. Morita, Y. Nishiuchi, K. Tada, J. Nagahara, T. Hata, Cleaning effect by fine bubbles generated with gas–liquid share method, J. Chem. Eng. Jpn. 51 (2) (2018) 170–174.

[107] T. Owerkowicz, R.M. Elsey, J.W. Hicks, Atmospheric oxygen level affects growth trajectory, cardiopulmonary allometry and metabolic rate in the American alligator (*Alligator mississippiensis*), J. Exp. Biol. 212 (9) (2009) 1237–1247.

[108] S.M. Yoshida, Water uptake and growth of cucumber plants (*Cucumis sativus* L.) under control of dissolved O_2 concentration in hydroponics, International Symposium on Plant Production in Closed Ecosystems 440 (1996).

[109] J.S. Park, K. Kurata, Application of microbubbles to hydroponics solution promotes lettuce growth, HortTechnology 19 (1) (2009) 212–215.

[110] K. Ebina, K. Shi, M. Hirao, J. Hashimoto, Y. Kawato, S. Kaneshiro, H. Yoshikawa, Oxygen and air nanobubble water solution promote the growth of plants, fishes, and mice, PLoS One 8 (6) (2013) e65339.

[111] S. Liu, Y. Kawagoe, Y. Makino, S. Oshita, Effects of nanobubbles on the physicochemical properties of water: the basis for peculiar properties of water containing nanobubbles, Chem. Eng. Sci. 93 (2013) 250–256.

[112] F.Y. Ushikubo, S. Oshita, T. Furukawa, Y. Makino, Y. Kawagoe, T. Shiina, A study of water containing micro and nano-bubbles and its possible effect on physiological activity, in: Central Theme, Technology for All: Sharing the Knowledge for Development. Proceedings of the International Conference of Agricultural Engineering, XXXVII Brazilian Congress of Agricultural Engineering, 2008.

[113] J. Dzubiella, Explicit and implicit modeling of nanobubbles in hydrophobic confinement, An Acad. Bras Ciências 82 (1) (2010) 3–12.

[114] J. Owen, C. McEwan, H. Nesbitt, P. Bovornchutichai, R. Averre, M. Borden, E. Stride, Reducing tumour hypoxia via oral administration of oxygen nanobubbles, PLoS One 11 (12) (2016) e0168088.

[115] M. Iijima, et al., Development of single nanometer-sized ultrafine oxygen bubbles to overcome the hypoxia-induced resistance to radiation therapy via the suppression of hypoxia-inducible factor-1α, Internet J. Oncol. 52 (3) (2018) 679–686.

[116] P.K. Sharma, et al., Influence of fluid shear and microbubbles on bacterial detachment from a surface, Appl. Environ. Microbiol. 71 (7) (2005a) 3668–3673.

[117] P.K. Sharma, et al., Microbubble-induced detachment of coadhering oral bacteria from salivary pellicles, Eur. J. Oral Sci. 113 (4) (2005b) 326–332.

[118] S. Hayakumo, et al., Effects of ozone nano-bubble water on periodontopathic bacteria and oral cells-in vitro studies, Sci. Technol. Adv. Mater. 15 (5) (2014) 055003.

[118a] R.M. Worden, M.D. Bredwell, Mass transfer properties of microbubbles 2, Analysis using a dynamic model, Biotechnol. Prog. 14 (1) (1998) 39–46.

[119] M.D. Bredwell, R.M. Worden, Mass-transfer properties of microbubbles. 1. Experimental studies, Biotechnol. Prog. 14 (1) (1998) 31–38.

Further reading

[1] H. Chen, H. Mao, L. Wu, J. Zhang, Y. Dong, Z. Wu, J. Hu, Defouling and cleaning using nanobubbles on stainless steel, Biofouling 25 (4) (2009) 353–357.

[2] K.K. Jyoti, A.B. Pandit, Hybrid cavitation methods for water disinfection: simultaneous use of chemicals with cavitation, Ultrason. Sonochem. 10 (4) (2003) 255–264.

[3] M. Takahashi, K. Chiba, P. Li, Free-radical generation from collapsing microbubbles in the absence of a dynamic stimulus, J. Phys. Chem. B 111 (6) (2007) 1343–1347.

[4] F.Y. Ushikubo, T. Furukawa, R. Nakagawa, M. Enari, Y. Makino, Y. Kawagoe, S. Oshita, Evidence of the existence and the stability of nano-bubbles in water, Colloids Surf., A 361 (1) (2010a) 31–37.

[5] Y. Wang, X. Li, Y. Zhou, P. Huang, Y. Xu, Preparation of nanobubbles for ultrasound imaging and intracellular drug delivery, Int. J. Pharm. 384 (1) (2010) 148–153.
[6] R.M. Worden, M.D. Bredwell, Mass-transfer properties of microbubbles. 2. Analysis using a dynamic model, Biotechnol. Prog. 14 (1) (1998) 39–46.

Chapter 5

Low-cost activated carbon from rice wastes in liquid-phase adsorption

Eleni A. Deliyanni
Laboratory of General and Environmental Chemistry, Department of Chemistry, Aristotle University of Thessaloniki, Greece

1. Introduction

Activated carbons (ACs) are important industrial materials due to their porous structure and their large surface area. Coal, biomass, and synthetic polymers are the precursors of ACs either through carbonization followed by activation or through a direct activation. During these processes, porosity and surface area are created as well as surface chemical functional groups over the carbon matrix. Due to their ability to adsorb gasses and aqueous pollutants [1] their major application is to be used as adsorbent materials for pollution control. Other important applications of AC include supercapacitors and catalyst support due to their high porosity that makes them able to be impregnated with a large amount of metals on their surface.

Different precursors for the preparation of ACs have been studied by many researchers. Various types of carbon sources have been used, such as glucose [2] and cellulose [3], with the biomass-derived carbon source to offer additional advantages of being renewable and available in abundance at extremely low cost [4]. Besides, ACs were prepared by different activating processes, which depend on the type of activating agent (chemical or physical), as well as the number of stages (direct and indirect activation). The preparation process, temperature, and time of activation affect dramatically the textural properties of the ACs produced, as well as the surface chemical functional groups [5].

Rice is an important source of carbohydrates with an average annual rice global production at 729 million tons [6]. After rice preparation a big part of rice biomass wastes is generated. Rice wastes have many applications such as

Advanced Low-Cost Separation Techniques in Interface Science. https://doi.org/10.1016/B978-0-12-814178-6.00005-4

bedding, compost, and feed. The bigger portion of these rice wastes are left or burnt in the field. This method of crop residue disposal is inexpensive, so it is practiced in many parts of the world to clear the surface biomass from land for faster crop rotation; to control undesirable weeds, pests, diseases; and to return some nutrients to the soil. This activity emits a large amount of toxic air pollutants (particles, and inorganic and/or organic gases). Toxic air pollutants include methane which can have detrimental effects on human health and atmospheric pollution [7]. Overall, emissions from field burning activities deteriorate local air quality and result in adverse health effects [8]. Elevated air pollution levels due to agricultural field burning activities have been reported in the United States, Europe, and Asia, e.g., Taiwan and Thailand [1]. In Asia, this air pollution leads to regional atmospheric phenomena such as Atmospheric Brown Clouds [9,10]. During the field burning of rice wastes, fine particles are also emitted that are harmful on human health and atmosphere [1]. Particulate matter is itself a complex mixture, and its potential environmental and health effects depend on particle size and composition.

Rice-derived wastes comprise two different materials: husks and straws (Fig. 5.1) [11]. Rice husk (Fig. 5.1C) is the outer surface of rice. Rice husks have been commonly investigated as potential precursors of ACs [1,11]. By contrast, straws have been candidates mostly, e.g., as precursors for production of paper pulp [12], artificial wood [13,14], compost [13], ethanol production, and as filler for rubber composites [14].

The annual global rice production implies more than 150 million tons of rice husk production, based on the data published by the Food and Agriculture Organization (FAO) of the United Nations in the "Rice Market Monitor" report [15]. Main components of rice husk are cellulose (38.3%), hemicellulose (31.6%), lignin (11.8%), and silica (18.3%) [16]. Rice husk grows in silicate-abundant soil. The upper and lower parts of rice husk contains SiO_2, for this reason the SiO_2 weight occupies 15–20 wt% of the total weight of rice husk. Due to the SiO_2 content, rice husk is different from most of the biomasses. The

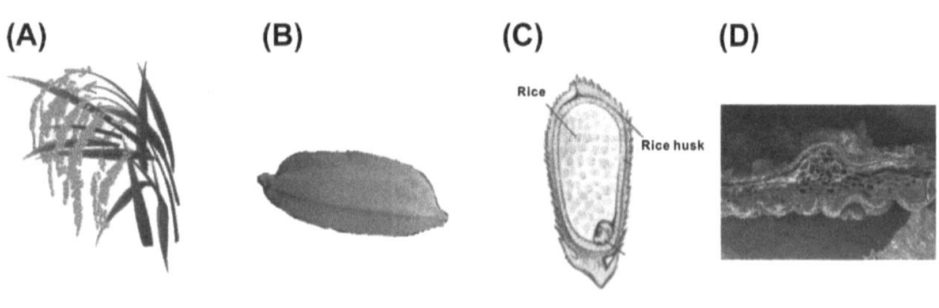

FIGURE 5.1 (A) Cartoon graph of rice husk, (B) digital graph of rice, (C) the schematic diagram of the cross section of rice, (D) SEM image of the cross section of rice husk. *Reprinted with permission, copyright Elsevier W. Zhang, N. Lin, D. Liu, J. Xu, J. Sha, J. Yin, X. Tan, H. Yang, H. Lu, H. Lin, Direct carbonization of rice husk to prepare porous carbon for supercapacitor applications, Energy 128 (2017) 618–625.*

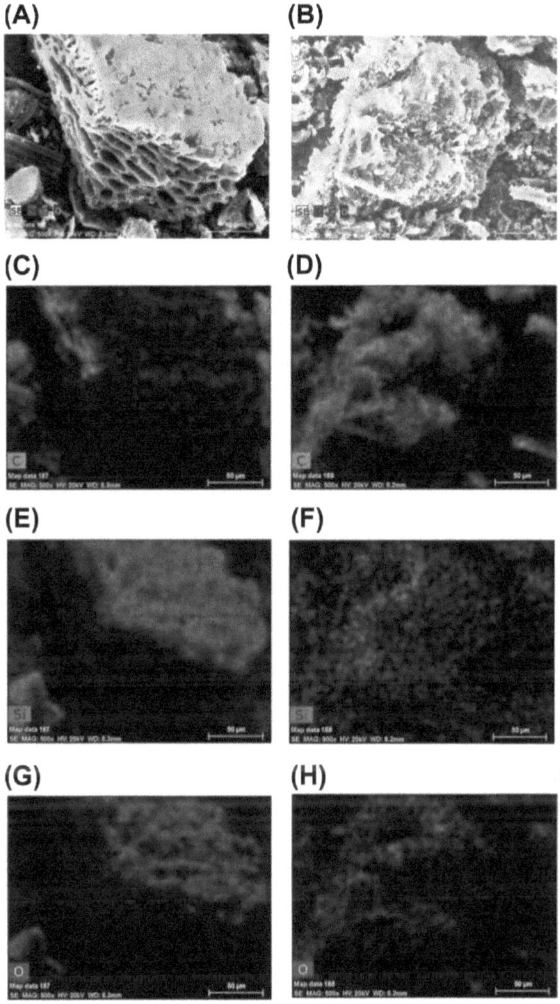

FIGURE 5.2 Element distribution of rice husk–activated carbon: (A) all, (C) C, (E) Si, (G) O; and rice husk–derived porous carbon: (B) all, (D) C, (F) Si, (H) O. *Reprinted with permission, copyright Elsevier W. Zhang, N. Lin, D. Liu, J. Xu, J. Sha, J. Yin, X. Tan, H. Yang, H. Lu, H. Lin, Direct carbonization of rice husk to prepare porous carbon for supercapacitor applications, Energy 128 (2017) 618–625.*

SiO_2 content is seen in the EDAX maps of Fig. 5.2E that represents the silica distribution on rice husk–derived AC. Given their silica content, they have been used in silica synthesis, synthesis of belite cement, and development of refractory ceramics materials [17]. Its use as animal fodder is restricted due to its low digestibility, they can be converted into high energy fuels and valuable chemicals, i.e., production of ethanol, methane, acetic acid, sugar (glucose and xylose), 5-hydroxymethylfurfural (HMF), and furfural [18].

2. Use of rice wastes for activated carbons

Agricultural wastes are used as precursors for ACs due to their low cost and the good quality of the carbons produced. Their synthesis steps include dehydration, conversion of organic matter to elemental carbon by carbonization of the noncarbon portion at high carbonization/pyrolysis temperatures (400–800°C), and activation that leads to pore enlargement [14]. The surface chemical properties of the carbon produced can be further modified after physical or chemical functionalization.

Rice husks have been commonly investigated as potential precursors of ACs by numerous researchers due to their availability [19–31]. In contrast, rice straws (RSs) have been less suggested for this application, whereas other routes of valorization have been considered in the past, e.g., pulp and paper, construction materials, compost, fuel, production of chemicals such as ethanol and bioadsorbents [23].

Some of the issues to be concerned and examined by many researchers during the conversion are the following:

1. The pretreatment of the waste biomass.
2. The better activation way: physical or chemical.
3. The choice of the activating agent.
4. The effect of temperature on chemical activation of biochar in order to obtain an AC with the highest surface area and microporosity.
5. The effectiveness of produced AC as water pollutant absorbent.

The production process and the activating agent significantly affect the properties of the materials.

3. Pretreatment of rice biomass

The basic route for AC preparation from waste RS and/or husk source is based on the burning of wastes in a special incinerator at elevated temperatures. The rice wastes (straw or husk) before activation have to be crushed, ground, and sieved; they have to be washed several times with distilled water and to get dried at 120°C. During the burning process, hydrocarbons are converted to methane, propane, carbon oxides, hydrogen, and other gases. The physical activation of rice waste biochar with steam and carbon dioxide are described by Eqs. (5.1) and (5.2), respectively [31].

$$C(s) + H_2O(g) \rightarrow CO(g) + H_2(g) \tag{5.1}$$

$$C(s) + CO_2(g) \rightarrow 2CO(g) \tag{5.2}$$

Due to the high ash content of rice biomass, especially rice husks, leaching of biomass by a base and/or an acid is needed in order to avoid the negative influence of ash content on carbon formation. Leaching may reduce the ash content to 92–98 wt% [32,33], having as result the specific surface area increase and the pore widening of the AC [32]. NaOH is the most important leaching agent for rice biomass, predominantly removing SiO_2 to form sodium silicate (Na_2SiO_3) [32]:

$$2NaOH + SiO \rightarrow Na_2SiO_3 + H_2O \tag{5.3}$$

Na_2SiO_3 is soluble in water and can be easily removed by washing with water [32]. Acid leaching of rice biomass may be also applied; it may remove the metallic impurities, via their dissolution in acid that can hinder the pore development [32]. The metallic impurities can be by rinsing with distilled water. Although the need for silica removal is pointed out as important in order to improve the properties of rice biomass–derived ACs, the low carbon yields is a disadvantage attributed to the removal of the lignin, that is a main constituent for char formation, after base leaching [33].

4. Activation

4.1 Physical activation

With the physical activation, char is produced after carbonization of rice wastes (husk and straw) at temperatures from 400 to 700°C in presence of N_2 [34]. The biochars prepared after physical activation exhibits a small porosity due to the pores tarry products formed during carbonization that block the pores of the biochars; for this reason, biochars present usually a small adsorption capacity. In the second stage, these tarry products are removed at elevated temperatures, between 600 and 1000°C, and with the aid of oxidizing gases such as steam, air, carbon dioxide, or their mixtures [34]. During this stage that is also known as "activation process" new porosity is created while the diameter of the existing pores is enlarged. After the physical activation process a product with a well-developed pore structure is prepared. During physical activation a weight loss of rice biochars occurs that increases linearly with activation temperature and time.

The commonly applied activating agent for physical activation of rice biochar is steam and carbon dioxide. Oxygen may cause burning of the biochar causing a great weight loss. Steam and carbon dioxide have a different reaction route during activation and lead to different physical properties of the ACs prepared. Specifically, steam presents a faster reaction with the biochars and results in widening of the micropores, while CO_2 is evolved in creation of new porosity and pore enlargement [35]. This can be attributed to the size of H_2O and CO_2 molecules; the H_2O molecule diffuses faster into the pores of the carbon than the CO_2 molecule due to its comparatively smaller size. Consequently, the reaction between char and steam is faster than that with CO_2. While steam causes widening of micropores from earlier activation stages, activation by CO_2 favors the development of new fine porosity followed by pore enlargement as seen in Table 5.1 [35–39]. Leaching of rice biomass as

TABLE 5.1 Textural parameters of rice straw (RS)–derived activated carbon (AC) by physical activation.

AC precursor	Carbon yield %	S_{BET} (m^2/g)	C_ S_{BET} (m^2/g)	$V_{0.99}$ (cm^3/g)	V_{DR} (cm^3/g)	V_{meso} (cm^3/g)	$V_{DR}/V_{0.99}$
RS	44.4	588	261	0.72	0.21	0.48	0.33
Fibrous RS	41	790	324	0.48	0.29	0.18	0.62
RS-MP	34.9	857	299	0.91	0.31	0.59	0.34
RS-MT	36.6	584	213	0.64	0.21	0.43	0.33
RS-NaP	34.7	368	128	0.52	0.13	0.39	0.26
RS-MT-NaP	37.2	783	291	0.62	0.29	0.34	0.46

Reprinted with permission A.H. Basta, V. Fierro, H. El-Saied, A. Celzard, Effect of deashing rice straws on their derived activated carbons produced by phosphoric acid activation, Biomass Bioenergy 35 (2011) 1954–1959.

well as pretreatment, pyrolysis temperature and conditions, activating agent are very important for the result on the surface area and porosity of the AC prepared.

Zhang et al. prepared rice husk carbon (RHC) after direct carbonization of rice husk biomass waste in a furnace without N_2 [30]. Rice husks were put into a tube furnace; N_2 flow excluded the air in the tube for half an hour, while during and after heat treatment air was no longer injected into the tube. The obtained carbon (RHC) was treated by 10 wt% NaOH solution (liquid to solid mass ratio of 10:1) at 100°C for 2 h to eliminate most of the residual SiO_2 in RHC. After drying, the rice husk–derived porous carbon (RHPC) was obtained. RHC presented high pore volume and specific surface area as seen in Table 5.2. The high specific surface area of RHC could be attributed to the fact that H_2O and CO_2 produced during the carbonization process leads to self-activation of rice husk. When RHC was further alkali treated, the AC prepared (RHPC) presented higher specific surface due to the SiO_2 removal during leaching.

4.2 Chemical activation

During chemical activation of rice wastes, it is possible that activation and carbonization stage to be performed in one stage, with the aid of chemical activating agents such as KOH at relatively low temperatures of 400–700°C [32,36,40], NaOH [20], H_3PO4 [41], and ZnCl2 [36]. This process is efficient only when acidic activating agents are used [32,36,40]. On the contrary, it is not successful when KOH and NaOH are used as activating agents [42] due to the fact that not enough activating agent can penetrate the biomass and create pores during carbonization, having as result AC of low surface area [36,40].

TABLE 5.2 Textural parameters of rice husk–derived activated carbon by physical activation.

Samples	S_{BET} m^2/g	St-plots m^2/g	St-plots/ S_{BET} %	V_{total} cm^3/g	V_{micro} cm^3/g	V_{meso} cm^3/g
Rice husk carbon	337.5	268.3	79.5	0.207	0.109	0.103
Rice husk–derived porous carbon	527.1	392.0	74.4	0.327	0.159	0.171

Reprinted with permission W. Zhang, N. Lin, D. Liu, J. Xu, J. Sha, J. Yin, X. Tan, H. Yang, H. Lu, H. Lin, Direct carbonization of rice husk to prepare porous carbon for supercapacitor applications, Energy 128 (2017) 618–625.

A two-stage chemical activation, with the precursor biomass to be precarbonized in the first stage, and chemically activated in the second stage, is more effective with KOH and NaOH as activating agents [20,36]. The two-stage chemical activation facilitates the reaction of the activating agent with the precarbonized material leading to a serious amount of porosity of the AC prepared [41]. Factors such as the precarbonization conditions, the type of activating agent, the impregnation ratio, the activation temperature, the heating rate, and the activation time play a crucial role in obtaining AC with the desired adsorption properties.

4.2.1 Chemical activation with KOH

The chemical activation with KOH solution is known as a very effective activation method. The carbons produced present a higher micropore volume and more micropores compared with those produced by physical activation methods [37,41]. The reaction of KOH with char derived from the pyrolysis of rice biomass resembles the one described with coal. The presence of a higher number of heteroatoms in biomass can enhance the catalytic role of KOH on the volatilization of the material, as seen with coals.

In the one-step process, KOH may catalyze the decomposition of biomass into volatile matters, leading this way to a low carbon yield. Since KOH is involved in the reactions with biomass constituents, not enough amounts are available in order to react with silica, leading this way to an AC with a high ash content from 29% to 67%. Various researchers activated rice husk wastes with KOH as activating agent examining one-step activation process [23,38,41,43,44]. In the one-step process, the precursor sources were impregnated with KOH solution at a ratio of precursor/KOH of 1/4 w/w, and then oven dried for 12 h at 110°C. The activation temperature, 800 °C, was maintained for 1 h. Some values from literature for specific surface of RS-derived ACs with chemical activation are presented in Table 5.3.

TABLE 5.3 Specific Surface Area of rice straw–derived activated carbon with chemical activation.

Activation agent	Activation temperature	Surface area, m^2/g	Vt, cm^3/g	Reference
KOH	800	2646	1.285 Vt	Linfeng Zhu et al. 2017 [45]
KOH	800	1917		Basta et al. 2009 [23]
KOH		143.3	0.175	Yakout et al. 2013 [44]
KOH		1304.8		Ken-Lin Chang et al. 2014 [46]
KOH		3333		Hong Jin et al. 2018 [43]
KOH		1330		Hyungseok Nam et al. 2018 [47]
H_3PO_4	450	857	0.91	Basta et al. 2011 [24]
H_3PO_4	450	500		Fierro et al. 2010 [31]
$(NH_4)_2PO_4$	700	1154	0.0670Vt	Pin Gao et al. 2011 [48]

In the two-step process, biomass is carbonized at the first stage, for a higher carbon yield. At the second stage, KOH reacts with carbon leading to higher pore volumes and surface areas. KOH may react with silica to soluble products that are leaching off during washing, hence ACs prepared present a low ash content from 6% to 29%, as presented in Table 5.4. Pores may be also unblocked and made open when soluble products are washed off, increasing the BET surface area. Since pretreatments usually lead to removal of ashes, the derived ACs present an increased BET surface area although their effect depends on the kind (one or two steps) of activation process.

Basta et al. produced AC with a two-step method, i.e., carbonization and activation with KOH from RS biomass [23]. The AC prepared presented a surface area of 2695 m^2/g at 850°C, low carbon yield, and high adsorption capacity for methylene blue, as seen in Table 5.4.

4.2.2 Chemical activation with H_3PO_4

Orthophosphoric acid (PA), H_3PO_4, is also an important and commonly used activating agent in ACs preparation from agricultural wastes [46,49–59]. Phosphoric acid plays a crucial role in the bond cleavage and in dehydration of the biopolymers at low temperatures [31]. The extensive cross-linking of the products formed has as result the binding of the volatile matter into the produced carbon matrix, increasing this way the carbon yield. As reported in

TABLE 5.4 Carbon yield and main pore texture parameters of activated carbons (ACs) derived after one- and two-step KOH activation from raw and pretreated rice straws.

AC precursor	Carbon yield, (C) (%)	Ash, ε(%)	S_{BET}, (m^2/g)	CxS_{BET}, (m^2/g)	$V_{0.99}$, (cm^3/g)	V_{DR}, (cm^3/g)	V_{meso}, (cm^3/g)	$V_{DR/0.99}$	L_0, (nm)
One-stage process									
Rice straw (RS)		66.71	420	73.7	0.31	0.15	0.16	0.49	0.88
RS–MP	8.5	61.31	430	36.4	0.51	0.16	0.35	0.31	0.9
RS–MT	8.8	38.43	866	72.9	0.57	0.32	0.25	0.57	0.82
RS–NaP1	13.9	28.74	553	76.7	0.41	0.21	0.2	0.51	0.66
RS–MT–NaP	10.1	38.36	657	66.2	0.4	0.25	0.15	0.63	0.66
Two-stage process									
RS	13.5	28.58	1554	209.5	0.93	0.59	0.34	0.64	1.14
RS–MP	18	21.06	1393	250.6	0.76	0.53	0.25	0.7	0.89
RS–MT	17	9.34	1444	244.9	0.67	0.54	0.13	0.81	0.84
RS–NaP	16.8	11.98	1762	295.3	0.91	0.65	0.26	0.72	1.08
RS–MT–NaP	24	5.72	1917	459.3	0.94	0.73	0.21	0.78	1.03

Reprinted with permission A.H. Basta, V. Fierro, H. El-Saied, A. Celzard, 2-Steps KOH activation of ricestraw: an efficient method for preparing high-performance activated carbons, Bioresour. Technol. 100 (2009) 3941–3947.

literature [60], the mechanism of phosphoric acid biomass activation starts with cellulose depolymerization followed by biopolymers dehydration, formation of aromatic rings, and finally the elimination of phosphate groups. This procedure leads to ACs with good percentage yields and high surface areas.

Adinaveen et al. [61] produced AC from pyrolyzed RS using H_3PO_4 as an activating agent. A surface area of 376 m^2/g and a maximum capacitance of 112 F/g were obtained for AC produced at 600°C.

The effect of physically and/or chemically deashing RS on the adsorption capacity of the ACs produced by chemical activation with H_3PO_4 was also studied [24]. RS was impregnated with an 85% phosphoric acid solution for 1 h using an acid to RS weight ratio of 1. The resultant mixture was introduced in a pyrolysis reactor, heated at 5 K/min up to the final activation temperature, which was held for 2 h under nitrogen flow. The derived carbon was cooled under nitrogen flow and washed with distilled water in a Soxhlet apparatus for 5 days. Surface area and pore volumes were determined by nitrogen adsorption at 77 K. Methylene blue adsorption isotherms onto ACs were carried out, and Langmuir and Freundlich models were applied to the adsorption data. ACs prepared by H_3PO_4 activation of raw and deashed RS have intermediate performances between those of ACs prepared by one- and two-step KOH activation, but are much cheaper to produce.

RS activation with carbon dioxide, potassium hydroxide, and phosphoric acid led to porous carbons; phosphoric acid ACs presented a high mesoporosity (94%–95%). RS precarbonization at 400°C prior to activation with KOH at 700°C produced carbon with a surface area of 1973 m^2/g. The precarbonized KOH AC was effective for the adsorption of methylene blue (cationic dye) and Congo red (anionic dye) presenting an adsorption capacity of 527.6 and 531.4 mg/g, respectively, as presented in Table 5.5.

4.2.3 Chemical activation with $ZnCl_2$

The BET surface area and specific capacitance of RHPC are further compared with previous porous carbon prepared from rice husk via carbonization–activation method. The porous carbon prepared from $ZnCl_2/CO_2$ activation method has a surface area of 480 m^2/g [62], lower than RHPC. Although the rice husk porous carbon prepared from SiO_2-leaching and CO_2 activation method has a much higher specific surface area (1357 m^2/g) [39] than RHPC, its capacitance only has a value of 106 F/g, comparable with RHPC.

4.2.4 Chemical activation with combined activation agents

It is interesting to note the reverse connection between surface area and yield of produced AC after chemical activating process. Whereas, the activating agents KOH and K_2CO_3 produce ACs with high surface areas, they present a

TABLE 5.5 Nitrogen adsorption results for activated carbons derived from rice straw.

Sample	S_{BET} (m^2/g)	S_{micro} (m^2/g)	Smeso (m^2/g)	V_{total} (cm^3/g)	V_{micro} (cm^3/g)	V_{meso} (cm^3/g)	V_{micro}/V_{total} (%)	V_{meso}/V_{total} (%)	Dp (nm)
N_2-700	158.1	101.5	56.6	0.129	0.051	0.078	39.70	60.30	3.269
CO_2-700	214.7	166.1	48.5	0.164	0.085	0.079	51.93	48.07	3.058
Pre-KOH	1973.0	1541.0	4321.5	1.131	0.769	0.362	67.97	32.03	2.292
Pre-H_3PO_4	392.6	47.3	345.2	0.463	0.021	0.443	4.54	95.46	4.727
RS-KOH	165.9	64.2	101.7	0.246	0.032	0.214	12.95	87.05	5.941
RS-H_3PO_4	629.1	111.0	518.2	0.935	0.054	0.881	5.81	94.19	5.949

Reprinted with permission S. Sangon, A.J. Hunt, T.M. Attard, Paweenuch Mengchang, Yuvarat Ngernyen, Nontipa Supanchaiyamat, Valorisation of waste rice straw for the production of highly effective carbon based adsorbents for dyes removal J. Clean. Prod. 172 (2018) 1128–1139.

TABLE 5.6 Rice husk chemical activation by combined activation methods (values from literature).

Chemical activator	Activation conditions	Surface area, m^2/g	Reference
NaOH and KOH	850	2681	Muniandy et al. 2014 [33]
NaOH and KOH		>2500	Cuo et al. 2002 [20]
NaOH and $ZnCl_2$	400 600	NaOH: 280 $ZnCl_2$: 1240	Isoda et al. 2014 [36]
NaOH and $ZnCl_2$		NaOH: 850 $ZnCl_2$: 900	Isoda et al. 2014 [36]
NaOH and H_3PO_4	700	750	Kalderis et al. [40]
H_3PO_4 and $ZnCl_2$	500	H_3PO_4: 1741 $ZnCl_2$: 2434	Liou et al. 2009 [32]

low yield; the reverse trend is noticed with the use of phosphoric acid (Table 5.6).

The precarbonization also affects the properties of the carbons produced. The precarbonization of RS at 400°C prior to activation at 700°C with KOH as activating agent resulted in AC with high specific surface area (1973 m^2/g) and a carbon content 55.5% C. The resulting advantages by precarbonization include avoiding the extensive pretreatments as well as the high mesoporosity presented by the ACs with H_3PO_4 as activator, either in direct pyrolysis or after precarbonization followed by activation [20,32,36,40,51].

4.2.5 Biological treatment

Activation with KOH and K_2CO_3 provide ACs with highest surface area, lowest yield, while phosphoric acid presented highest yield and lower surface area. Besides the above described activation methods, recently biological methods appeared to present sufficient results and gain increasing interest in the utilization of rice by-products as precursors for efficient ACs production [1].

Basta et al. [23] and Fierro et al. [31] compared for the first time the biological pretreatment of rice husk and RS to chemical pretreatment using H_3PO_4 as activating agent. They prepared ACs with high specific surface area as well as high yield. The ACs produced were successfully tested as adsorbents for their removal efficiency of MB.

Biological treatments were performed by cellulase (Cl) and peroxidase (Pe) producer organisms: *Rhizopus oryzae* and *Aspergillus awamori*. The rice waste fibers were exposed to cellulase (Cl) and peroxidase (Pe); after their exposure they were incubated for 14 days at 27 C. The RS fibers exposed to cellulase (Cl) and peroxidase (Pe) were labeled (RS-Cl) and (RS-Pe) and the rice husk fibers (RH-Cl) and (RH-Pe), respectively.

The chemical treatment (AT) of rice wastes was carried out with 12% NaOH at 60°C for 1 h. The rice waste biomass sources, after biological and chemical treatments, were dried at 105°C. The un- and treated rice waste fibers after their activation with phosphoric acid (ratio phosphoric acid: fiber 3:1) were pyrolized in nitrogen atmosphere at 450°C for 60 min. Finally the resulted ACs were washed with distilled water till neutrality and oven dried at 105°C.

Biological treatment on rice husk wastes, specifically treatment with cellulase, revealed beneficial effect for the surface area (1000 m^2/g) and yield x S_{BET} (777 m^2/g) of the AC prepared as seen in Table 5.7 [62]. The ACs from RS and RH, pretreated by cellulase provided higher maximum adsorption capacities (211–214 mg/g) for methylene blue than that presented by RS-based ACs carbonized by H_3PO_4 and commercial Aquasorb 1000 AC (~88.7 mg/g and 81.3 mg/g, respectively). Results are presented in Table 5.7.

FTIR measurements for ACs prepared after biological and chemical treatment indicated a blue shift and also an increase in intensity of absorption bands related to the stretching vibration of –OH groups. The higher alterations were noticed for chemically treated biomass precursors attributed to the removal not only of the undesirable components (silica and wax) but also parts of hemicelluloses, lignin, and lower chain of celluloses. For biologically treated RS and RH biomass, by cellulase and peroxidase, a selective removal of components occurred.

4.2.6 Hydrothermal treatment

Recently, a new method for ACs preparation has been developed; carbonization takes place in hot water at temperatures between 180–250°C under self-generated pressure [63]. The hydrothermal carbon (HTC) prepared this way could be functionalized for catalytic activities. Although the hydrothermal carbonization process resembles the process of biomass conversion to HMF, furfural, glucose, levulinic acid, and formic acid, it demands longer time [63]. Biomass residues, i.e., amino acids [64], phenolic compounds [65], etc., could be successfully converted to ACs with HTC. Defatted rice bran comprising cellulose (25.48 wt.%), hemicellulose (9.35 wt.%), lignin (25.63 wt.%), ash (9.89 wt.%), and others (29.65 wt.%) has been successfully used for the preparation of ACs by hydrothermal treatment [66]. The HTC was prepared according to a typical procedure; an amount of carbon was added in a closed stainless steel batch reactor in specific volume of distilled water. The reactor was then heated to the desired temperature (180–250°C), and after a period of 1–8 h at the desired reaction temperature, the reactor was cooled to room temperature. The HTC prepared was separated from the liquid carbonization

TABLE 5.7 The textural properties of activated carbons (ACs) prepared from different alkaline and microbial treatments of rice husk and rice straw.

Sample code	S_{BET} (m^2/g)	CxSBET (m^2/g)	S_{mic} (m^2/g)	VT(0.95) (cm^3/g	V_{mic} (cm^3/g)	V_{mes} (cm^3/g)	V_{mic}/V_T (%)	Iodine value (mg/g)
RS-cont.	543	329	328	0.304	0.154	0.15	50.7	658
RS-AT	814	460	393	0.612	0.176	0.436	28.8	1272
RS-Cl	538	404	282	0.375	0.129	0.246	34.4	601
RS-Pe	601	345	242	0.304	0.111	0.193	36.5	627
RH-cont.	426	299	228	0.278	0.106	0.172	38.1	687
RH-AT	504	337	303	0.267	0.139	0.128	52.1	949
RH-Cl	1000	777	565	0.607	0.261	0.346	43.0	744
RH-Pe	503	412	327	0.236	0.153	0.083	64.8	904

Reprinted with permission A.H. Basta, V.F. Lotfy, M.S. Hasanin, P. Trens, H., El-Saied, Efficient treatment of rice byproducts for preparing high-performance activated carbons, J. Clean. Prod. 207 (2019) 284–295.

portion by filtration, washed with deionized water until neutral pH, and dried overnight at 110°C (HTCDRB).

5. Activated carbons by rice husk and rice straw as adsorbents

ACs are important adsorbent materials and have been widely used for removing different pollutants from water and air. ACs prepared by rice husk and RS have been tested as adsorbents for dyes as well as for metals and pharmaceutical products. The main parameters examined for a study of pollutant adsorption include the pH of the solution, contact time, initial concentration, adsorbent dose, and temperature.

Luna et al. [67] conducted the removal of Eriochrome Black T dye using rice husk–activated carbon (RHAC) to find the optimum conditions based on initial concentration, adsorbent dose, and pH. The highest removal achieved was 95.9%.

Nama et al. [47] evaluated the performance of AC as adsorbent. Rice straw biochar was chemically activated with KOH solution. The largest surface area and micropores for the AC they prepared were found for the activation temperature of 750°C (1330 m^2/g). The AC was evaluated as an adsorbent material to remove pharmaceutical pollutants.

Up to 95% and 95.3% of acetaminophen and ibuprofen pollutants, respectively, in an aqueous solution were removed after a 24 h contact time [38], similar to the results reported, i.e., 95.9% of Eriochrome Black T dye by rice straw AC by Luna et al. [67]. Table 5.8 shows the percent removal of acetaminophen and ibuprofen from the initial concentration.

Low-cost RS-based carbons were prepared with a surface area and pore volume of 67.4 m^2/g and 0.134 cm^3/g, respectively. The adsorption of azo-coumarin dye was sufficient with respect to initial dye concentration, contact time, solution pH, and temperature.

Porous ACs [38] have been produced by pyrolysis from waste RS. AC prepared from RS after precarbonization at 400 °C followed by activation with potassium hydroxide at 700°C, presented a surface area of 1973 m^2/g and

TABLE 5.8 Adsorptive removal of acetaminophen and ibuprofen using rice straw–activated carbon by RSAC750.

Compounds	Initial concentration, mg/L	Final concentration, mg/L	% Removal
Acetaminophen	10.0	0.50	95.0
Ibuprofen	10.0	0.47	95.3

TABLE 5.9 Adsorption capacities for rice straw–activated carbons (ACs).

ACs	Adsorbent surface area (m^2/g)	Adsorption capacity (mg/g)	
		MB	CR
N_2-700	151.8	105.1	245.1
CO_2-700	214.7	44.2	253.9
Pre-KOH	1973.0	527.6	531.4
Pre-H_3PO_4	392.6	34.7	34.7
RS-KOH	165.9	392.4	178.4
RS-H_3PO_4	629.1	629.1	208.4

Reprinted with permission S. Sangon, A.J. Hunt, T.M. Attard, Paweenuch Mengchang, Yuvarat Ngernyen, Nontipa Supanchaiyamat, Valorisation of waste rice straw for the production of highly effective carbon based adsorbents for dyes removal J. Clean. Prod. 172 (2018) 1128–1139.

proved to be effective for the adsorption of methylene blue (cationic dye) and Congo red (anionic dye) with a maximum adsorption capacity of 527.6 and 531.4 mg/g, respectively, as presented in Table 5.9. Adsorption isotherm studies revealed that adsorption was mainly physical while the pre-KOH material found to be the most efficient dye adsorbent for MB and CR as well.

Methylene blue adsorption isotherms onto ACs were carried out [24], and Langmuir and Freundlich models were applied to the adsorption data. ACs prepared with H_3PO_4 activation presented an intermediate adsorption capacity between these obtained by one-step and two-step KOH ACs, but their production was much cheaper. Although H_3PO_4 activation compared to two-step KOH activation presented a lower cost and comparatively lower performances of the resultant adsorbents, the resulted ACs may be successfully used as inexpensive yet effective adsorbents for dye removal in water treatment applications.

Biological treatment on rice husk wastes, specifically treatment with cellulase, revealed beneficial effect for the surface area (1000 m^2/g) and yield x SBET (777 m^2/g) of the AC prepared [62]. The ACs from RS and RH, pretreated by cellulase provided higher maximum adsorption capacities (211–214 mg/g) for methylene blue than that presented by RS-based ACs carbonized by H_3PO_4 and commercial Aquasorb 1000 AC ($\sim$88.7 mg/g and 81.3 mg/g, respectively). However, alkali treatment of RS amply surpasses methylene blue adsorption capacity (247 mg/g) as presented in Table 5.10, while in Table 5.10 it is presented the adsorption capacity for different pollutants, of the rice wastes–derived ACs along with the activation method and the specific surface area it was developed during their activation.

TABLE 5.10 Adsorption capacities for rice straw– and rice husk–activated carbons (values from literature).

	Preactivation	Activation	Surface area, m^2/g	Pollutant	Q_{max}, mg/g	Reference
Rice straw (RS)	Preoxidation at 200°C	$(NH_4)_2HPO_4$ 700°C	1154	Methylene blue	129.5	Pin Gao et al. 2011 [48]
RS	NaOH pulping	Two-step KOH 800°C	1917	Methylene blue		Basta et al. 2009 [23]
RS RH		Cellulase treatment Alkali treatment Peroxidase treatment	1000	Methylene blue	247 214 196	Basta et al. 2019 [1]
RS	Precarbonization 400°C	Carbon dioxide, potassium hydroxide, and phosphoric acid 700°C	1973	Methylene blue Congo red	527.6 531.4	Suwiwat Sangon et al. 2018 [38]
RS		H_3PO_4 500°C	857	Methylene blue	103.2	Basta et al. 2011 [24]
				Methylene blue		Fierro et al. 2010 [31]
RS		Mixture of HNO_3 and H_2O_2		Cd	93.2	Huiyan Zhang et al. 2018 [68]
RS		1200°C	67.4	Azocoumarin dye	43.69 25 °C 156.34 60°C	Ashraf A. El-Bindary et al. 2014 [39]

Continued

TABLE 5.10 Adsorption capacities for rice straw– and rice husk–activated carbons (values from literature).—cont'd

	Preactivation	Activation	Surface area, m^2/g	Pollutant	Q_{max}, mg/g	Reference
RS		KOH	1304.8	Carbofuran	296.52	Ken-Lin Chang et al. 2014 [46]
RS		Microwave-assisted H_2SO_4 activation		Hg (II)	500	Somaye Mashhadi et al. 2016 [69]
RS		Pyrolysis at 600°C	162.6	Cu Zn	56.5 38.6	Jong-Hwan Parka et al. 2017 [70]
RS	Burned in the air	KCl	249	Cr(VI)	100% (in 50 mg)	Nai-Hua Hsua et al. 2009 [71]
RS	Biochar		71.53	Sr	175.95	Jiseon Jang et al. 2018 [72]
RS	Steam 550°C	KOH	143.3	Uranium (VI) and humic acid (HA)	100 mg/g 21.1 mg/g	Yakout et al. 2013 [44]
RS RH	110°C in air	NaCl	282.1 281.7	Methylene blue	90.9 72.4	Jatinder Kumar Ratan et al. 2018 [73]

From the adsorption results presented in literature for AC adsorbents prepared from rice wastes, it is concluded that although KOH activator provided, even with lowest yield, ACs with highest surface area than phosphoric acid activator (that presented highest yield and lower surface area), the ACs prepared by both activators shown importantly high adsorption results. Besides, ACs activated with biological methods appeared to present sufficient adsorption results, as well. Adsorption capacities estimated for ACs from rice wastes exceeded in many times than those reported in literature for other absorbents that require less pretreatment. From the economic point of view, these highly porous carbons from waste rice biomass are cost competitive with other commercially available ACs. They can be utilized as low-cost yet very effective adsorbents for pollution control as well as in other applications including catalysis and chromatography.

6. Conclusions

Rice waste biomass, e.g., rice husk and RS proved to be a sufficient source for AC preparation. ACs can be produced after different activation routes with or without pretreatment, at different activation temperatures and with different activators. In all cases, the carbons derived present efficient specific surface area and pore structure. The surface area found to reach a higher value when two-step activation with KOH follows pretreatment while mesoporosity was developed when H_3PO_4 was the activating agent. Biological treatment, with cellulase and peroxidase presented sufficient results in surface area and porosity while hydrothermal methods was less effective. RS and rice husk−derived ACs presented high maximum adsorption capacity when tested as adsorbents for dyes, pharmaceuticals, and metals.

References

[1] A.H. Basta, V.F. Lotfy, M.S. Hasanin, P. Trens, H. El-Saied, Efficient treatment of rice byproducts for preparing high-performance activated carbons, J. Clean. Prod. 207 (2019) 284−295.

[2] M.H. Zong, Z.Q. Duan, W.Y. Lou, T.J. Smith, H. Wu, Preparation of a sugar catalyst and its use for highly efficient production of biodiesel, Green Chem. 9 (5) (2007) 434−437.

[3] S. Suganuma, K. Nakajima, M. Kitano, D. Yamaguchi, H. Kato, S. Hayashi, M. Hara, Synthesis and acid catalysis of cellulose-derived carbon-based solid acid, Solid State Sci. 12 (6) (2010) 1029−1034.

[4] P. Wataniyakula, P. Boonnounb, A.T. Quitainc, T. Kidac, N. Laosiripojanad, A. Shotipruka, Preparation of hydrothermal carbon acid catalyst from defatted rice bran, Ind. Crops Prod. 117 (2018) [286]−[294].

[5] R.C. Bansal, M. Goyal, Activated Carbon Adsorption, CRC Press, Boca Raton, FL, 2005.

[6] FAO, Rice Market monitor, in: Food and Agricultural Commodities Production, 2017, vol. XVIII (4), Food and Agriculture Organisation of the United Nations, 2017, pp. 1−34.

[7] N.T. Kim Oanh, T.L. Bich, D. Tipayarom, B.R. Manandhar, P. Prapat, C.D. Simpson, L.J. Sally Liu, Characterization of particulate matter emission from open burning of rice straw atmos, Environ. 45 (2) (2011) 493–502.
[8] K. Torigoe, S. Hasegaw, O. Numata, S. Yazaki, M. Matsumaga, N. Boku, M. Hiura, H. Ino, Influence of emission from rice straw burning on bronchial asthma in children, Pediatr. Int. 42 (2000) 143–150.
[9] UNEP, 2002.
[10] United Nations Environment Programme, Annual Report, 2008.
[11] W. Zhang, N. Lin, D. Liu, J. Xu, J. Sha, J. Yin, X. Tan, H. Yang, H. Lu, H. Lin, Direct carbonization of rice husk to prepare porous carbon for supercapacitor applications, Energy 128 (2017) 618–625.
[12] D. Suhardy, S. Mohd Nazry, K. Farizul Hafiz, S. Saiful Azher, Characterization of sugarcane bagasse, rice straw and rice husk and their suitability for paper production, in: International Conference of Sustainable Materials (ICoSM), Universiti Malaysia Perlis, 2007.
[13] M. Yasin, A. Bhutto, A.A. Bazmi, S. Karim, Efficient utilization of rice-wheat straw to produce value added composite products, Int. J. Chem. Environ. Eng. 1 (2) (2010) 136–143.
[14] Y. Li, X. Ding, Y. Guo, L. Wang, C. Rong, Y. Qu, Z. Wang, A simple and highly effective process for the preparation of activated carbons with high surface area, Mater. Chem. Phys. 127 (3) (2011) 495–500.
[15] FAO Rice Market Monitor, November, 2012, vol. XV, issue no. 4
[16] M.M. Hessien, M.M. Rashad, R.R. Zaky, E.A. Abdel-Aal, K.A. El-Barawy, J. Mater. Sci. Eng. B 162 (1) (2009) 14–21.
[17] A. Álvaro Guzmán, S. Marisol Gordillo, A. Silvio Delvasto, V. María Francisca Quereda, V. Enrique Sánchez, Optimization of the technological properties of porcelain tile bodies containing rice straw ash using the design of experiments methodology, Ceram. Int. 42 (2016) 15383–15396.
[18] I. Demiral, A. Eryazıcı, A. Sensoz, Bio-oil production from pyrolysis of corn-cob (*Zea mays* L.), Biomass Bioenergy 36 (2012) 43–49.
[19] N. Yalcin, V. Sevinç, Studies of the surface area and porosity of activated carbons prepared from rice husks, Carbon 38 (14) (2000) 1943–1945.
[20] Y. Guo, S. Yang, K. Yu, J. Zhao, Z. Wang, H. Xu, The preparation and mechanism studies of rice husk based porous carbon, Mater. Chem. Phys. 74 (3) (2002) 320–323.
[21] G.T.K. Fey, C.L. Chen, High-capacity carbons for lithium-ion batteries prepared from rice husk, J. Power Sources 97–98 (2001) 47–51.
[22] H. El-Saied, A.H. Basta, A.I. Waly, O.A. El-Hady, C.Y. El-Dewiny, Evaluating the grafting approaches for utilizing the rice straw as environmental friendly and potential low cost hydrogels, Emir. J. Food Agric. 25 (3) (2013) 211–224.
[23] A.H. Basta, V. Fierro, H. El-Saied, A. Celzard, 2-Steps KOH activation of ricestraw: an efficient method for preparing high-performance activated carbons, Bioresour. Technol. 100 (2009) 3941–3947.
[24] A.H. Basta, V. Fierro, H. El-Saied, A. Celzard, Effect of deashing rice straws on their derived activated carbons produced by phosphoric acid activation, Biomass Bioenergy 35 (2011) 1954–1959.
[25] A.H. Basta, H. El-Saied, O.A. El-Hadi, C.Y. El-Dewiny, Evaluation of rice straw based hydrogels for purification of wastewater, Polym. Plast. Technol. Eng. 52 (2013) 1074–1080.
[26] A.H. Basta, H. El-Saied, V.F. Lofty, Performance assessment of deashed and dewaxed rice straw on improving the quality of RS-based composites, RSC Adv. 4 (42) (2014a) 21794–21801.

[27] A.H. Basta, H. El-Saied, V.F. Lofty, Performance of rice straw-based composites using environmentally friendly polyalcoholic polymers-based adhesive system, Pigment Resin Technol. 42 (1) (2014b) 24–33.

[28] G.H. Oh, C.R. Park, Preparation and characterization of rice straw based porous carbons with high adsorption capacity, Fuel 81 (2002) 327–336.

[29] Y. Guo, J. Zhao, H. Zhang, S. Yang, J. Qi, Z. Wang, H. Xu, Use of rice husk-based porous carbon for adsorption of Rhodamine B from aqueous solutions, Dyes Pigments 66 (2) (2005) 123–128.

[30] Y. Zhang, Y.H. Guo, X. Liu, Y.H. Zao, Z.C. Deng, L. Wang, Adsorption of creatinine on porous carbons prepared from rice husks, Acta Physico-Chim. Sin. 23 (2007) 825–828.

[31] V. Fierro, G. Muniz, A.H. Basta, H. El-Saied, A. Celzard, Rice straw as precursor of activated carbons: activation with ortho-phosphoric acid, Hazard Mater 181 (2010) 27–33.

[32] T.H. Liou, S.J. Wu, Characteristics of microporous/mesoporous carbons prepared from rice husk under base- and acid-treated conditions, J. Hazard Mater. 171 (1–3) (2009) 693–703.

[33] L. Muniandy, F. Adam, A.R. Mohamed, E.P. Ng, The synthesis and characterization of high purity mixed microporous/mesoporous activated carbon from rice husk using chemical activation with NaOH and KOH, Microporous Mesoporous Mater. 197 (2014) 316–323.

[34] D. Kalderis, S. Bethanis, P. Paraskeva, E. Diamadopoulos, Production of activated carbon from bagasse and rice husk by a single-stage chemical activation method at low retention times, Bioresour. Technol. 99 (15) (2008) 6809–6816.

[35] F. Rodriguez-Reinoso, M. Molina-Sabio, M.T. Gonzalez, The use of steam and CO_2 as activating agents in the preparation of activated carbons, Carbon 33 (1) (1995) 15–23.

[36] N. Isoda, R. Rodrigues, A. Silva, M. Gonçalves, D. Mandelli, F.C.A. Figueiredo, W.A. Carvalho, Optimization of preparation conditions of activated carbon from agriculture waste utilizing factorial design, Powder Technol. 256 (2014) 175–181.

[37] J. Pallarés, A. González-Cencerrado, Inmaculada Arauzo, Production and characterization of activated carbon from barley straw by physical activation with carbon dioxide and steam, Biomass Bioenergy 115 (2018) 64–73.

[38] S. Sangon, A.J. Hunt, T.M. Attard, Paweenuch Mengchang, Yuvarat Ngernyen, Nontipa Supanchaiyamat, Valorisation of waste rice straw for the production of highly effective carbon based adsorbents for dyes removal, J. Clean. Prod. 172 (2018) 1128–1139.

[39] A.A. El-Bindary, A.Z. El-Sonbati, A.A. Al-Sarawy, K.S. Mohamed, M.A. Farid, Adsorption and thermodynamic studies of hazardous azocoumarin dye from an aqueous solution onto low cost rice straw based carbons, J. Mol. Liq. 199 (2014) 71–78.

[40] D. Kalderis, D. Koutoulakis, P. Paraskeva, E. Diamadopoulos, E. Otal, J.O. Valle, del, C. Fernández-Pereira, Adsorption of polluting substances on activated carbons prepared from rice husk and sugarcane bagasse, Chem. Eng. J. 144 (1) (2008) 42–49.

[41] S. Somasundaram, K. Sekar, V.K. Gupta, S. Ganesan, Synthesis and characterization of mesoporous activated carbon from rice husk for adsorption of glycine from alcohol-aqueous mixture, J. Mol. Liq. 177 (2013) 416–425.

[42] Z.Z. Chowdhury, S.B.A. Hamid, R. Das, M.R. Hasan, S.M. Zain, K. Khalid, M.N. Uddin, Preparation of carbonaceous adsorbents from lignocellulosic biomass and their use in removal of contaminants from aqueous solution, BioResources 8 (4) (2013) 6523–6555.

[43] H. Jin, J. Hua, S. Wua, X. Wanga, Z. Hui, X. Hui, K. Lian, Three-dimensional interconnected porous graphitic carbon derived fromrice straw for high performance supercapacitors, J. Power Sources 384 (2018) 270–277.

[44] S.M. Yakout, S.S. Metwally, T. El-Zakla, Uranium sorption onto activated carbon prepared from rice straw: competition with humic acids, Appl. Surf. Sci. 280 (2013) 745–750.

[45] L. Zhu, F. Shen, L. Richard, Smith Jr., L. Yan, L. Li, Xinhua Qi, Black liquor-derived porous carbons from rice straw for high-performance supercapacitors, Chem. Eng. J. 316 (2017) 770–777.
[46] K.-L. Chang, C.-C. Chen, J.-H. Lin, J.-F. Hsien, Y. Wang, F. Zhao, Y.-H. Shih, Z.-J. Xing1, S.-T. Chen, Rice straw-derived activated carbons for the removal of carbofuran from an aqueous solution, N. Carbon Mater. 29 (1) (2014).
[47] H. Nama, W. Choic, D.A. Genuinoa, S.C. Capareda, Development of rice straw activated carbon and its utilizations, J. Environ. Chem. Eng. 6 (2018) 5221–5229.
[48] P. Gao, Z.-hong Liu, G. Xue, B. Han, Z. Mei-hu, Preparation and characterization of activated carbon produced from rice straw by (NH4)2HPO4 activation, Bioresour. Technol. 102 (2011) 3645–3648.
[49] B.S. Girgis, A.A. El-Hendawy, Porosity development in activated carbons obtained from date pits under chemical activation with phosphoric acid, Micropor. Mesopor.Mater. 52 (2002) 105–117.
[50] T. Vernersson, P.R. Bonelli, E.G. Cerrella, A.L. Cukierman, Arundodonax cane as a precursor for activated carbons preparation by phosphoric acid activation, Bioresour. Technol. 83 (2002) 95–104.
[51] J. Guo, A.C. Lua, Adsorption of sulphur dioxide onto activated carbon prepared from oil-palm shells with and without pre-impregnation, Separ. Purif. Technol. 30 (2003) 265–273.
[52] R.A. Shawabkeh, D.A. Rockstraw, R.K. Bhada, Copper and strontium adsorption by a novel carbon material manufactured from pecan shells, Carbon 40 (2002) [781]–[786].
[53] Y. Diao, W.P. Walawender, L.T. Fan, Activated carbons prepared from phosphoric acid activation of grain sorghum, Bioresour. Technol. 81 (2002) [45]–[52].
[54] H. Benaddi, T.J. Bandosz, J. Jagiello, J.A. Schwarz, J.N. Rouzaud, D. Legras, F. Béguin, Surface functionality and porosity of activated carbons obtained from chemical activation of wood, Carbon 38 (2000) 669–674.
[55] M. Jagtoyen, F. Derbyshire, Activated carbons from yellow poplar and white oak by H_3PO_4 activation, Carbon 36 (1998) 1085–1097.
[56] C. Toles, S. Rimmer, J.C. Hower, Production of activated carbons from a Washington lignite using phosphoric acid activation, Carbon 34 (1996) 1419–1426.
[57] H. Teng, T.S. Yeh, L.Y. Hsu, Preparation of activated carbon from bituminous coal with phosphoric acid activation, Carbon 36 (1998) 1387–1398.
[58] A.M. Puziy, O.I. Poddubnaya, B. Gawdzik, M. Sobiesiak, D. Dziadko, Heterogeneity of synthetic carbons obtained from polyimides, Appl. Surf. Sci. 196 (2002) 89–97.
[59] A.M. Puziy, O.I. Poddubnaya, A. Martínez-Alonso, F. Suárez-García, J.M.D. Tascón, Characterization of synthetic carbons activated with phosphoric acid, Appl. Surf. Sci. 200 (2002) 196–202.
[60] H. Benaddi, D. Legras, J.N. Rouzaud, F. Béguin, Influence of the atmosphere in the chemical activation of wood by phosphoric acid, Carbon 36 (1998) 306–309.
[61] T. Adinaveen, L.J. Kennedy, J.J. Vijaya, G. Sekaran, Surface and porous characterization of activated carbon prepared from pyrolysis of biomass (rice straw) by two-stage procedure and its applications in supercapacitor electrodes, J. Mater. Cycles Waste Manag. 17 (4) (2015) 736–747.
[62] S. Karthikeyan, P. Sivakumar, P.N. Palanisamy, Novel activated carbons from agricultural wastes and their characterization, J. Chem. 5 (2008) 409–426.

[63] J.A. Libra, K.S. Ro, C. Kammannn, A. Funke, N.D. Berge, Y. Neubauer, Hydrothermal carbonization of biomass residuals: a comparative review of chemistry, processes and applications of wet and dry pyrolysis, Biofuels 2 (1) (2011) 71–106.

[64] I. Sereewatthanawut, S. Prapintip, K. Watchiraruji, M. Goto, M. Sasaki, A. Shotipruk, Extraction of protein and amino acids from deoiled rice bran by subcritical water hydrolysis, Bioresour. Technol. 99 (2008) 555–561.

[65] P. Wataniyakul, P. Pavasant, M. Goto, A. Shotipruk, Microwave pretreatment of defatted rice bran for enhanced recovery of total phenolic compounds extracted by subcritical water, Bioresour. Technol. 124 (2012) 18–22.

[66] K. Yua, J. Lia, Q. Hui, C. Liang, High-capacity activated carbon anode material for lithium-ion batteriesprepared from rice husk by a facile method, Diam. Relat. Mater. 86 (2018) 139–145.

[67] M.D.G. de Luna, E.D. Flores, D.A.D. Genuino, C.M. Futalan, M.-W. Wan, Adsorption of Eriochrome Black T (EBT) dye using activated carbon prepared from waste rice hulls—optimization, isotherm and kinetic studies, J. Taiwan Ins. Chem. Eng. 44 (2013) 646–653.

[68] H. Zhang, X. Yue, F. Li, R. Xiao, Y. Zhang, D. Gu, Preparation of rice straw-derived biochar for efficient cadmium removal by modification of oxygen-containing functional groups, Sci. Total Environ. 631–632 (2018) 795–802.

[69] S. Mashhadi, R. Sohrabi, H. Javadian, M. Ghasemi, I. Tyagi, S. Agarwal, V.K. Gupta, Rapid removal of Hg (II) from aqueous solution by rice straw activated carbon prepared by microwave-assisted H_2SO_4 activation: kinetic, isotherm and thermodynamic studies, J. Mol. Liq. 215 (2016) 144–153.

[70] J.-H. Parka, J.J. Wanga, S.-H. Kimb, J.-S. Choc, S.-W. Kangc, R.D. Delauned, K.-J. Hana, D.-C. Seo, Recycling of rice straw through pyrolysis and its adsorption behaviors for Cu and Zn ions in aqueous solution, Colloids Surf. A 533 (2017) [330]–[337].

[71] N.-H. Hsua, S.-L. Wanga, Y.-H. Liaoa, S.-T. Huangb, Y.-M. Tzoua, Y.-M. Huanga, Removal of hexavalent chromium from acidic aqueous solutions using rice straw-derived carbon, J. Hazard Mater. 171 (2009) 1066–1070.

[72] J. Jang, W. Miran, S.D. Divine, M. Nawaz, Asif shahzad, Seung Han Woo, Dae sung lee, Rice straw-based biochar beads for the removal of radioactive strontium from aqueous solutions, Sci. Total Environ. 615 (2018) 698–707.

[73] J. Kumar Ratan, M. Kaur, B. Adiraju, Synthesis of activated carbon from agricultural waste using a simple method: characterization, parametric and isotherms study, Mater. Today: Proceedings 5 (2018) 3334–3345.

Chapter 6

Low-cost materials in gas-phase adsorption

Ramonna I. Kosheleva,* George Z. Kyzas and Athanasios C. Mitropoulos*
Hephaestus Advanced Laboratory, Department of Chemistry, International Hellenic University, Kavala, Greece
**Corresponding author*

1. Introduction

Adsorption technology is broadly applied in many fields. Industrial application of adsorbents is mainly focused on wastewater treatment and water purification [1,2], namely for removal of dyes [3,4], heavy metal ions [5,6], and other compounds, i.e., pharmaceuticals [7]. Comparing available literature for liquid-phase adsorption with that of gases adsorption, there is a gap for the latter case, especially concerning the application of low-cost and environmental friendly materials. A low-cost material is best determined by the precursor used for its fabrication; some of the criteria considered are its abundance/availability, the ratio of material obtained per amount of raw material needed, adsorption capacity, etc. [8] On the other hand, eco-friendly materials are those that conform with green technology in respect to their synthesis process as well as their reusability [9]. Many of the aforementioned aspects are reached by the implementation of carbon-based materials derived from various sources: biomass [10], wastes of many different kinds (food waste [11,12], tires [13,14], agro-wastes [15,16], etc.), and others [17].

Application of adsorbents for gas-phase adsorption is concentrated on toxic and hazardous gases produced by industrial processes and other sources, involving human activity. Greenhouse gases are those of high concern due to the formation of greenhouse effect. According to Intergovernmental Panel on Climate Change (IPCC), greenhouse effect is the warming process of the planet's surface due to radiation retainment by its atmosphere. Climate change and environmental protection is the number one priority worldwide, while

Advanced Low-Cost Separation Techniques in Interface Science. https://doi.org/10.1016/B978-0-12-814178-6.00006-6

technologies toward mitigation of greenhouse gas emissions are developing and upgrading constantly to reach the desired balance [18].

As it was already mentioned, carbonaceous materials are the most common adsorbents because of their physical and, when modified, physicochemical properties. Materials such as activated carbon, biochar, and their composites are of big interest through older and recent history. The presented work has six sections; an introduction part where the most common carbonaceous materials (activated carbon and biochar) are briefly presented; the second section includes adsorbent trends for CO_2 capture; in the third section the reduction technologies of acidifying gases (NOx and Sox) are discussed; the fourth section presents adsorbents used for CH_4; the fifth for volatile organic compounds (VOCs), and in the last sections there are conclusions based on the context of this chapter. Although there are a sufficient number of released reviews on gas-phase adsorption, most of the works select to discuss adsorbents applied for adsorption of one gas at a time, with CO_2 holding a leading role. Finally, this chapter provides a summarized overview of updated studies in respect to carbonaceous materials for gas-phase adsorption with focus on the main greenhouse gases. Materials for VOCs adsorption are discussed briefly as well, while at the end of each section there is a table for a quick overview of recently developed materials.

1.1 Activated carbon

Activated carbon is a well-known material, presenting high surface area, mechanical stability, and reusability. The common procedure for activated carbon synthesis is a two-step process: (1) carbonization of the precursor and (2) activation of the resulted char [19]. There are many cases, that activation step is completed along with carbonization, so we have only one-step process. Most of the cases of the latter, activation that takes place is a physical one with thermal treatment at a selected gas atmosphere (i.e., activation with CO_2), while when chemical activation is selected, the process performed is usually a two-step one [20]. The obtained properties of the final material are related to many parameters. For example, both nature of the precursor and synthesis conditions have a crucial impact on the resulted characteristics of adsorbent [6,21]. One of the most important factors that affect surface properties of the activated carbon were found to be carbonization and/or activation temperature like literature suggests. In Fig. 6.1, one can realize the role of carbonization temperature on the evolution of pores at wood-based activated carbon surface. The pore structure of activated carbon is usually micropore-dominant, a fact that makes it a strong candidate for gas-phase adsorption [22].

Modification of activated carbon surface can lead to improvement of adsorption, mechanical, and other properties of the material. Modification is accomplished by introducing functional groups onto the surface resulting in

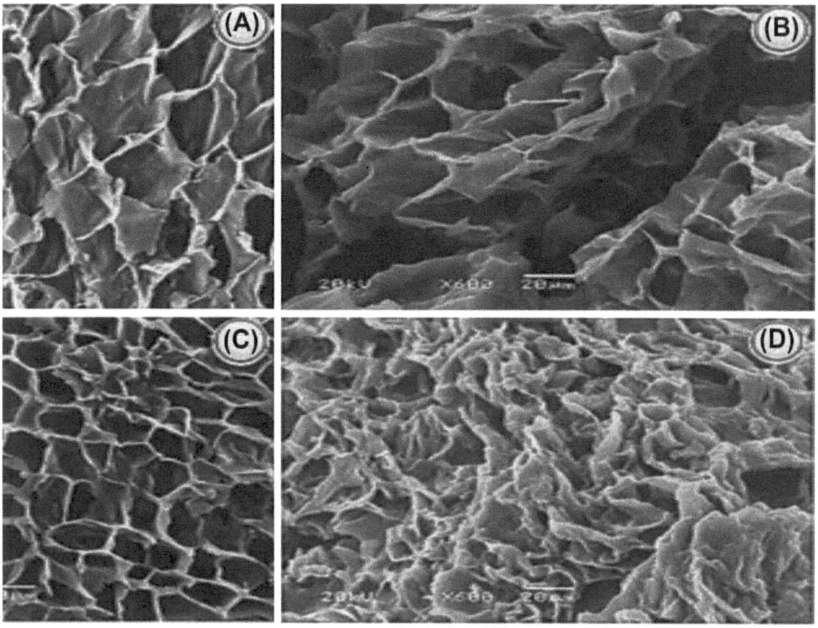

FIGURE 6.1 SEM micrographs of wood-based activated carbon at different carbonization temperature with H_3PO_4 impregnation ratio of 4 at (A) 300°C, (B) 400°C, (C) 500°C and (D) 600°C [23].

chemical properties alteration, hence achieving higher adsorption capacity of the target adsorbate [24,25].

1.2 Biochar

In contrast to activated carbon, biochar is commonly produced after slow pyrolysis usually at low temperature and total absence of oxygen [26,27]. Additionally, a recent trend is hydrothermal treatment of biomass, concluding in the so-called hydrochar. Hydrochar production process offers some significant advantages over the former; no necessary drying of biomass, no emissions, and by-products compared with pyrolysis as well as higher carbon yield are some of them [28]. As it is shown in Fig. 6.2, structural characteristics of biochar differ from those of hydrochar.

In fact, biochar used as adsorbent gains popularity among other carbonaceous materials. Biochar is considered to cost less than activated carbon while having superior performance even for gas-phase adsorption [29−31]. As it is common for all pyrogenic materials, synthesis parameters are those that have the key role of the overall adsorbent performance. Like activated carbon, biochar can be modified by various chemicals in order to reach ultimate quality and/or specific application requirements [32,33].

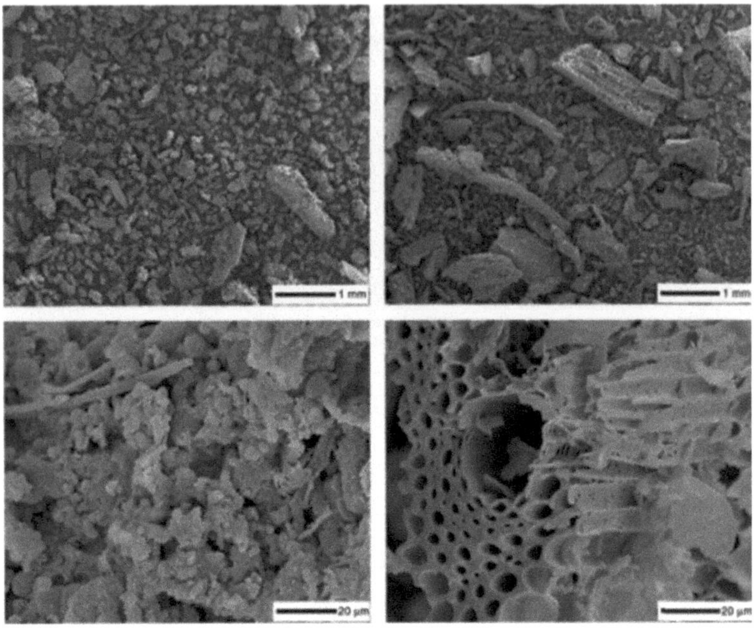

FIGURE 6.2 SEM micrographs of (left) biochar and (right) hydrochar. *Reprinted with permission by Elsevier from J. Fang, L. Zhan, Y.S. Ok, B. Gao, Minireview of potential applications of hydrochar derived from hydrothermal carbonization of biomass, J. Ind. Eng. Chem. 57 (2018) 15–21. doi: https://doi.org/10.1016/j.jiec.2017.08.026.*

2. Carbon dioxide (CO_2) capture

Carbon dioxide is a greenhouse gas that comes first regarding its production rate and total emission amounts worldwide. According to Environmental Protection Agency (EPA) and reports from other related sources, CO_2 holds 76% of the greenhouse gas emissions. The majority of CO_2 is emitted from the agricultural sector and industrial processes such as power production. Due to global warming and its ever increasing trend, there have been developed CO_2 capture and sequestration technologies [34], involving trapping, transportation, and its long-term storage. Adsorption is one of the most common methods for both trapping and storage of the emitted CO_2. Although CO_2 storage refers to storage in geological formations, trapping of CO_2 is achieved by adsorption onto synthesized materials such as MOFs, modified zeolites, and various membranes. The particular step of CO_2 capture (trapping) can take place at different stages of CO_2 production/emission cycle; such options include (1) precombustion, (2) postcombustion, (3) oxy-fuel, and (4) CO_2 capture from the air stream prior to combustion process.

As was mentioned before, many materials have the potential to be an effective adsorbent with high CO_2 uptake. However, the implementation of some of them has restrictions regarding performance under specific operation conditions [35], the vapor content of the stream [36,37] and the overall cost estimation [38]. Adsorption of CO_2 by carbon-based materials is the most promising approach especially in terms of cost-effectiveness and adsorption efficiency. Numerous studies have been conducted regarding carbon-based materials for the specific purpose, suggesting new synthesis processes [39,40] and materials or enhancing material properties by surface modification [41,42]. One of the most important cost reduction parameters of any kind of adsorbents is the abundance of their precursors. Carbonaceous materials can be derived from many sources including waste materials, resins, used tires [43], and many others, presenting good stability [44,45], fast kinetics [35], as well as reusability [46].

From the very early stage of industrialization, the most common adsorbent for industrial use was activated carbon. Until recently, activated carbon was preferred for processes such as wastewater treatment [47,48]. Although activated carbon has a poor adsorption performance of gaseous phase, modification of its surface in numerous ways is currently been studied. The interest to this approach comes from the fact that activated carbon has a low cost, and it can be manipulated easily in order to reach the required properties also in a relatively low cost [49].

Sea mango was the precursor of the synthesized activated carbon for CO_2 adsorption studied by Ali et al. [50]. The produced material was activated chemically by phosphoric acid (H_3PO_4), while a portion of the resulted material was treated further with various amines following Box–Behnken Design and the surface area was modified. During the synthesis, factors such as temperature for sea mango carbonization and H_3PO_4 concentration of activation process were investigated as well. After optimization of the mentioned conditions, both cases (modified and virgin activated carbon) were tested for CO_2 uptake capacity. After amine impregnation, materials were evaluated regarding common adsorption parameters and their performance in a bed column. As it was expected, amine groups blocked some of the pores of the activated carbon, leaving functional groups proper not only for efficient CO_2 adsorption but also for regeneration ability. The activated carbon that had the best performance in the bed column was modified with AMP amine resulting in adsorption bed capacity of 23.05 mg CO_2/g of the material.

Another emerging technology for efficient CO_2 uptake is the modification of activated carbon surface by developing nitrogen active sites [51]. Nitrogen-enriched carbonaceous materials present superior properties of adsorption capacity due to added functional groups that contribute to the polarization of CO_2 molecules [52]. In a study, conducted by Yami et al., of carbon material from coconut shell, nitrogen doping of the derived activated carbon provided remarkable textural and surface properties to the final material [46]. The material was activated chemically by KOH, and its performance was investigated for a fixed-bed column. As it is reported, adsorption properties were enhanced due to surface modification reaching a CO_2 ultimate uptake of 4.23 mmol/g of the material under mild operation conditions. Parameters that affected the CO_2 adsorption capacity found to be the initial concentration of CO_2, the temperature as well as the flow rate of the stream. Besides the good performance, the final material showed good stability after 20 adsorption–desorption cycles, indicating its reusability. A recent work of Manmuanpom et al. [53], investigated the CO_2 capacity of activated and nonactivated carbons. Specifically, both sample series were produced by carbonization of polybenzoxanine at a temperature of 800°C, while for activated samples a further physical activation under CO_2 at 900°C was followed. As it was revealed, activated carbon samples presented a higher specific surface area compared with that of nonactivated ones. In addition, XPS analysis showed N-sites onto the structure of the material. The adsorption capacity of the samples was investigated at different temperatures (from 30 to 50°C) and pressures (from 1 to 7 bars). The highest CO_2 uptake (3.59 mmol/g) was obtained for samples examined at a combination of 30°C and 7 bar, where the samples had the highest surface area and micropore volume. The overall good performance of the nitrogen-enriched carbons was attributed to both textural properties due to activation process and chemisorption mechanism induced by formed N-sites.

Coconut shell, in a study conducted by Yue et al., was used as precursor for porous carbon production by chemical activation [54]. In this work, coconut shell was urea-modified and activated by K_2CO_3 after the carbonization process. Again, samples were examined regarding their CO_2 capacity at two different temperatures (0 and 25°C) keeping the pressure at 1 bar. Results of the experiment showed that the decrease in temperature leads to increase of CO_2 uptake, hence at 0°C 1 g of sample could adsorb up to 5.12 mmol compared with 3.71 mmol/g at 25°C. Interestingly, findings of simulation-based investigation of CO_2 adsorption capacity revealed that besides nitrogen content and microporosity of the adsorbent, the distribution of the micropores also plays an important role in material's CO_2 capture capacity at mild conditions.

Similarly, a detailed characterization of KOH- and NaOH-activated carbons [55] support the outcomes of the aforementioned study. Argan fruit shell, this time, was carbonized and then chemically activated with the mentioned acids. The conditions of CO_2 uptake measurements were 1 bar and 25°C as previously. Once more, the high N content along with large surface area of narrow micropores domain reached a CO_2 uptake of 5.63 mmol/g, a value close to the previously presented one at same conditions. The comparison of the two cases can lead to the conclusion that micropore structure and significantly high nitrogen content are indeed the two major factors of carbonaceous material CO_2 capture performance.

It is widely accepted that, except activation conditions and surface modification, the performance of a material is also attributed to its form. For example, there are comparative studies upon powdered and activated fiber forms, suggesting that activated carbon fibers have advantage over the powder form for reasons of packing density [56]. Activated carbon fibers can be produced from any fabric waste or synthesized from various precursors; woven cotton [57] and denim [58] waste, poly(acrylonitrile) (PAN), and other synthetic polymers [59] can be the raw material for carbon fibers manufacturing.

Carrott et al. conducted a study investigating the effect of a preoxidation of industrial acrylic fibers on structural properties of the final activated carbon fibers [60]. In this work, high carbon yield during carbonization of acrylic fibers was achieved by prior oxidation of the precursor. This approach resulted in maximization of carbon content at the stage of carbonization and minimizing the reaction of the char with CO_2 at activation stage. A BET surface area of 2064 m^2/g and total pore volume of 1.15 cm^3/g were achieved under a controlled range of activation temperature and time. The temperature range as well as the activation time influenced dramatically the microstructure of the material providing a wide range of porosities, from ultramicropores to pores of 4 nm in size. However, N_2 adsorption and consequent CO_2 adsorption analysis revealed that the samples with ultrafine micropores which are well developed and distributed at lower activation temperatures had the highest CO_2 uptake (2.1 mmol/g) reaching even CO_2 capacities obtained from chemically activated carbons measured at same conditions. Chemically activated industrial sisal fiber was used for investigating the effect of carbonization parameters on the surface characteristics of the final carbonaceous material [61]. Although this study was not conducted for examining CO_2 capacity, it is interesting to notice that parameters such as N2 flow rate and carbonization temperature and time have the opposite effect; higher temperature and longer hold time develop higher BET surface area and pore volume, whereas higher N_2 flow rate has negative impact on the aforementioned properties.

Enhancement of activated carbon fibers properties over CO_2 capacity was done by nitrogen-enriched carbon nanotube (CNT) growth onto PAN-derived fibers. This attempt conducted by Chiang and his coworkers [52] resulted in spaghetti-like and randomly oriented CNT grafted to activated carbon fibers. For comparison reasons, a portion of used fibers were immersed to a cobalt acetate solution prior CVD process for CNT growth. Fig. 6.3 shows the results of each treatment.

The above figure, Fig. 6.3, presents clearly the effect of preimmersion of activated carbon fiber on the resulted CNTs sizes. According to authors, images b and c reveals the remaining particles of the used catalyst, a fact that indicates the low reaction of the catalyst with the surface of the fibers while the open pore end is attributed to catalyst particle falling-off during CVD process.

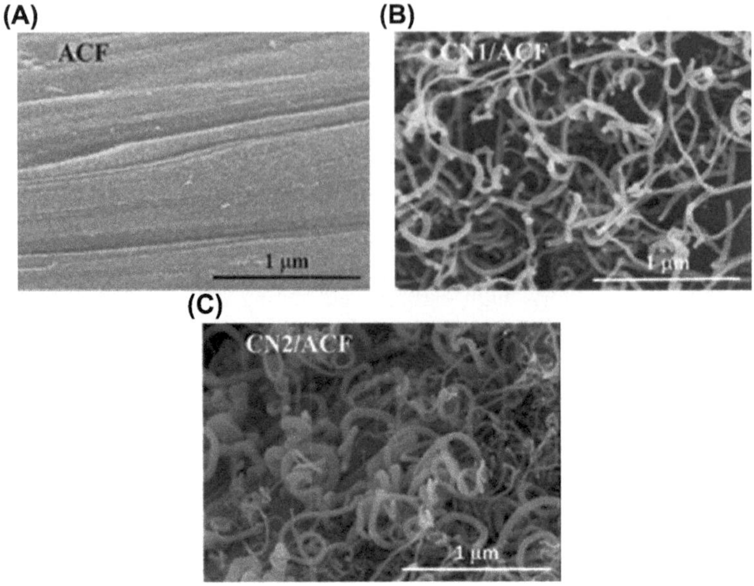

FIGURE 6.3 FESEM micrographs of (A) activated carbon fibers as-received, (B) carbon nanotube (CNT)-grafted activated carbon fibers, and (C) immersed to cobalt/acetate solution CNT-grafted activated carbon fibers. *Reprinted with permission by Elsevier from Y.-C. Chiang, R.-S. Juang, Surface modifications of carbonaceous materials for carbon dioxide adsorption: a review, J. Taiwan Inst. Chem. Eng. 71 (2017) 214–234.*

Spherical form of activated carbon has shown superior properties over various applications: high mechanical and thermal stability, better packing, surface smoothness, and pore manipulation are some of them [62].

Nitrogen- and sulfur-enriched spherical activated carbon were prepared by Sun et al. as CO_2 adsorbents [62]. Spheres of some millimeters in size were synthesized by a three-step process via final chemical activation. Resin of poly (styrene-vinylimidazole-divinylbenzene) was first H_2SO_4-sulfonicated, carbonized, and finally activated with KOH. The sterene and nitric components of resin at different ratios were the sources for carbon and nitrogen, while H_2SO_4 and KOH were responsible for simultaneous cross-linking and sulfur-enriching the surface thermally. The as-prepared sample with the higher sterene/N-vinylimidazole ratio (1:0.75) had the best performance at different operation conditions for CO_2 adsorption affected by the concentration of N-sites. Remarkably, obtained comparative results at different operation pressure revealed that at lower pressure the efficiency of adsorption was due to surface chemistry, while at increased pressure the mechanism was better attributed to microporosity. Finally, authors concluded that the suggested adsorbent has good potentials for CO_2 capture due to enhanced properties of regeneration stability and fast kinetics at various operation conditions (Table 6.1).

TABLE 6.1 The most recent studies of adsorbents for CO_2 capture.

Adsorbent	Precursor	Carbonization	Activation	Uptake	References
Porous carbon xerogels	Resorcinol-formaldehyde-phloroglucinol gels	900°C	–	2.89 mmol/g	[63]
Three-dimensional porous carbon frameworks	Mangosteen peel	700 °C	KOH	6.93 and 4.77 mmol/g (0 and 25 °C)	[64]
Heteroatom-doped hierarchically porous carbons	Glucose hydrogels	600–1000°C	KOH	4.4 mmol/g	[65]
N-doped porous carbon	Imidazole-based hyper-cross-linked polymers (hcps)	–	KOH	180–258 mg/g (273 K, 1.0 bar) 4173 mg	[66]
Activated carbon	Date seeds	600–900°C	Physically activated	141.14 mg	[67]
Activated carbon	Wood sawdust	400°C	H_3PO_4, NaOH, and KOH	0.96 mol/kg	[68]
Porous carbons	EFB biomass	600 or 800 C	KOH	5.184, 3.712, and 2.343 mmol/g	[68]
N-doped nanoporous carbon	Walnut shell	600°C	KOH	7.42 mmol/g 14.03 mmol/g (1 and 10 bar, respectively)	[69]
N-doped active carbon	Chitosan	550 C	KOH	1.86 mmol/g	[70]
Nanoporous carbons	Cigarette butts	600 C	KOH	6.0 mmol/g	[71]

3. Acidifying pollutants (NOx, SOx)

Nitrous oxides, sulfur oxides, and their derivatives are indirect greenhouse gases. Although NOx comprises only 6% of the total greenhouse gases emissions, they are responsible for phenomena such as acid rain and smog [9,72]. In addition, photo chemical reaction of NOx in the atmosphere leads to the production of tropospheric ozone [12]. The main sources of all acidifying pollutant emissions are internal combustion engines (road transport) with 13% [11], energy production 28%, while the agriculture sector have the 27% of the total share. Emissions of NOx and SOx from on-road sources have been reduced significantly by the implementation of clearer fuels and more efficient catalysts [10,15]. However, there is still space for emerging technologies or improvement of existing ones for NOx and Sox reduction by adsorption in other sectors. Nitrogen oxide emissions, and especially that of N_2O resulting from agricultural activities, are suppressed by soil enrichment with various substances known as amendment [73]. In general, soil amendment is a substance added to soil in order to improve textural properties [74], fertilization of soil [75], etc.

Biochar is the most common soil amendment for N_2O reduction and soil chemical stabilization [76]. Biochar is produced by pyrolysis of biomass at total or partial absence of oxygen [27]. Biochars present excellent physical and chemical properties such as high surface area, well developed porosity, high sorption capacities, etc. Surface structure properties are influenced massively by conditions of the synthesis process; Fig. 6.2 shows the effect of carbonization temperature on porosity development of wood biomass after slow pyrolysis [77].

Many studies have been conducted for examining the adsorption capacity of produced biochars for the specific purpose. Although biochar is extensively used in soil with great benefits, regarding its role as a good adsorbent of N_2O very few works have been presented. For determination of the importance of biochar adsorption capacity, adsorption measurements of N_2O onto wood-derived biochar were conducted by Xiao et al. [78]. At this work, wood biomass–based samples were prepared by anaerobic pyrolysis at a temperature range from 300 to 700°C. After carbonization, samples were further heated in the presence of air at 400°C. Adsorption measurements of CO_2 were conducted, and surface area was determined correlating well with Langmuir capacity. Furthermore, simulated pressure conditions in soil were used in order to investigate if adsorption–desorption properties of chars are influenced by char coating with soil matter. The results revealed that coating does not affect sorption characteristics of the biochar which take place in the developed micropores; therefore a biochar concentration of 1% in the soil could have great potentials to retain N_2O.

Nitrogen oxide and sulfur oxide adsorption by activated carbon have been reviewed with focus on surface properties and consequently required char treatment processes to reach the desired adsorption capacities [79–81]. One example of surface chemistry importance is provided by Ghouma et al. [82]. This study investigates three samples of activated carbons derived from agro-industrial wastes and prepared by both chemical and physical activation. Nitrogen oxide (N_2O) capacity onto the as-prepared samples was investigated at ambient temperature in a fix-bed column with low N_2O flow rate. The findings of this study revealed that chemically activated carbon has greater N_2O uptake at those conditions than the physically activated one. This result indicates that functional groups present on the surface of the material have a key role in N_2O adsorption mechanism. Furthermore, authors suggest that pore structure has also a crucial role regarding N_2O diffusion rates. Therefore, as activated carbon prepared by H_3PO_4 activation presents a mesoporous surface structure, N_2O is diffused more efficiently compared with CO_2-activated samples.

Adsorption of NOx and SOx, from flue gas at the same time, onto activated carbon monoliths supporting metal oxide group (CO_3O_4) was investigated [83]. The scope of Silas and his coworkers' study was to investigate the contribution of the deposition precipitation, pore volume impregnation, and hydrothermal methods to the efficiency of as-synthesized materials. The outcome of this research was that hydrothermally prepared samples showed higher adsorption capacities of the two target gases than those prepared differently; with an uptake of 123.1 and 130.2 mg/g of SO_2 and NOx, respectively. According to the obtained results of breakthrough curves (Fig. 6.4), material performs higher efficiency for NOx adsorption attributed to its physicochemical properties as well as experimental conditions (Fig. 6.5).

Additionally to the previous analysis, reusability of adsorbents was evaluated after two cycles of adsorption/regeneration. The reduction of adsorption capacity both for NOx and SOx was minor, while the regeneration efficiency was kept above 94% for both samples after the total two cycles.

4. Methane separation

Although CO_2 is considered the main culprit gas for global warming, CH_4 is also a potent greenhouse gas. According to EPA, CH_4 has a 14% of greenhouse gas emissions with the majority derived from agricultural and energy sectors. Separation technologies are usually studied in respect to CH_4 separation from a mixed gas stream in natural gas refining processes. The most common technologies applied for this purpose are pressure swing adsorption, temperature swing adsorption, vacuum swing adsorption, or the vacuum pressure swing adsorption [84]. Various adsorbents are used in the referred methods including MOFs [85], zeolites [86], silica-based, and others. Carbonaceous materials are often utilized in the aforementioned processes for efficient gas mixture separation [85,87,88].

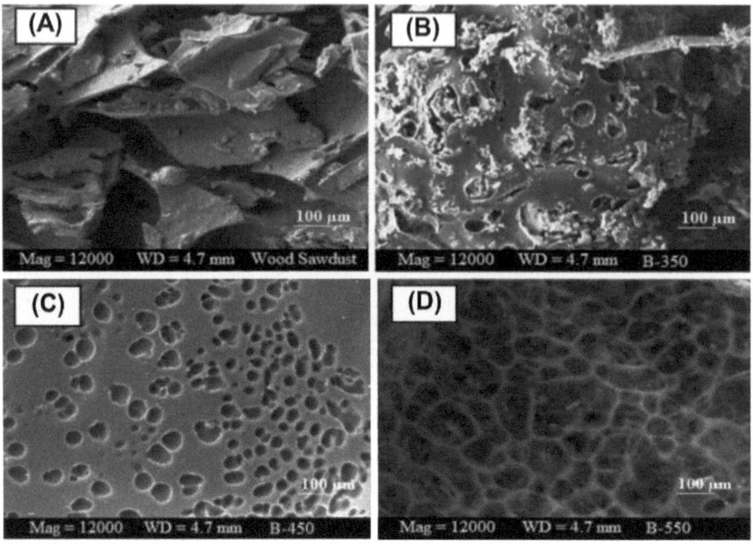

FIGURE 6.4 SEM micrographs of (A) wood biomass and (B) carbonization at 350°C, (C), at 450°C, and (D) at 550°C. *Reprinted with permission by Elsevier from Z.Z. Chowdhury, M.Z. Karim, M.A. Ashraf, K. Khalid, Influence of carbonization temperature on physicochemical properties of biochar derived from slow pyrolysis of durian wood (Durio zibethinus) Sawdust, Bio Resour. 11 (2016).*

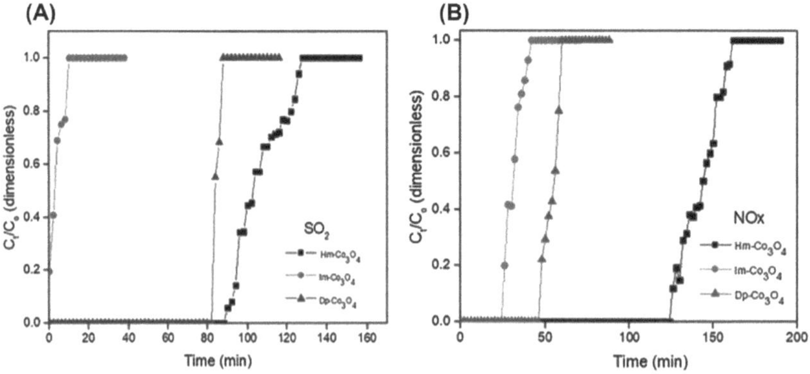

FIGURE 6.5 Breakthrough curves of simultaneous (A) NOx and (B) SOx adsorption onto hydrothermally derived adsorbent. *Reprinted with permission by Elsevier from K. Silas, W.A.W. A. K. Ghani, T. S.Y. Choong, U. Rashid, Breakthrough studies of Co3O4 supported activated carbon monolith for simultaneous SO2/NOx removal from flue gas, Fuel Process. Technol. 180 (2018) 155–165. doi: https://doi.org/10.1016/j.fuproc.2018.08.018.*

4.1 CH_4/CO_2

The selection of the adsorbent in use is a matter of the (1) gas nature, (2) composition of the gas mixture, (3) operation conditions, and (4) thermodynamics of the system (kinetics, equilibrium, time, etc.). Four types of activated carbon were compared with zeolites regarding CO_2 removal from different gas mixtures. Experiments were conducted in order to evaluate adsorption performance of both types and their efficiency under a range of operation conditions. The degradation of material was examined as well. Activated carbon examined showed better characteristics over zeolites especially for reusability.

Gas binary mixtures of CO_2, CH_4, and N_2 selective adsorption on carbon materials derived from starch sugar were studied by Wang et al. Four glucose-based carbons were synthesized differentiating in KOH/C ratio (2:1, 3:1, 4:1, and 5:1). Isosteric heat of adsorption for each gas was obtained by several techniques. Nitrogen adsorption revealed that the optimal from the as-prepared material was that with KOH/C ratio of 4:1; high surface area, and high CO_2 uptake reaching capacities even of best MOFs (3153 m_2/g and 22.4 mmol/g, respectively). In addition, the KOH concentration seems to play a highly important role on surface chemistry of the resulted materials. Given the example of the presented study, at KOH/C from 2 to 5, pore structure was developing accordingly as it is depicted in Fig. 6.6. The selectivity of the adsorbents was estimated by ideal absorbed solution theory, predicting high selectivity rates at 30 operation pressure of binary gas mixtures. The outcome of this study promotes the implementation of this kind of materials as efficient adsorbents for CH_4 separation from gas mixtures.

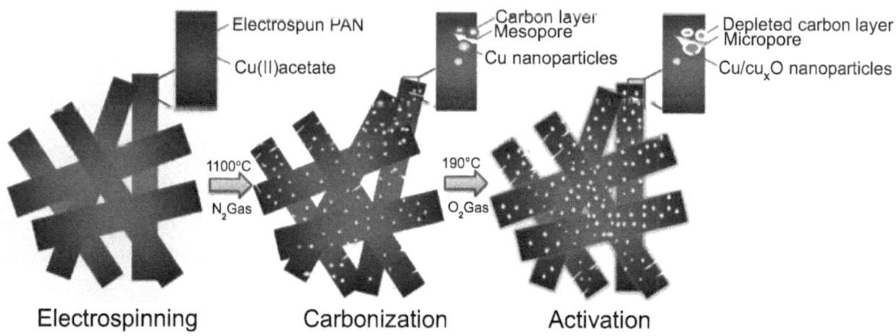

FIGURE 6.6 Schematic representation of Cu-/CuxO-modified activated carbon fiber composites synthesis process. *PAN*, poly(acrylonitrile). *Reprinted with permission by Elsevier from B. Bajaj, H.-I. Joh, S. M. Jo, J. H. Park, K. B. Yi, S. Lee, Enhanced reactive H2S adsorption using carbon nanofibers supported with Cu/CuxO nanoparticles, Appl. Surf. Sci. 429 (2018) 253–257. doi: https://doi.org/10.1016/j.apsusc.2017.06.280.*

4.2 CH_4/H_2S

Hydrogen sulfide (H_2S) causes major health problems when one is exposed even to relevantly low concentration. Eye and skin irritation, breathing difficulties, and loss of odor are some of the symptoms after the exposure. According to Occupational Safety and Health Administration (OSHA), H_2S causes asphyxiation and has serious effect on human nervous system, that is why H_2S removal/adsorption technologies are in the prime line of investigation. Furthermore, H_2S concentration in a natural gas stream may cause significant damage to piping network and natural gas treating units [89].

Selectivity along with adsorption capacity of an adsorbent is paramount for optimal efficacy of any chemical process. Regarding H_2S, the most common applied materials for adsorption methods are zeolites [84], MOFs [90], and various carbon materials. Because H_2S has low affinity to carbon, modification of the derived carbon material is promoted by researchers of the field [91].

In the previous section of this chapter, it was mentioned that carbonaceous materials in fiber gain an increasing interest due to its advantages. Also, it was mentioned that further modification of adsorbents surface is preferred in most of the cases. As an example for these two, utilized by a recently developed methodology of electrospinning, the work of Bajaj et al. is discussed [92]. This team used activated carbon fibers and modified the surface with nanoparticles of Cu/CuxO. As source material, electrospun PAN/copper fibers were used. After carbonization process, the produced composite was activated at mild conditions. Schematic representation of the synthesis steps is clear and comprehensive as it is seen in Fig. 6.7.

The as-synthesized material was a result of electrospinning PAN stabilized under air at 280°C. The following steps include carbonization in nitrogen-rich environment at elevated temperature (1100°C) and activation at 190°C in O_2 as the final step. Surface elemental characterization was conducted by XRD and XPS, while SEM and TEM were used for structural properties. Characterization analysis indicates the presence of Cu/CuxO onto the carbon fibers. Measurements of H_2S uptake were conducted at 300°C in a fixed column bed. Pure activated carbon fibers were subjected to the breakthrough time measurements as well. The amount of nanocomposite used was half of the amount of pure fibers. The experiment stopped when H_2S concentration reached 50ppm (from 761ppm). Finally, the presence of Cu enhanced the adsorption capacity by many times compared with pure activated carbon fibers, suggesting their potential for H_2S adsorption applications.

5. Volatile organic compounds

VOCs is an umbrella term for any compounds, according to the European Union, that has an initial boiling point below 250°C. Emissions of VOCs, besides industrial processes, are produced from many every day activities.

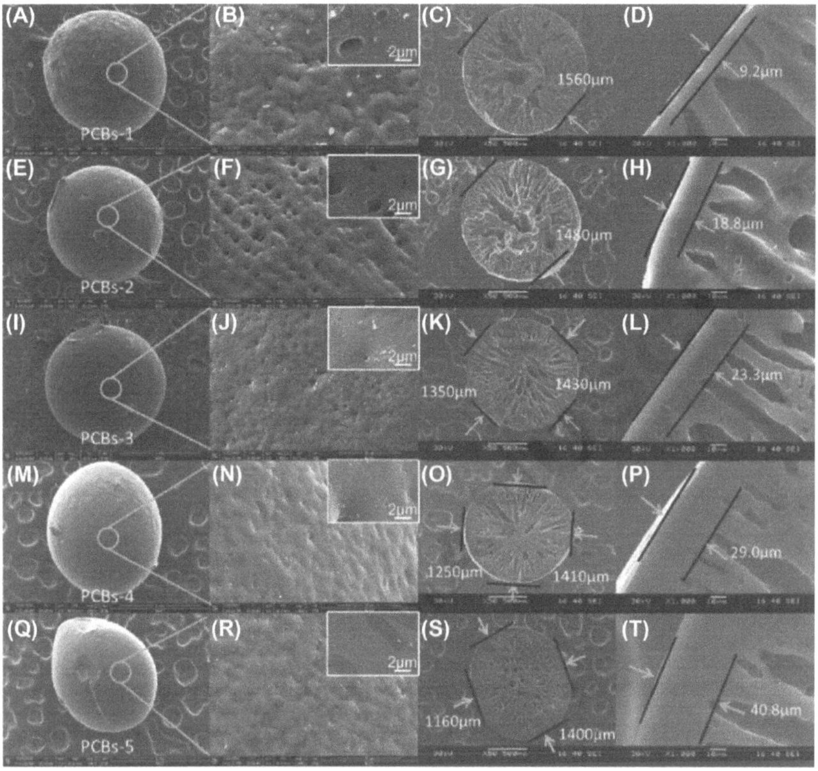

FIGURE 6.7 SEM micrographs of PCBs, from top to bottom; PCB-1 to PCB-5. The presentation of each sample includes surface and cross sections at different magnifications (from left-hand to right-hand side). *Reprinted with permission by Elsevier from J. Qi, J. Li, Y. Li, X. Fang, X. Sun, J. Shen, L. Wang, Synthesis of porous carbon beads with controllable pore structure for volatile organic compounds removal, Chem. Eng. J. 307 (2017) 989–998. doi: https://doi.org/10.1016/j.cej.2016.09.022.*

Compounds of this category cause harm not only to plants and animals but to human in the long term [93]. Moreover, VOCs contribute to climate change significantly through ozone formation [94] and smog [72] as NOx do. Technologies for VOCs limitation and management can be divided in two categories: destruction and recycling. Although the efficiency of the former is at a satisfying level, it requires massive energy consumption. The latter option includes absorption, adsorption, separation, and condensation. Photocatalysis is one of the most studied cases concerning air purification especially for indoor application, while adsorption onto materials for large-scale units is among the emerging technologies [95,96].

In many cases regarding industrial applications, spherical shaped adsorbents are preferred. Superior mechanical and surface properties of carbonaceous spheres have attracted recently the interest of scientific community. In an experimental work of Qi and coworkers, beads of porous surface structure were synthesized by phase inversion method [97]. After phase inversion, the as-produced polymeric samples were carbonized. Beads were derived from four different polymers, namely PES, PSF, PAN, and PVDF. For achieving the desired shape, the resulted solution was introduced via a syringe to a coagulation bath with deionized water and iso-propanol mixture at different ratios (from pure water to 40:60, with a composition change step of 10). The different water/iso-propanol ratio contributed both to the size and to the microporosity of the beads. After this, oven drying was accomplished as an intermediate step prior to carbonization at 350°C and further at 950°C at nitrogen-rich environment. The as-prepared beads were designated regarding the order of ratio increase, so we have PCB (porous carbon beads)-1, PCB-2, PCB-3, PCB-4, and PCB-5 (where 1 is for 0:100 and so on). As it is observed in Fig. 6.6, the best structure was developed on samples of 10:90 (PCB-2) water/iso-propanol.

The optimized coagulation bath ratio provides materials with high surface area of 1166 m^2/g as well as good adsorption capacity of benzene (1.467 g/g), toluene (1.229 g/g), n-hexane (0.600 g/g), and acetone (0.770 g/g). The almost unchanged performance of the material after six cycles of sorption process promotes its application for VOCs treatment.

Regeneration efficiency of an adsorbent plays a crucial role in the overall performance estimation. Especially for industrial applications, high thermal and mechanical stability are the main criteria. To this direction, studies of desorption method applied for regenerating a specific adsorbent, after having adsorbed a specific compound, are very important for process optimization [86,98]. Among other properties, desorption of benzene and toluene from reduced graphene oxide was investigated [99]. In specific, graphene oxide and its reduced form was synthesized from graphite following a slightly altered Hummer's method. Although the interlayer space of the sheetlike structure of the graphene oxide was found to be higher, reduced graphene oxide possesses a higher surface area of 292.6 m^2/g over 236.4 m^2/g of the former. Experiments conducted for benzene and toluene uptake capacities of the reduced graphene oxide showed values of 276.4 and 304.4 mg/g, respectively. Finally, the regeneration efficiency of the adsorbent was examined, and it was determined that reduced graphene oxide can be fully regenerated at low temperatures (150°C).

In general, separation techniques require more strong chemical bonds between the adsorbate and the adsorbent. Surface modification and/or chemical activation are the most common approach to improve original material's chemical properties. Table 6.2 summarizes carbon-based materials for (1) CH_4 separation from gas mixtures, (2) H_2S rementation and VOCs reduction by adsorption.

TABLE 6.2 Summary of carbon-based materials for various greenhouse gases (with exception of CO_2).

Adsorbent	Precursor	Carbonization	Activation	Uptake	Application	Ref.
N-doped activated carbon	Chlorella, isochrysis, and *Platymonas subcordiformis*	500°C	–	7.27 mmol/g	CO_2/CH_4	[100]
Activated carbon	Lignin	800°C	KOH	600 mg/g	VOCs	[96]
Activated carbon fiber	Mesophase pitch fiber, acf textile kyrol acf-15 (nippon kynol)	800–1000°C	CO_2/H_2O	Toluene 59.6 g/100 g Ethanol/methanol 54.4/49.2 g/100 g	VOCs	[101]
Monolithic tubular activated carbon	Cotton stalks	Ac-p 400°C Ac-z 600°C	Phosphorous acid and zinc chloride	328 mg/g wet 258 mg/g dry 221 mg/g wet	VOCs	[102]
Activated carbon	Corncob wastes	350, 400, 450, 550, and 650°C	$ZnCl_2$	414.6 mg/g 3000 mg/m (25 °C)	VOCs	[103]
Activated carbon	Gingko leaf and peanut shell	800°C	N_2	CO_2, CH_4, and N_2 2.13/0.64/0.2 mmol/g, 3.63/1.38/0.45 mmol/g, 2.26/0.89/0.31 mmol/g, 1.62/0.45/0.17 mmol/g	CO_2, CH_4, N_2	[88]
N-doped carbon	Crab shell	600–900°C	N_2/CO_2	6.8 mmol/g 2.49 mmol/g	CH_4/N_2, CO_2/N_2	[100]
Activated carbon	Cherry stones	–	CO_2	–	CO_2/CH_4	[87]
Activated carbon cloth	Viscose rayon cloth	630°C	CO_2	2.68 mmol/g or 10.55 wt% 0.99 mmol/g or 1.56 wt%	CO_2/CH_4	[104]
Activated carbon	Coconut shell	1173K	CO_2	6.0 mmol/g 4 mmol/g	CO_2/CH_4	[105]

6. Conclusions

This chapter reviews all the recent literature on carbonaceous materials for gas-phase adsorption. The increased concern on climate change and its irreversible impact on the planet induced the scientific research toward environmental protection technologies. Aspects that accord with eco-friendly processes utilization involve the implementation of technologies for gas purification and clean energy consumption, toxic and hazardous gas reduction, greenhouse effect mitigation, and so on. Up to the very recent history, adsorption was linked with decontamination technologies of liquid phase. Nowadays, adsorption seems to have a key role in the mentioned emerging technologies with high level of efficiency for gas-phase applications. However, the selection of the proper adsorbent is a multicriteria decision: cost efficiency, physicochemical properties, and mechanical properties are some of the most important. In the perspective of all mentioned parameters, and especially cost, carbon-based materials have gained popularity. Carbonaceous materials can be derived from a wide range of precursors resulting with high surface and good stability properties. The performance of each adsorbent depends mainly on factors such as the source, the synthesis process, and surface chemistry. The latter is easily adjusted by surface modification techniques. The present work encloses studies of gas-phase adsorbents with potentials for large-scale applications. In addition, to best of our knowledge, this is one of the few works that reviews the performance and properties of carbon-based material for several different greenhouse gases. At this point, it should be noticed that the number of published studies upon CO_2 is disproportionate regarding the existing literature for other harmful gases, and further research is suggested toward this direction.

References

[1] K. Bennaceur, Chapter 26 - CO_2 capture and sequestration, in: T.M. Letcher (Ed.), Future Energy, second ed., Elsevier, Boston, 2014, pp. 583–611.

[2] J. Jiang, Z. Lu, M. Zhang, J. Duan, W. Zhang, Y. Pan, J. Bai, Higher Symmetry multinuclear clusters of metal–organic frameworks for highly selective CO_2 capture, J. Am. Chem. Soc. (2018).

[3] X. He, The latest development on membrane materials and processes for post-combustion CO_2 capture: a review, SF J Material Chem Eng. 1 (1) (2018) 1009.

[4] G.N. Nikolaidis, E.S. Kikkinides, M.C. Georgiadis, A model-based approach for the evaluation of new zeolite 13X-based adsorbents for the efficient post-combustion CO_2 capture using P/VSA processes, Chem. Eng. Res. Des. 131 (2018) 362–374.

[5] C. Chen, H. Huang, Y. Yu, J. Shi, C. He, R. Albilali, H. Pan, Template-free synthesis of hierarchical porous carbon with controlled morphology for CO_2 efficient capture, Chem. Eng. J. 353 (2018) 584–594.

[6] P. Zhang, Y. Zhong, J. Ding, J. Wang, M. Xu, Q. Deng, S. Deng, A new choice of polymer precursor for solvent-free method: preparation of N-enriched porous carbons for highly selective CO_2 capture, Chem. Eng. J. 355 (2019) 963–973.

[7] M. Sevilla, A.S.M. Al-Jumialy, A.B. Fuertes, R. Mokaya, Optimization of the pore structure of biomass-based carbons in relation to their use for CO_2 capture under low-and high-pressure regimes, ACS Appl. Mater. Interfaces 10 (2) (2018) 1623–1633.

[8] D. Yu, J. Hu, L. Zhou, J. Li, J. Tang, C. Peng, H. Liu, Nitrogen-Doped coal tar pitch based microporous carbons with superior CO_2 capture performance, Energy & Fuels 32 (3) (2018) 3726–3732.

[9] J. Wettestad, Clearing the Air: European Advances in Tackling Acid Rain and Atmospheric Pollution, Routledge, 2018.

[10] M.T. Chaichan, Q.A. Abass, Study of NOx emissions of SI engine fueled with different kinds of hydrocarbon fuels and hydrogen, Al-Khwarizmi Eng. J. 6 (2) (2018) 11–20.

[11] A.A. Abbas, Monitoring of air pollution by road transport with purpose of protection of public health, in: ка» Џредставлено вь Змгдц Внодрмзвус Змтдрматзнмак ЋЗВЗЏэ НЛБ дкзбрарчћ руэ Онкзсж Свжнкаркч Бзбкзнёраожчэ Рдсдарвжќ Бзбэ Турйзсж Дгуватзнм З̇мгдц, vol. 12, Научная Џериодика Украины, 2018.

[12] S. Tiwari, M. Agrawal, Tropospheric Ozone Budget: Formation, Depletion and Climate Change, in: Tropospheric Ozone and its Impacts on Crop Plants, Springer, 2018, pp. 31–64.

[13] D. Štreimikienė, Review of internalization of externalities and dynamics of atmospheric emissions in energy sector of Baltic States, Renew. Sustain. Energy Rev. 70 (2017) 1131–1141.

[14] R. Suarez-Bertoa, P. Mendoza-Villafuerte, F. Riccobono, M. Vojtisek, M. Pechout, A. Perujo, C. Astorga, On-road measurement of NH_3 emissions from gasoline and diesel passenger cars during real world driving conditions, Atmos. Environ. 166 (2017) 488–497.

[15] J.E.D.A. Goes, L. Olsson, M. Berggrund, A. Kristoffersson, L. Gustafson, M. Hicks, Performance studies and correlation between vehicle-and rapid-aged commercial lean NOx trap catalysts, SAE Int. J. Eng. 10 (2017) 1613–1626, 2017-01-0940.

[16] S. Ramalingam, S. Rajendran, P. Ganesan, Performance improvement and exhaust emissions reduction in biodiesel operated diesel engine through the use of operating parameters and catalytic converter: a review, Renew. Sustain. Energy Rev. 81 (2018) 3215–3222.

[17] S.N. Nandeshwar, A.S. Mahakalakar, R.R. Gupta, G.Z. Kyzas, Green activated carbons from different waste materials for the removal of iron from real wastewater samples of Nag River, India, J. Mol. Liq. 216 (2016) 688–692. https://doi.org/10.1016/j.molliq.2015.12.065.

[18] P. Arnell, Greening trade and investment: environmental protection without protectionism by Eric Neumayer, 2001. Earthscan, 228 pp, ISBN 1-85383-788-1, Corp. Soc. Responsib. Environ. Manag. 10 (2) (2003) 111–112. https://doi.org/10.1002/csr.34.

[19] N. Mohamad Nor, L.C. Lau, K.T. Lee, A.R. Mohamed, Synthesis of activated carbon from lignocellulosic biomass and its applications in air pollution control—a review, J. Environ. Chem. Eng. 1 (4) (2013) 658–666. https://doi.org/10.1016/j.jece.2013.09.017.

[20] A. Zubrik, M. Matik, S. Hredzák, M. Lovás, Z. Danková, M. Kováčová, J. Briančin, Preparation of chemically activated carbon from waste biomass by single-stage and two-stage pyrolysis, J. Clean. Prod. 143 (2017) 643–653. https://doi.org/10.1016/j.jclepro.2016.12.061.

[21] E. Köseoğlu, C. Akmil-Basar, Preparation, structural evaluation and adsorptive properties of activated carbon from agricultural waste biomass, Adv. Powder Technol. 26 (3) (2015) 811–818. https://doi.org/10.1016/j.apt.2015.02.006.

[22] G. Singh, I.Y. Kim, K.S. Lakhi, P. Srivastava, R. Naidu, A. Vinu, Single step synthesis of activated bio-carbons with a high surface area and their excellent CO_2 adsorption capacity, Carbon 116 (2017) 448–455. https://doi.org/10.1016/j.carbon.2017.02.015.

[23] S. Yorgun, D. Yıldız, Preparation and characterization of activated carbons from Paulownia wood by chemical activation with H_3PO_4, J. Taiwan Inst. Chem. Eng. 53 (2015) 122–131. https://doi.org/10.1016/j.jtice.2015.02.032.

[24] M.J. Rwiza, S.-Y. Oh, K.-W. Kim, S.D. Kim, Comparative sorption isotherms and removal studies for Pb(II) by physical and thermochemical modification of low-cost agro-wastes from Tanzania, Chemosphere 195 (2018) 135–145. https://doi.org/10.1016/j.chemosphere.2017.12.043.

[25] C.Y. Yin, M.K. Aroua, W.M.A.W. Daud, Review of modifications of activated carbon for enhancing contaminant uptakes from aqueous solutions, Separ. Purif. Technol. 52 (3) (2007) 403–415. https://doi.org/10.1016/j.seppur.2006.06.009.

[26] L. Leng, H. Huang, An overview of the effect of pyrolysis process parameters on biochar stability, Bioresour. Technol. 270 (2018) 627–642. https://doi.org/10.1016/j.biortech.2018.09.030.

[27] M. Tripathi, J.N. Sahu, P. Ganesan, Effect of process parameters on production of biochar from biomass waste through pyrolysis: a review, Renew. Sustain. Energy Rev. 55 (2016) 467–481. https://doi.org/10.1016/j.rser.2015.10.122.

[28] J. Fang, L. Zhan, Y.S. Ok, B. Gao, Minireview of potential applications of hydrochar derived from hydrothermal carbonization of biomass, J. Ind. Eng. Chem. 57 (2018) 15–21. https://doi.org/10.1016/j.jiec.2017.08.026.

[29] H. Bamdad, K. Hawboldt, S. MacQuarrie, A review on common adsorbents for acid gases removal: focus on biochar, Renew. Sustain. Energy Rev. 81 (2018) 1705–1720. https://doi.org/10.1016/j.rser.2017.05.261.

[30] K.A. Thompson, K.K. Shimabuku, J.P. Kearns, D.R.U. Knappe, R.S. Summers, S.M. Cook, Environmental comparison of biochar and activated carbon for tertiary wastewater treatment, Environ. Sci. Technol. 50 (20) (2016) 11253–11262. https://doi.org/10.1021/acs.est.6b03239.

[31] L. Zhou, C. Richard, C. Ferronato, J.-M. Chovelon, M. Sleiman, Investigating the performance of biomass-derived biochars for the removal of gaseous ozone, adsorbed nitrate and aqueous bisphenol A, Chem. Eng. J. 334 (2018) 2098–2104. https://doi.org/10.1016/j.cej.2017.11.145.

[32] V. Chemerys, E. Baltrėnaitė, A review of lignocellulosic biochar modification towards enhanced biochar selectivity and adsorption capacity of potentially toxic elements 8 (1) (2018) 12. https://doi.org/10.15421/2018_183.

[33] T. Sizmur, T. Fresno, G. Akgül, H. Frost, E. Moreno-Jiménez, Biochar modification to enhance sorption of inorganics from water, Bioresour. Technol. 246 (2017) 34–47. https://doi.org/10.1016/j.biortech.2017.07.082.

[34] V.K. Singh, E.A. Kumar, Measurement of CO_2 adsorption kinetics on activated carbons suitable for gas storage systems, Greenhouse Gases: Sci. Technol. 7 (1) (2017) 182–201. https://doi.org/10.1002/ghg.1641.

[35] J. Yu, L.-H. Xie, J.-R. Li, Y. Ma, J.M. Seminario, P.B. Balbuena, CO_2 capture and separations using MOFs: computational and experimental studies, Chem. Rev. 117 (14) (2017) 9674–9754.

[36] N. Chanut, S. Bourrelly, B. Kuchta, C. Serre, J.S. Chang, P.A. Wright, P.L. Llewellyn, Screening the effect of water vapour on gas adsorption performance: application to CO_2 capture from flue gas in metal–organic frameworks, ChemSusChem 10 (7) (2017) 1543–1553.

[37] J. Hu, Y. Liu, J. Liu, C. Gu, Effects of water vapor and trace gas impurities in flue gas on CO_2 capture in zeolitic imidazolate frameworks: the significant role of functional groups, Fuel 200 (2017) 244–251.

[38] T.P. Senftle, E.A. Carter, The holy grail: chemistry enabling an economically viable CO_2 capture, utilization, and storage strategy, Accounts Chem. Res. 50 (3) (2017) 472–475.

[39] W. Chan, M. Mazlee, Z.A. Ahmad, M. Ishak, J. Shamsul, The development of low cost adsorbents from clay and waste materials: a review, J. Mater. Cycles Waste Manag. 19 (1) (2017) 1–14.

[40] R. Kosheleva, A.C. Mitropoulos, G.Z. Kyzas, Activated Carbon from Food Waste, in: Green Adsorbents for Pollutant Removal, Springer, 2018, pp. 159–182.

[41] H. Ebrahimi, H. Roghani-Mamaqani, M. Salami-Kalajahi, S. Shahi, A. Abdollahi, Preparation of furfuryl alcohol-functionalized carbon nanotube and epoxidized novolac resin composites with high char yield, Polym. Compos. 39 (S2) (2018) E1231–E1236.

[42] R. Heimböckel, S. Kraas, F. Hoffmann, M. Fröba, Increase of porosity by combining semi-carbonization and KOH activation of formaldehyde resins to prepare high surface area carbons for supercapacitor applications, Appl. Surf. Sci. 427 (2018) 1055–1064.

[43] K.M. Rambau, N.M. Musyoka, N. Manyala, J. Ren, H.W. Langmi, M.K. Mathe, Preparation of carbon nanofibers/tubes using waste tyres pyrolysis oil and coal fly ash derived catalyst, J. Environ. Sci. Health Part A (2018) 1–8.

[44] W. Liu, S. Xu, S. Guan, R. Liang, M. Wei, D.G. Evans, X. Duan, Confined synthesis of carbon nitride in a layered host matrix with unprecedented solid-State quantum yield and stability, Adv. Mater. 30 (2) (2018) 1704376.

[45] J.N. Schwämmlein, G.S. Harzer, P. Pfändner, A. Blankenship, H.A. El-Sayed, H.A. Gasteiger, Activity and stability of carbon supported PtxY alloys for the ORR determined by RDE and single-cell PEMFC measurements, J. Electrochem. Soc. 165 (15) (2018) J3173–J3185.

[46] A. Yaumi, M.A. Bakar, B. Hameed, Reusable nitrogen-doped mesoporous carbon adsorbent for carbon dioxide adsorption in fixed-bed, Energy 138 (2017b) 776–784.

[47] Y. Jiang, Preparation of activated carbon from walnut shell and its application in industrial wastewater, in: Paper Presented at the AIP Conference Proceedings, 2017.

[48] G.Z. Kyzas, E.A. Deliyanni, K.A. Matis, Activated carbons produced by pyrolysis of waste potato peels: cobalt ions removal by adsorption, Colloid. Surf. Physicochem. Eng. Asp. 490 (2016) 74–83. https://doi.org/10.1016/j.colsurfa.2015.11.038.

[49] Z. Guo, J. Zhang, H. Liu, Y. Kang, J. Yu, C. Zhang, Optimization of the green and low-cost ammoniation-activation method to produce biomass-based activated carbon for Ni (II) removal from aqueous solutions, J. Clean. Prod. 159 (2017) 38–46.

[50] U.F.M. Ali, N.H. Azmi, K.M. Isa, M.K. Aroua, T.R. Shien, M.H. Khamidun, Optimization study on preparation of amine functionalized sea mango (cerbera odollam) activated carbon for Carbon Dioxide (CO_2) adsorption, Combust. Sci. Technol. 190 (7) (2018) 1259–1282.

[51] A. Yaumi, M.A. Bakar, B. Hameed, Recent advances in functionalized composite solid materials for carbon dioxide capture, Energy 124 (2017a) 461–480.

[52] Y.-C. Chiang, R.-S. Juang, Surface modifications of carbonaceous materials for carbon dioxide adsorption: a review, J. Taiwan Inst. Chem. Eng. 71 (2017) 214–234.

[53] N. Manmuanpom, U. Thubsuang, S.T. Dubas, S. Wongkasemjit, T. Chaisuwan, Enhanced CO_2 capturing over ultra-microporous carbon with nitrogen-active species prepared using one-step carbonization of polybenzoxazine for a sustainable environment, J. Environ. Manag. 223 (2018) 779–786.

[54] L. Yue, Q. Xia, L. Wang, L. Wang, H. DaCosta, J. Yang, X. Hu, CO_2 adsorption at nitrogen-doped carbons prepared by K_2CO_3 activation of urea-modified coconut shell, J. Colloid Interface Sci. 511 (2018) 259–267.

[55] O. Boujibar, A. Souikny, F. Ghamouss, O. Achak, M. Dahbi, T. Chafik, CO_2 capture using N-containing nanoporous activated carbon obtained from argan fruit shells, J. Environ. Chem. Eng. 6 (2) (2018) 1995–2002.

[56] D. Lozano-Castello, D. Cazorla-Amoros, A. Linares-Solano, Powdered activated carbons and activated carbon fibers for methane storage: a comparative study, Energy & Fuels 16 (5) (2002) 1321–1328.

[57] J. Zheng, Q. Zhao, Z. Ye, Preparation and characterization of activated carbon fiber (ACF) from cotton woven waste, Appl. Surf. Sci. 299 (2014) 86–91.

[58] T.L. Silva, A.L. Cazetta, P.S. Souza, T. Zhang, T. Asefa, V.C. Almeida, Mesoporous activated carbon fibers synthesized from denim fabric waste: efficient adsorbents for removal of textile dye from aqueous solutions, J. Clean. Prod. 171 (2018) 482–490.

[59] E. Frank, F. Hermanutz, M.R. Buchmeiser, Carbon fibers: precursors, manufacturing, and properties, Macromol. Mater. Eng. 297 (6) (2012) 493–501.

[60] P. Carrott, M.R. Carrott, P. Correia, Evolution of porosity of activated carbon fibres prepared from pre-oxidized acrylic fibres, Microporous Mesoporous Mater. 264 (2018) 176–180.

[61] M. Dizbay-Onat, U.K. Vaidya, C.T. Lungu, Preparation of industrial sisal fiber waste derived activated carbon by chemical activation and effects of carbonization parameters on surface characteristics, Ind. Crops Prod. 95 (2017) 583–590.

[62] Y. Sun, K. Li, J. Zhao, J. Wang, N. Tang, D. Zhang, Z. Jin, Nitrogen and sulfur Co-doped microporous activated carbon macro-spheres for CO_2 capture, J. Colloid Interface Sci. 526 (2018) 174–183. https://doi.org/10.1016/j.jcis.2018.04.101.

[63] S. Wang, Y. Xu, M. Yan, L. Zhang, Z. Liu, The effect of surfactants on carbon xerogel structure and CO_2 capture, J. Non-Cryst. Solids 499 (2018) 101–106. https://doi.org/10.1016/j.jnoncrysol.2018.07.023.

[64] Y. Li, X. Wang, M. Cao, Three-dimensional porous carbon frameworks derived from mangosteen peel waste as promising materials for CO_2 capture and supercapacitors, J. CO2 Util. 27 (2018) 204–216. https://doi.org/10.1016/j.jcou.2018.07.019.

[65] L. Liang, M. Zhou, C. Tan, X. Tian, K. Li, Easily tunable hydrogel-derived heteroatom-doped hierarchically porous carbons as multifunctional materials for supercapacitors, CO_2 capture and dye removal, Microporous Mesoporous Mater. 271 (2018) 92–99. https://doi.org/10.1016/j.micromeso.2018.05.050.

[66] L. Shao, Y. Sang, J. Huang, Imidazole-based hyper-cross-linked polymers derived porous carbons for CO_2 capture, Microporous Mesoporous Mater. 275 (2019) 131–138. https://doi.org/10.1016/j.micromeso.2018.08.025.

[67] A.E. Ogungbenro, D.V. Quang, K. Al-Ali, M.R.M. Abu-Zahra, Activated carbon from date Seeds for CO_2 capture applications, Energy Procedia 114 (2017) 2313–2321. https://doi.org/10.1016/j.egypro.2017.03.1370.

[68] J.B. Mathangi, M. Sadeesh Sharma, B. Mercy Jacquline, M. Helen Kalavathy, Development of carbon-based material from biomass for the removal of Ni^{2+} and CO_2 from fluid phase, Vacuum 158 (2018) 236–248. https://doi.org/10.1016/j.vacuum.2018.09.056.

[69] Z. Rouzitalab, D. Mohammady Maklavany, A. Rashidi, S. Jafarinejad, Synthesis of N-doped nanoporous carbon from walnut shell for enhancing CO_2 adsorption capacity and separation, J. Environ. Chem. Eng. 6 (5) (2018) 6653−6663. https://doi.org/10.1016/j.jece.2018.10.035.

[70] D. Li, J. Zhou, Z. Zhang, L. Li, Y. Tian, Y. Lu, L. Wen, Improving low-pressure CO_2 capture performance of N-doped active carbons by adjusting flow rate of protective gas during alkali activation, Carbon 114 (2017) 496−503. https://doi.org/10.1016/j.carbon.2016.12.039.

[71] H. Sun, P. La, R. Yang, Z. Zhu, W. Liang, B. Yang, W. Deng, Innovative nanoporous carbons with ultrahigh uptakes for capture and reversible storage of CO_2 and volatile iodine, J. Hazard Mater. 321 (2017) 210−217. https://doi.org/10.1016/j.jhazmat.2016.09.015.

[72] A. Ashraf, A. Butt, I. Khalid, R.U. Alam, S.R. Ahmad, Smog analysis and its effect on reported ocular surface diseases: a case study of 2016 smog event of Lahore, Atmos. Environ. 198 (2019) 257−264. https://doi.org/10.1016/j.atmosenv.2018.10.029.

[73] A. Hyodo, S. Malghani, Y. Zhou, R.M. Mushinski, S. Toyoda, N. Yoshida, J.B. West, Biochar amendment suppresses N_2O emissions but has no impact on 15N site preference in an anaerobic soil, Rapid Commun. Mass Spectrom. (2018). https://doi.org/10.1002/rcm.8305.

[74] E. Lapied, J. Nahmani, G.X. Rousseau, Influence of texture and amendments on soil properties and earthworm communities, Appl. Soil Ecol. 43 (2) (2009) 241−249. https://doi.org/10.1016/j.apsoil.2009.08.004.

[75] M. Diacono, F. Montemurro, Effectiveness of organic wastes as fertilizers and amendments in Salt-affected soils, Agriculture 5 (2) (2015) 221.

[76] R. Van Poucke, J. Ainsworth, M. Maeseele, Y.S. Ok, E. Meers, F.M.G. Tack, Chemical stabilization of Cd-contaminated soil using biochar, Appl. Geochem. 88 (2018) 122−130. https://doi.org/10.1016/j.apgeochem.2017.09.001.

[77] Z.Z. Chowdhury, M.Z. Karim, M.A. Ashraf, K. Khalid, Influence of carbonization temperature on physicochemical properties of biochar derived from slow pyrolysis of durian wood (Durio zibethinus) Sawdust, Bio Resour. 11 (2016).

[78] F. Xiao, B. Gámiz, J.J. Pignatello, Adsorption and desorption of nitrous oxide by raw and thermally air-oxidized chars, Sci. Total Environ. 643 (2018) 1436−1445. https://doi.org/10.1016/j.scitotenv.2018.06.280.

[79] M. Belhachemi, M. Jeguirim, L. Limousy, F. Addoun, Comparison of NO_2 removal using date pits activated carbon and modified commercialized activated carbon via different preparation methods: effect of porosity and surface chemistry, Chem. Eng. J. 253 (2014) 121−129. https://doi.org/10.1016/j.cej.2014.05.004.

[80] P. González-García, Activated carbon from lignocellulosics precursors: a review of the synthesis methods, characterization techniques and applications, Renew. Sustain. Energy Rev. 82 (2018) 1393−1414. https://doi.org/10.1016/j.rser.2017.04.117.

[81] M. Jeguirim, M. Belhachemi, L. Limousy, S. Bennici, Adsorption/reduction of nitrogen dioxide on activated carbons: textural properties versus surface chemistry − a review, Chem. Eng. J. 347 (2018) 493−504. https://doi.org/10.1016/j.cej.2018.04.063.

[82] I. Ghouma, M. Jeguirim, U. Sager, L. Limousy, S. Bennici, E. Däuber, A. Ouederni, The potential of activated carbon made of agro-industrial residues in NOx immissions abatement, Energies 10 (10) (2017) 1508.

[83] K. Silas, W.A.W.A.K. Ghani, T.S.Y. Choong, U. Rashid, Breakthrough studies of Co_3O_4 supported activated carbon monolith for simultaneous SO_2/NO_x removal from flue gas,

Fuel Process. Technol. 180 (2018) 155–165, in: https://doi.org/10.1016/j.fuproc.2018.08.018.

[84] M. Kacem, M. Pellerano, A. Delebarre, Pressure swing adsorption for CO_2/N_2 and CO_2/CH_4 separation: comparison between activated carbons and zeolites performances, Fuel Process. Technol. 138 (2015) 271–283, in: https://doi.org/10.1016/j.fuproc.2015.04.032.

[85] X. Wang, B. Yuan, X. Zhou, Q. Xia, Y. Li, D. An, Z. Li, Novel glucose-based adsorbents (Glc-Cs) with high CO_2 capacity and excellent $CO_2/CH_4/N_2$ adsorption selectivity, Chem. Eng. J. 327 (2017) 51–59. https://doi.org/10.1016/j.cej.2017.06.074.

[86] E. Gabruś, K. Witkiewicz, J. Nastaj, Modeling of regeneration stage of 3A and 4A zeolite molecular sieves in TSA process used for dewatering of aliphatic alcohols, Chem. Eng. J. 337 (2018) 416–427. https://doi.org/10.1016/j.cej.2017.12.112.

[87] N. Álvarez-Gutiérrez, M.V. Gil, F. Rubiera, C. Pevida, Adsorption performance indicators for the CO_2/CH_4 separation: application to biomass-based activated carbons, Fuel Process. Technol. 142 (2016) 361–369, in: https://doi.org/10.1016/j.fuproc.2015.10.038.

[88] J. Park, N.F. Attia, M. Jung, M.E. Lee, K. Lee, J. Chung, H. Oh, Sustainable nanoporous carbon for CO_2, CH_4, N_2, H_2 adsorption and CO_2/CH_4 and CO_2/N_2 separation, Energy 158 (2018) 9–16. https://doi.org/10.1016/j.energy.2018.06.010.

[89] E. Latosov, M. Loorits, B. Maaten, A. Volkova, S. Soosaar, Corrosive effects of H_2S and NH_3 on natural gas piping systems manufactured of carbon steel, Energy Procedia 128 (2017) 316–323. https://doi.org/10.1016/j.egypro.2017.08.319.

[90] G. Liu, A. Cadiau, Y. Liu, K. Adil, V. Chernikova, I.-D. Carja, W.J. Koros, Enabling fluorinated MOF-based membranes for simultaneous removal of H_2S and CO_2 from natural gas, Angew. Chem. Int. Ed. 57 (45) (2018) 14811–14816. https://doi.org/10.1002/anie.201808991.

[91] T.J. Bandosz, Effect of pore structure and surface chemistry of virgin activated carbons on removal of hydrogen sulfide, Carbon 37 (3) (1999) 483–491. https://doi.org/10.1016/S0008-6223(98)00217-6.

[92] B. Bajaj, H.-I. Joh, S.M. Jo, J.H. Park, K.B. Yi, S. Lee, Enhanced reactive H_2S adsorption using carbon nanofibers supported with Cu/CuxO nanoparticles, Appl. Surf. Sci. 429 (2018) 253–257. https://doi.org/10.1016/j.apsusc.2017.06.280.

[93] C.-W. Lee, Y.-T. Dai, C.-H. Chien, D.-J. Hsu, Characteristics and health impacts of volatile organic compounds in photocopy centers, Environ. Res. 100 (2) (2006) 139–149. https://doi.org/10.1016/j.envres.2005.05.003.

[94] L. Hui, X. Liu, Q. Tan, M. Feng, J. An, Y. Qu, M. Jiang, Characteristics, source apportionment and contribution of VOCs to ozone formation in Wuhan, Central China, Atmos. Environ. 192 (2018) 55–71. https://doi.org/10.1016/j.atmosenv.2018.08.042.

[95] H. Ren, P. Koshy, W.-F. Chen, S. Qi, C.C. Sorrell, Photocatalytic materials and technologies for air purification, J. Hazard Mater. 325 (2017) 340–366. https://doi.org/10.1016/j.jhazmat.2016.08.072.

[96] D. Saha, N. Mirando, A. Levchenko, Liquid and vapor phase adsorption of BTX in lignin derived activated carbon: equilibrium and kinetics study, J. Clean. Prod. 182 (2018) 372–378. https://doi.org/10.1016/j.jclepro.2018.02.076.

[97] J. Qi, J. Li, Y. Li, X. Fang, X. Sun, J. Shen, L. Wang, Synthesis of porous carbon beads with controllable pore structure for volatile organic compounds removal, Chem. Eng. J. 307 (2017) 989–998. https://doi.org/10.1016/j.cej.2016.09.022.

[98] R. Cherbański, Regeneration of granular activated carbon loaded with toluene – comparison of microwave and conductive heating at the same active powers, Chem. Eng. Process 123 (2018) 148–157. https://doi.org/10.1016/j.cep.2017.11.008.

[99] L. Yu, L. Wang, W. Xu, L. Chen, M. Fu, J. Wu, D. Ye, Adsorption of VOCs on reduced graphene oxide, J. Environ. Sci. 67 (2018) 171–178. https://doi.org/10.1016/j.jes.2017.08.022.

[100] H.S. Kim, M.S. Kang, S. Lee, Y.-W. Lee, W.C. Yoo, N-doping and ultramicroporosity-controlled crab shell derived carbons for enhanced CO_2 and CH_4 sorption, Microporous Mesoporous Mater. 272 (2018) 92–100. https://doi.org/10.1016/j.micromeso.2018.06.021.

[101] Z. Yue, A. Vakili, J. Wang, Activated carbon fibers from meltblown isotropic pitch fiber webs for vapor phase adsorption of volatile organic compounds, Chem. Eng. J. 330 (2017) 183–190. https://doi.org/10.1016/j.cej.2017.07.150.

[102] L. Hu, Y. Peng, F. Wu, S. Peng, J. Li, Z. Liu, Tubular activated carbons made from cotton stalk for dynamic adsorption of airborne toluene, J. Taiwan Inst. Chem. Eng. 80 (2017) 399–405. https://doi.org/10.1016/j.jtice.2017.07.029.

[103] J. Zhu, Y. Li, L. Xu, Z. Liu, Removal of toluene from waste gas by adsorption-desorption process using corncob-based activated carbons as adsorbents, Ecotoxicol. Environ. Saf. 165 (2018) 115–125. https://doi.org/10.1016/j.ecoenv.2018.08.105.

[104] N. Kostoglou, C. Koczwara, C. Prehal, V. Terziyska, B. Babic, B. Matovic, C. Rebholz, Nanoporous activated carbon cloth as a versatile material for hydrogen adsorption, selective gas separation and electrochemical energy storage, Nano Energy 40 (2017) 49–64. https://doi.org/10.1016/j.nanoen.2017.07.056.

[105] P.C. Vilella, J.A. Lira, D.C.S. Azevedo, M. Bastos-Neto, R. Stefanutti, Preparation of biomass-based activated carbons and their evaluation for biogas upgrading purposes, Ind. Crops Prod. 109 (2017) 134–140. https://doi.org/10.1016/j.indcrop.2017.08.017.

Chapter 7

New trends in molecular imprinting techniques

Ramonna I. Kosheleva, Athanasios C. Mitropoulos and George Z. Kyzas*
Hephaestus Advanced Laboratory, Department of Chemistry, International Hellenic University, Kavala, Greece
**Corresponding author*

1. Introduction

Molecular imprinting technology (MIT) is an umbrella term for all the methods and approaches involved in molecularly imprinted polymer (MIP) synthesis process [1]. The as-synthesized MIPs can be implemented in many fields including biological and electrochemical sensors [2,3], separation and adsorption technologies [4] as well as sample preparation applications mainly in analytical chemistry for solid-phase extraction (SPE), microextraction, and others. First in 1931 by Polyakov and then in 1949 by Dickey [5], the effect of participating solvent in polymerization process and the resulting affinity of the synthesized polymer to the "pattern molecule" led to a more elaborative investigation on the topic by researchers globally. The MIP technology gained popularity during the last two decades, acquiring its present sense through an enormous number of studies and review articles. In fact, a brief research though scientific publications database (ScienceDirect and Scopus) of related literature showed that a great amount of work done is within a time frame of 2015 to present, with 2017 being the year of the most published studies on the topic. Also, a closer review of the available literature can lead to estimations regarding the trends both in synthesis and evaluation of MIP. In this respect, the specific chapter aims to cover the recent development within MIP technology, providing examples from recently published works.

Advanced Low-Cost Separation Techniques in Interface Science. https://doi.org/10.1016/B978-0-12-814178-6.00007-8

2. Synthesis of molecularly imprinted polymers

The synthesis of MIP is considered an easy and effective process requiring (1) a template which is essentially the target molecule, (2) one or more functional monomer(s), (3) cross-linker(s), (4) initiator(s), (5) a porogenic solvent, and (6) a solvent for template extraction after polymerization of the complex (Fig. 7.1). The resulted material is a three dimensional polymer with formed cavities in shape and size that can bind and take in, via functional groups of the polymer, only a molecule with the same specifications. Thus, MIPs are characterized by high degree of selectivity for a predetermined compound, rapid mass transfer and kinetics, low-cost and easy synthesis procedure for trace analytes binding. Detailed description of the common synthesis protocol is given in multiple literature reviews [7], textbooks [8,9], and studies [10], therefore it is omitted from the specific work.

A successful metaphor to describe this mechanism is lock and key hypothesis that was postulated by Emil Fisher in 1894 in his work upon enzymes selectivity [11]. Actually, MIPs are artificial materials that mimic such behavior of natural occurring processes (i.e., antibodies) being though superior due to the fact that they provide [9] the ability to bind even compounds that are not present in nature while the number of binding sites is

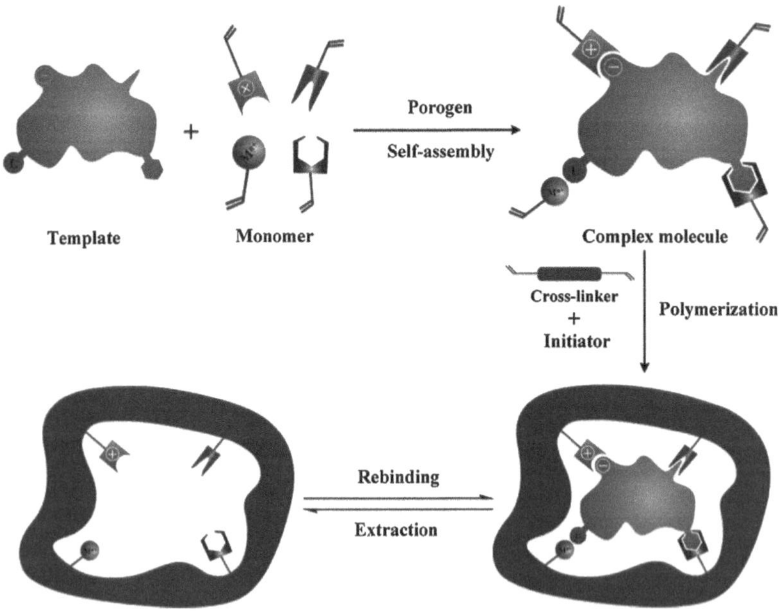

FIGURE 7.1 Schematic representation of molecularly imprinted polymer synthesis [6].

increased by many orders of magnitude. Fig. 7.1 represents schematically the whole process and outcomes of each step.

In MIP technology, many subgroups can be formed under various criteria; a classification could be done, for example, according to the MIP synthesis approach (solid-phase method, polymerization, surface imprinting, etc.) or the resulted MIP type (inorganic–organic molecularly imprinted polymer (IMIP), magnetic molecularly imprinted polymer (MMIP), multitemplate MIP, nanoparticles composite molecularly imprinted polymer (MIP-NP), etc.), the list can be endless due to versatile nature of MIPs. Another way for MIP categorization could be by application or target molecule. So ever, MIP technology is one of the emerged technologies with rapid development. The amount of studies conducted so far indicates the maturity level of MIT and hence MIP. The present work gives an overview of hot topics of the moment and the conducted studies upon the latest trends. Table 7.1 presents some of the new trends of this technology.

3. Inorganic molecularly imprinted polymer

According to the recent literature, a small yet portion of scientists have shown interest in using hybrids of organic–inorganic compounds in MIP synthesis instead of pure organic ones. This can be attributed to the fact that such hybrid materials can overcome some drawbacks. The as-synthesized MIPs, with traditional organic monomers, suffer from low performance in aqueous media, and many times there are difficulties of obtaining uniform particle size, some of the most important aspects considered especially in biological applications such as drug delivery [25] and therapeutic/diagnostic procedures [26]. For the latter case, "shell-" and "core-" like MIPs are implemented where "shell" is referred to imprinting the surface layer of nanosphere-shaped polymer [27] while "core" usually constitutes from a silica-based [28] or Fe_2O_3 [29] nanoparticles.

The embedment of inorganic particles into organic matrices leads to high-quality materials regarding mechanical strength, thermal and chemical stability, high selectivity performance, and other properties prerequisite for adsorption applications. Hence, this new generation of MIP materials is not restricted only to biological applications, but they are used in extend as adsorbents for effective decontamination of aqueous media. A large number of studies has been conducted upon investigation of IMIPs efficacy as adsorbents in many aspects including environmental [30], food processing [31], drug [32], and valuable metals recovery technology [33] and others.

Immobilization of template combined by solid-phase method for MIP nanoparticles synthesis has been first introduced in 2013 by Poma and associates [34]. The team achieved to fully automate the synthesis process in a device producing reuseable immobilized templates eliminating extreme operation conditions. Since then, the specific method has been utilized

TABLE 7.1 Few examples of new trends in molecular imprinting technology (MIT).

MIP type	Approach	Target molecule	Monomer/cross-linker/initiator	Chain transfer agent	Porogen	Solvent	References
IMIP-NP	Solid phase	Vancomycin	MAA/EGDMA, TRIM/DABE	PETMP			[12]
IMIP-NP	Solid phase	IBA					[13]
IMIP-NP	Solid phase	4-Ethylphenol	4VP, FER/EDMA, RIM/INEFERTER	Pentaerythritol tetrakis (3-mercaptopropionate)	Acetonitrile, tetrahydrofuran	–	[14]
IMIP-NP	Solid phase	Trypsin	Nipam, BIS, TBA, NAPMA, AA/BIS, EGDMA, TRIM, NAPMA, AAm/ammonium persulfate aqueous solution, TEMED	–	–	–	[15]
IMIP-NP	Solid phase	THC	BIS, EGDMA, TRIM, NAPMA, AAm/PETMP, INIFERTER			DMF, ACN	[15]
MIP	Bulk polymerization	Ketoprofen	2-Vinylpyridine/ethylene glycol dimethacrylate, 1,1′/azobis (cyclohexanecarbonitrile)	–	Toluene/acetonitrile	–	[16]
Multi-template MIP	Bulk polymerization	Naproxen, ibuprofen and diclofenac	2-Vinylpyridine/ethylene glycol dimethacrylate	–	Toluene	–	[17]
SPI-MIPs	Bulk polymerization	Spiramycin	MAA/EGDMA/AIBN	–	–	Methanol	[18]

SBA-15@MIP	Surface molecular imprinting	dicyandiamide	DCD/EGDMA/AIBN	—	Tetrahydrofuran-ethanol	Methanol	[19]
SiO_2@MIPs	Surface molecular imprinting	17β-Estradiol	APTES/CTAB/BTEB	—	Ammonia	Na_2CO_3	[20]
Sol-gel MIPs	Surface molecular imprinting	Gossypol	APTES/TEOS/-	—	Acetic acid	Methanol/HCL	[21]
MMIPs	Precipitation polymerization	CDNB	EGDMA/MDI/-	Phloroglucinol	Tetrahydrofuran	Methanol/acetic acid	[22]
MMIPs	RAFT polymerization	BPA	4-Vinylpyridine and β-CD/TRIM/AIBN	—	—	Acetic acid/ methanol	[23]
MMIPs	Surface polymerization	Folic acid	Acrylonitrile/ EGDMA/AIBN	—	Ethanol	Methanol/acetic acid	[24]

IMIPs, inorganic–organic molecularly imprinted polymers; *MIPs*, molecularly imprinted polymers; *MMIPs*, magnetic molecularly imprinted polymers.

successfully by many other researchers for various purposes. Commonly, immobilization of the template is achieved by grafting covalently organic molecules on an inorganic substrate either heterogeneously or homogenously [35].

A silica-based hybrid was evaluated for Hippuric acid (HA) extraction by solid-phase method [36]. The inorganic part of this composite material was a silica-based compound, and the target molecule recognition site was formed by 3-aminopropyl trimethoxysilane as the functional monomer cross-linked by tetraethoxysilane while methacryloxypropyltrimethoxysilane provided the ability of MIP surface acrylamide coating. The resulted material is characterized as restricted access material (RAM) due to the formation of hydrophilic barriers-like network. Furthermore, Arabi et al. [36], performed routine MIP characterization techniques for morphological evaluation including transmission electron microscope (TEM), scanning electron microscope (SEM), and FTIR as well as investigations of MIP performance by applying SPE of HA in a cartridge packed with the resulted material. After optimization of the methodology, the template to monomer ratio was found to be 1:5. Chromatography analysis of simple MIP was compared with that of RAM-MIP which demonstrated higher performance and repeatability while its reusability was determined to be high after a number of SPE cycles.

Enantiomers of albendazole sulfoxide (ABZSO), a drug usually used in veterinary as an antihelminthic for sheep, were successfully extracted from water by a poly(methacrylic-acid)silica-based IMIP. According to the study of Anacleto et al. [30], the so-named MIP-poly(MAA)-SiO_2 was synthesized and evaluated and compared with nonimprinted polymer (NIP) for its adsorption capacity, kinetics, and the effect of pH for the optimal performance. The results of the analysis showed that adsorption equilibrium was reached after 20 min of contact in aqueous solution of pH 9.0, while pseudo-second-order equation was that best fitted to kinetics analysis giving the same R value in both cases. This outcome leads to estimations that both examined materials adsorbed the target compound onto their external surfaces although at equilibrium MIP-poly-(MAA)-SiO_2 adsorbed up to 70% of enantiomers while NIP adsorbed only 50%. Additionally, the agreement with Langmuir isotherm indicates that monolayer molecule adsorption was performed giving information on adsorption capacity which was 140 and 35 mg/g for MIP-poly(MAA)–SiO_2 and NIP-poly(MAA)-SiO_2, respectively. Finally the specific work concluded that such MIPs can be implemented efficiently for separation and removal purposes of pharmaceuticals in water, providing reusability of the material as well.

In another work, organic–inorganic MIPs were synthesized on selenium (Se) nanoparticles for cholesterol binding. Specifically, two different monomers were polymerized namely 2-hydroxyethyl methacrylate (HEMA) and ethylene glycol dimethacrylate (EGDMA) and stabilized by poly(vinyl pyrrolidone) (PVP) by Polyakova et al. [37]. Pickering microemulsion was used as synthesis method; the resulted sorbent was evaluated by chemical and morphological evaluation techniques. In order to compare, same examination was done for NIP as well. SEM micrographs revealed that synthesized MIPs had better porous microstructure while adsorption properties investigation indicated that the initial concentration of cholesterol plays a significant role in material's overall performance. Langmuir isotherm model describes better the adsorption process which seems to be monolayer deposition of the target compound, hence material's imprinted surface can be considered as advantage for cholesterol extraction from blood plasma.

4. Magnetic molecularly imprinted polymer

MMIPs are a new class of materials attracting an increased attention during past years mainly due to their easier after-treatment processes which are much more convenient and effective than those of conventional MIPs. An additional factor for their preference is the selectivity, for the target molecule, that they perform [38]. As it is rational, MMIPs are synthesized following surface imprinting technique where the polymeric surface layer of coated magnets is imprinted by the target analogue. Specifically, as a common methodology, emulsion polymerization take place in order to obtain core particles covered with imprinted shells [39]. A schematic representation of a common utilized methodology for MMIP synthesis is illustrated in Fig. 7.2 [39]. Besides many other advantages of core–shell structured materials, the obtained uniformity regarding the size and shape of the resulted particles as well as the greater overall efficiency makes this approach highly attractive in many aspects.

High internal phase emulsion polymerization by Pickering method was the methodology of choice for porous MMIP tailoring used for λ-cyhalothrin (LC) adsorption [40]. After its synthesis, material was evaluated regarding kinetics, adsorption capacity, and selectivity by competitive recognition test. The adsorption capacity was investigated under no vibration and although the equilibrium was reached after 2 h, the amount that could be adsorbed was roughly determined at 1 h of the contact with the pesticide. Isothermal sorption measurements at ambient temperature estimated the adsorption capacity to 404.4 μmol/g, while from the thermodynamic point of view, the one-layer molecular deposition occurred spontaneously. Finally, in order to qualitatively define the MMIP selectivity, one similar to λ-cyhalothrin compound, namely fenvalerate (FL), and a nonanalogue such as diethyl phthalate (DEP) were examined as well. The test revealed that the MMIP presented higher affinity to LC and its analogue while no selectivity to DEP was observed. The authors attribute this behavior to the fact that LC and FL is bound to MMIP by hydrogen bonds while DEP lacks them, limiting its ability to get into the formed cavities despite its small molecule.

Zhang et al. [41] managed to successfully bind salicylic acid (SA) onto core–shell MMIP. In fact, comparative experiments of synthesized MMIP by

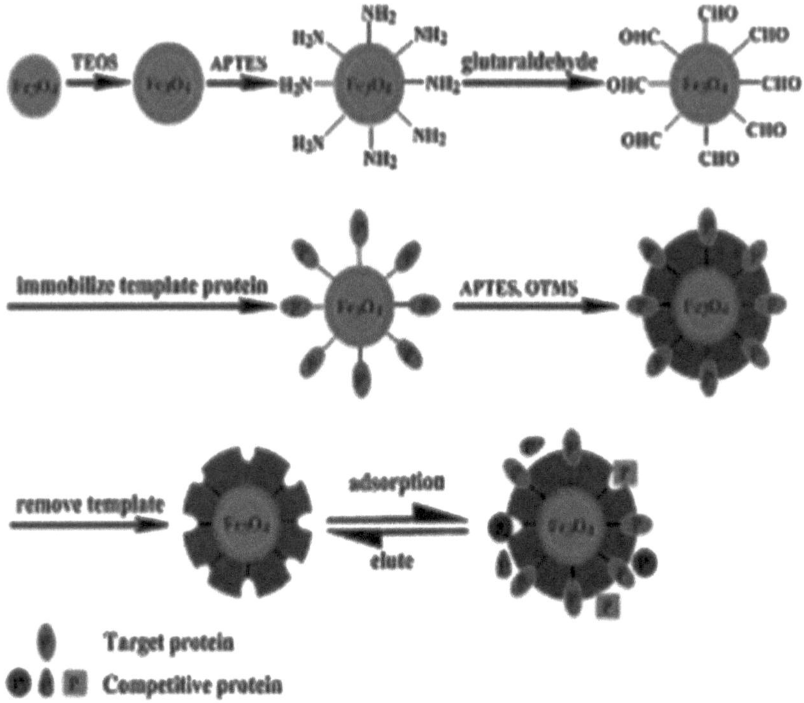

FIGURE 7.2 A schematic illustration of core–shell magnetic molecularly imprinted polymer synthesis. Fe_3O_4 nanoparticle is used as core and coated with a silica shell. After processing with tetraethoxysilane (TEOS), $Fe_3O_4@SiO_2$ is produced. Further, reaction with 3-aminopropyltriethoxysilane (APTES) take place in order to obtain amino groups on its surface, while further reaction with glutaraldehyde (GA) leads to surface modification by creating aldehydes. Then, a 3D polymeric structure is formed under siloxane copolymerization by APTES and octyltrimethoxysilane after target compound had been immobilized via covalent approach onto the modified surface. The created network constitutes protein complex in which protein is removed leaving cavities of specific characteristics concerning shape, size, and functionality [39].

sol–gel polymerization and the same not imprinted material demonstrated a greater binding capacity by many times while its recovery was up to 108%. Therefore the particular synthesis strategy is suggested by the research team as a multifaceted method for production of materials for harsh and complex environments.

Another study on core–shell structured MIP nanoparticles, conducted by Wu and his team [42], proposes the synthesized MMIP for tetrabromobisphenol A (TBBPA) extraction from water. Fe_3O_4 nanoparticles were coated by imprinted SiO_2, and evaluation of the material revealed its high efficiency toward TBBPA removal. Specifically, MMIP adsorbed 2.3 times more of the target compound than MNIP at a minimal equilibrium time of 40 min presenting also high reusability, while its stability did not show to decrease even after seven cycles of TBBPA extraction by adsorption.

5. Dummy-template molecularly imprinted polymer

Environmental safety is one of the most significant concerns in many fields of science. Green technology and all the subsequent aspects have been developed in recent years and the number of firm supporters is ever increasing. In the field of chemistry, some of the major issues arise are the logical use of reactants, the minimization of energy consumption, the reusability of the produced material, and the optimization of the whole process; so it could be considered eco-friendly and safe to be implemented [43]. In this sense, the participation of computational chemistry in MIP synthesis process is essential. During the past years many works have been conducted and published supporting computational approach for synthesis optimization. In most of the cases, computational tools such as density functional theory and molecular dynamics [10] are used for an efficient screening of monomers in order to find the suitable for the intent use, thus avoiding the generation of unusable waste like it happens when trial-and-error approach is implemented. MIPs synthesis is considered easy to do, but there are some steps of the process that are crucial for the final result; template manipulation is one of them. The main advantage of an MIP is its capability to recognize even traces of a particular compound, that is why it is preferred for sample preparation in already mentioned methods. In many cases, the target compound may be toxic or valuable enough so any leakage from the MIP matrix is undesirable. To solve this problem, dummy templates are used instead of original compounds. As a dummy template for MIP synthesis (DMIP) is an analogue to the target molecule; another molecule of similar shape, size, and functionality but not toxic or valuable.

In order to overcome low selectivity for SA of MIPs due to its weak hydrogen bonding ability with conventional monomers, Xiang et al. [44] used ionic liquid as functional monomer for polymerization of proposed MIP in aqueous medium. As template they replaced SA with a dummy compound which was benzoic acid while MCM-48 was the porous substrate. During this study, parameters such as the kind of template, the analogy of template—monomer—cross-linker, and porogen selection were optimized for synthesizing an MIP of high adsorption capacity and improved selectivity. The resulted material was a hollow porous MIP presenting better properties than any reported MIP fabricated traditionally or by surface imprinting technology. The affinity of fabricated MIP to SA was determined by means of adsorption capacity and imprinting factor. Imprinting factor was the ratio of adsorption capacity of MIP and an NIP as reference material, and it was found 5.61 with adsorption capacity of MIP being 29.75 mg/g. BET analysis provided information about surface area which was 543.9 m^2/g, while kinetics test showed that reaction equilibrium was reached after 25 min of contact following Freundlich equation and described by pseudo-second-order model. Finally, the recovery rate was found to be up to 94.5%, implying the specific MIP can be applied successfully for SA extraction.

One additional challenge in analytical chemistry is the ability for a material to be multifunctional, in other words, to possess the ability to recognize and bind more than one compounds simultaneously. MIP offers this opportunity by suitable design and optimization. For example, Song et al. [45] synthesized a dual template MIP for detection of 16 compounds, 8 fluoroquinolones (FQs), and 8 sulfonamides (SAs). As dummy templates, pipemidic acid (PA) and nalidixic acid (NA) were selected for FQs while sulfanilamide (SA) and sulfabenzamide (SB) were found to be suitable for SAs. Theoretical results of simulation modeling tool and experiments were compared in order to examine the performance of the fabricated MIP for FQs and SAs affinity. Then the optimized materials were used for mentioned drugs detection in pork and chicken meat by SPE column as sample preparation method for ultra performance liquid chromatography. The results obtained indicated the good potentials of the specific MIP with maximum drug adsorption at 74.2 μg and recovery up to 99% even after eight times of column usage. As it is implied in Fig. 7.3, the prepared MIPs present high affinity to specific analytes that can be observed from high recovery of the target compounds while other competitor compounds are not recognized by any of the MIPs.

In some cases, synthetic template preparation is restricted by high cost and complexity, hence other solutions should be provided. One clever way to overcome such issues is to structurally divide the target compound producing fragments that are same or similar to it like He et al. [46] achieved in order to selectively extract gonyautoxins from seawater by SPE. The research team followed a multitemplate MIP method for dummy template preparation using fragments of dummy analogues (2,4,6-triaminopyrimidine, 4-hydroxy-2-butanone, and imidazole). Bulk polymerization was the selected approach

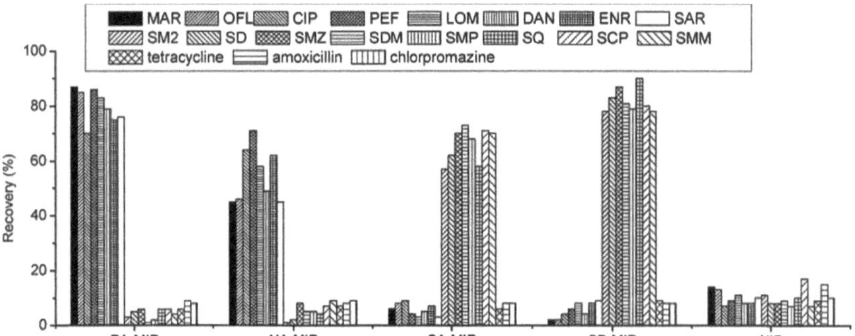

FIGURE 7.3 Recoveries of fluoroquinolones and SAs by four prepared molecularly imprinted polymers compared with blank nonimprinted polymer material [45].

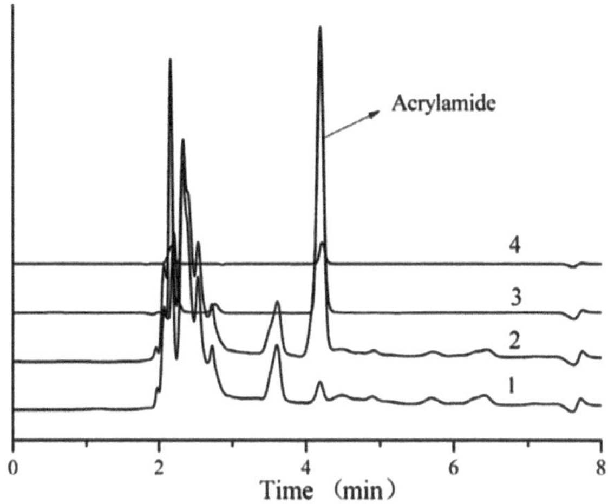

FIGURE 7.4 HPLC–UV chromatograms of food samples. The collected spectra shows (1) not spiked, (2) spiked with 4 ng/g acrylamide and without extraction, (3) spiked with 4 ng/g acrylamide and with AMDSMIPsGOFe$_3$O$_4$ extraction, and finally, (4) spiked with 4 ng/g acrylamide and the remaining solution after extraction [47].

for MIP synthesis involving prepolymerization by sonication for template and monomers dissolution. After this process, cross-linkers were added and mixed by stirring. The template was removed by a mixture of methanol/acetic acid of 9:1 in a Soxhlet after the polymerization was completed. The most significant remarks of this study was the results obtained by comparison of the fabricated multitemplate MIP with that of monotemplate, prepared in the previous work of the same team for 1 and 4 gonyautoxins isolation. It was revealed that MIP prepared by multitemplate method had lower adsorption capacity than that prepared by monotemplate method, authors attribute this outcome to incorrect combination of template fragments into the MIP synthesis leading to binding sites not suitable for the specific analyte.

The above spectra (Fig. 7.4) make clear the high selectivity of an MIP for acrylamide detection and removal from food. Specifically, Ning et al. [47], fabricated for the first time MIP selective to acrylamide. The difficulty in such case is that, usually, acrylamide is used a cross-linker making restrictions to its manipulation as template. Moreover, acrylamide is toxic and cancerogenic, hence its usage is preferred to be avoided. Instead acrylamide, the mentioned group introduced an analog to acrylamide; the so-named propionamide. The most remarkable of the work, was that the as-synthesized magnetic graphene oxide MIP (AMDSMIPsGOFe3O4) presented high selective recognition of the target compound even from fried foods, with high recovery yield.

6. MIP performance

6.1 Characterization and performance evaluation methods

Materials such as MIPs can be characterized by means of many methods including morphological and chemical evaluation. The most common techniques for morphological characterization of MIPs are TEM and SEM, while chemical analysis is provided by methods of liquid chromatography, FTIR, mass spectroscopy, and others. Binding efficiency can be determined by sorption test of the examined material, providing information about its kinetics and adsorption capacity of the target pollutant onto its surface, while surface properties can be further estimated by methods such as N_2 porosimetry in order to obtain the surface area. However, MIP properties are significantly affected by some factors both form the perspective of selected reagents and the followed synthesis approach. Following, effect of porogenic solvent, template removal method, and synthesis process on the resulted MIPs are presented by some examples of the recent literature.

6.2 Effect of the porogenic solvent on MIP efficiency

Porogen plays an important role within MIP synthesis process. Porogen is responsible not only for dissolving properly all the agents during polymerization but also for the development of porous structure of the MIP. Additionally, porogen must not interact with the formed template–monomer complex retaining its stability [49]. Thus, reagent used as porogen must be selected thoughtfully for MIP synthesis regarding the desired results and MIPs usability.

As it was mentioned in a previous section, green chemistry perspectives are appealing to chemists worldwide, therefore a lot of focus has been put on finding more eco-friendly, safer, and lower cost solutions whenever it is possible. Such an example is the usage of ionic liquids instead of conventional reagents in MIP synthesis process. Due to their nature, ionic liquids among other properties demonstrate nonvolatility, nonflammability, and also they are highly dispersible in inorganic/organic solvents, thus they have been implemented in recent studies substituting conventional monomers or/and crosslinker [50] in bulk polymerization or as functional substrate in surface MIT [51]. Ionic liquid can also be successfully used as porogen, adding special characteristics to MIPs. As an example, Booker and her team have conducted a series of studies upon the performance of ionic liquids as porogen within MIP synthesis for trans-aconitic acid 1 and cocaine 2. Investigations of porogen volumes and process temperature conditions as well as polymerization method were included. Comparisons with MIP prepared in volatile organic carbon VOC solvents such as CH_3CN were performed as well. In the initial study, acetonitrile and chloroform porogen performance for MIP preparation of aforementioned target compounds was compared with that of (BMIM) (BF4) and (BMIM) (PF6) ionic liquids for the same purpose. Two different polymerization methods were implemented, namely bulk and precipitation polymerization at 5 and 60 C. Remarkably, it was revealed that conventional porogens failed to develop 3D polymer complexion at low temperature in both polymerization methods. In contrast, ionic liquids as porogen achieved to form MIP and by temperature increase the polymerization rate was found to be significantly greater than the respective of conventional ones. Subsequent study of the same research team [48] generated additional MIPs characterized by SEM as well. Cocaine MIPs were prepared as monolithic when 5 mL of porogen was used and by precipitation method (25 mL of porogen). Obtained MIP characterization by SEM (Fig. 7.5) supports the results concluding that porogen nature impacts significantly MIP morphology.

As it was mentioned, MIPs are widely used in sensor technology for trace detection and monitoring as they are a good alternative of biological receptor–based methods whose usage is restricted by their stability and cost. A carbon paste electrode with appropriate MIP was proposed as sensor for histamine monitoring by Akhoundian et al. [52]. In their study, two different porogen solvents were used in MIP synthesis; acetonitrile and methacrylic acid following bulk polymerization method. For determination of porogen influence of MIP selectivity, optical rebinding test was performed by UV-VIS spectroscopy, after the MIP/histamine (at concentrations from 0 to 1 mM for 4 h of contact at ambient temperature) suspension was centrifuged, for both MIPs and their respective NIPs (Fig. 7.6).

The obtained results reveal that MIP synthesized in chloroform as porogen solvent performs better than MIP prepared in MeCN. This is attributed to the different polarities of the two porogens and to the fact that histamine binding is based mainly on hydrogen bonds in the presence of the methacrylic acid (as monomer) and chloroform (as porogen) combination. The latter is confirmed

FIGURE 7.5 SEM images of cocaine-imprinted polymers prepared in 5 mL porogen at 60 and 0 °C [48]. (A) CHCl3-60-5; (B) [bmim]BF4-60-5; (C) [bmim]PF6-60-5; (D) [bmim]HSO4-60-5; (E) [bmim]BF4-0-5; (F) [bmim]PF6-0-5; (G) [bmim]HSO4-0-5.

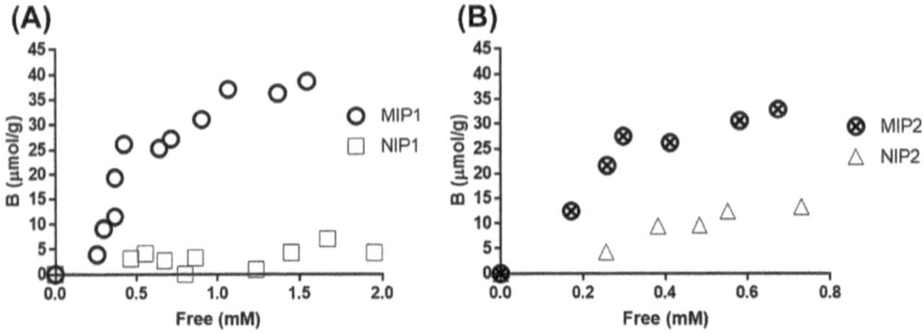

FIGURE 7.6 Rebinding isotherm of histamine examined by UV-VIS spectroscopy for (A) MIP1, NIP1; (B) MIP2, NIP2 in 50 mM PBS buffer (pH 7.4). MIP1 and MIP2 (corresponding NIPs) were prepared in $CHCl_3$ and MeCN, respectively [52].

by Trikka and her team [54] investigating binding and recognition properties of MIPs by conducting combinational NMR and UV-VIS spectroscopy analysis.

Supercritical fluids implementation is an emerged technology and can be considered as another route for meeting green technology requirements. To this direction, supercritical fluid such as $scCO_2$ has recently been used as porogen solvent for MIP prepolymerization phase. Due to its inherent characteristics, $scCO_2$ has been reported as a promising solution for sustainable, clean and with low environmental impact solvent. Moreover, $scCO_2$ apolar nature makes it perfect for porogen solvent while it does not interact with template–monomer complex resulting in MIP of controlled structural properties [55]. In order to fabricate a bisphenol A selective MIP, Rebocho and group synthesized a ferrocenyl-based MIP [56]. The polymerization process was occurred in a specially designed cell were all reactants were mixed at CO_2 environment with the assistance of sonication bath to ensure solids' dissolution (for detailed methodology refer to specific work).

6.3 Template removal methods

The method of template desorption is crucial for the optimal performance of the fabricated MIP. Many methods have been developed to efficiently remove template from the formed 3D network leaving the desired cavities for target molecule capturing. Lorenzo and Carro [57] conducted a replete review of MIP template extraction methods providing information about technologies up to 2011. Although MIT is a rapidly developed field with new materials and composition methods emerging every now, template extraction seems not to follow the same tempo. As it is well known, a successful template removal contributes to overall success of MIP synthesis. If template remainings are still present when prepared MIP is applied, then the adsorption capacity is decreased due to lower number of available binding sites (some of the binding sites are occupied with template). Additionally and especially for applications that requires high accuracy such as analytical methods (SPE, MSPE, etc.), a phenomenon called "template bleeding" may cause many problems leading to inaccurate estimations. Common strategies for this last step of MIP synthesis procedure include solvent extraction, microwave and ultrasound assisted extraction, pressurized-liquid extraction, and others. Template removal is considered as the least cost-effective process of MIP synthesis, thus optimization is mandatory for achieving good results at as low cost as possible but also with limited environmental impact. In fact, there are cases that, in order to remove the template, extreme conditions are applied leading most of the times to disintegration of formed cavities.

One of the latest work upon the specific topic has been conducted, and pressurized hot water extraction (PHWE) was proposed as the optimal template removal method by Batlokwa [53]. In the specific study, PHWE was compared with other common methods, namely soxhlet and ultrasonic extraction. Three MIPs were prepared and colored for distinguishing purposes. The optimal operation conditions of purge water were 220 °C at 50 bars with a flow rate of 2 mL/min. Comparative experiment for extraction efficiency (EE) was performed for all three methods revealing that PHWE not only had the highest EE percentage (99.6%) for all MIPs but also template wash-off was achieved in less time. The most interesting and worth to mention outcome of the specific study is the results from template bleeding test presented in Fig. 7.7.

As it is shown, PHWE presented no bleeding at all of chlorophyll and quercetin MIP and only a small percentage for Phthalocyanine of 0.02% compared to the rest two methods. The conclusion of this study is that water for template removal is the optimal solution while it is cheaper, readily available, and does not cause environmental damage as organic solvents do.

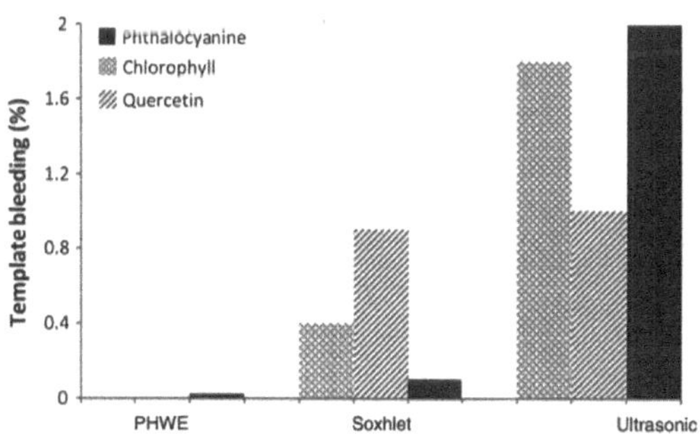

FIGURE 7.7 Percentage of template bleeding after the initial wash-off by all three template extraction methods [53].

6.4 Other parameters

An interesting perspective about the effect of MIP particle shape has been postulated by Roy et al. [58] with results that indicate the importance of MIP shape in selectivity and adsorption capacity. Silver nanoparticles (AgNPs) were synthesized acquiring spherical, hexagonal, rod, and flower shape and further modified by surface molecularly imprinting for phenformin recognition. AgNPs were subjected to characterization by UV-VIS spectroscopy, X-ray diffraction (XRD), SEM (Fig. 7.8), TEM (Fig. 7.9), while surface-modified nanoparticles were examined additionally for adsorption capacity, selectivity of the target molecule, surface area, and electrocatalytic activity.

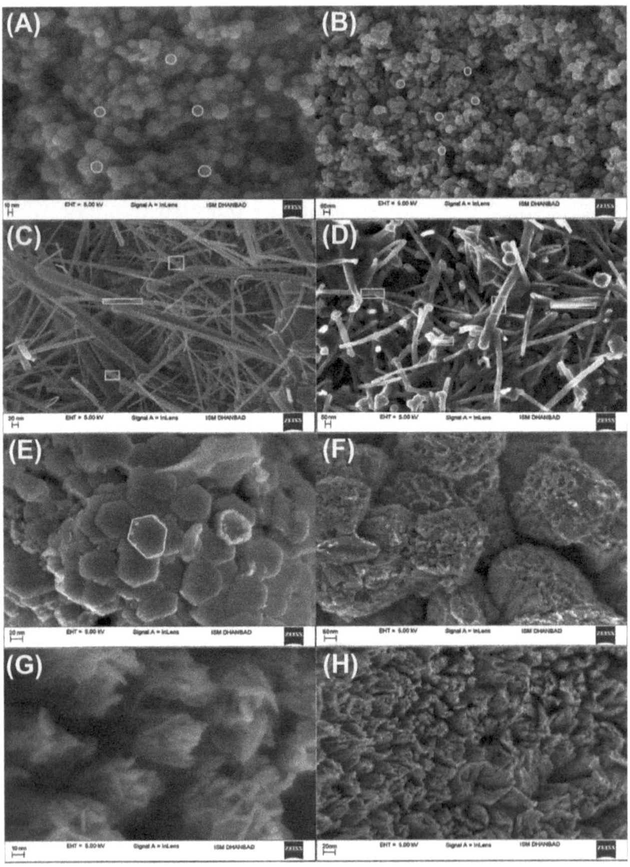

FIGURE 7.8 SEM micrographs of silver nanoparticles and the resulted molecularly imprinted polymers: (A) AgNSs, (B) MIPAgNSs, (C) AgNRs, (D) MIPAgNRs, (E) AgHPs, (F) MIPAgHPs, (G) AgFLs, and (H) MIPAgFLs.

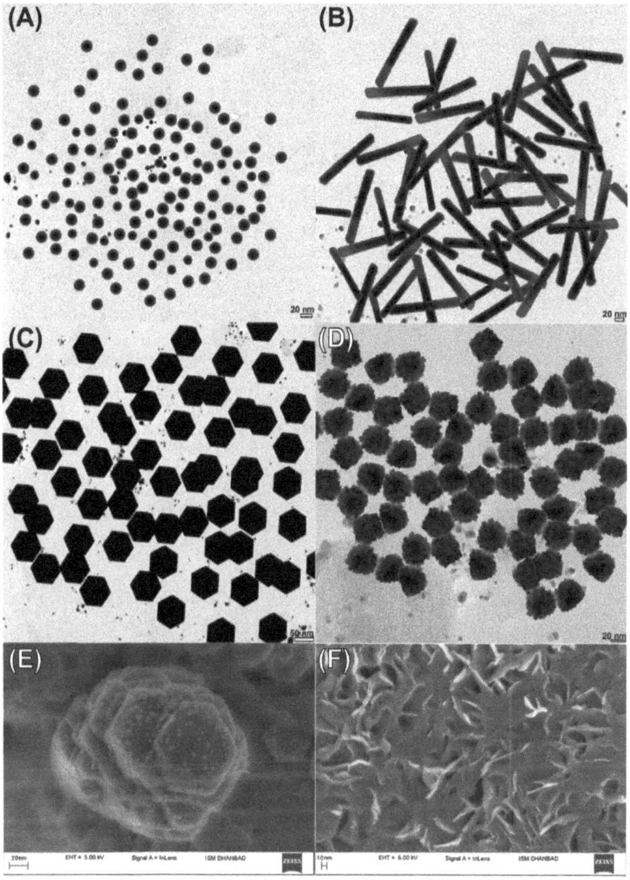

FIGURE 7.9 TEM imaging of silver nanoparticles and the resulted molecularly imprinted polymers: (A) AgNSs, (B) AgNRs, (C) AgHPs, and (D) AgFLs. High-resolution FE-SEM images of (E) MIPAgHPs and (F) MIPAgFL [58].

In another work of the same team, AgNPs of three different shapes (hexagonal, cubical, and spherical) were modified with vinyl group and used as one of the functional monomers in MIP synthesis for tramadol binding after they were embedded into the shell of Fe_3O_4 nanoparticle (NP) with high magnetism. As it was in the previous work, the scope of the study was to investigate the shape-related efficiency of the resulted MIP-NP for the morphological characterization SEM, TEM, and FE-SEM. Fig. 7.10 illustrates the whole synthesis process.

In both studies, results indicate that the shape of used MIP-NPs is crucial for their performance and this can be attributed to the fact that shape at nanoscale range alters physicochemical properties of the material as it has been reported elsewhere [60,61].

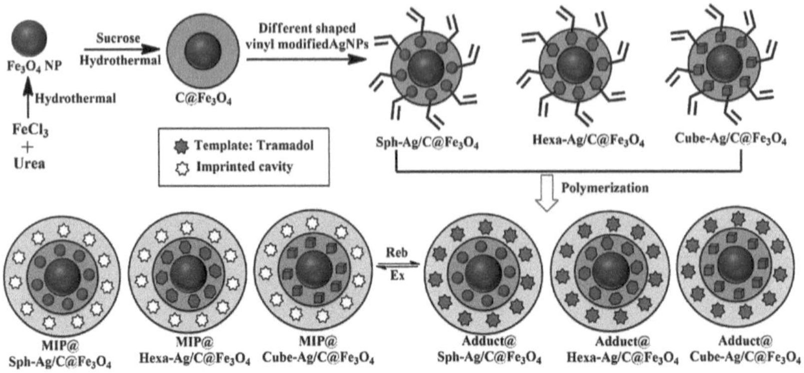

FIGURE 7.10 Synthesis process graphical illustration of resulted molecularly imprinted polymer (MIP) with different shaped AgNPs; Sph-Ag (spherical), Hexa-Ag (hexagonal), cube-Ag (cubical) [59].

7. Conclusions

Recent published studies have been reviewed and discussed. An attempt to present new trends in MIP technology and further subcategorized according to the nature of reagents and MIP utility, namely into IMIP, MIPs as magnetic particles composites (MMIP), and dummy template (DMIP) ones, was conducted. In addition, factors of major importance in MIP overall performance, such as porogenic mixture and template removal methods, as well as and other less common parameters, were highlighted. From the literature review, it was found that although there is a growing trend for MIP synthesis improvement, some involving methods (i.e., template removal) lack the appropriate attention. The latter observation indicates a misbalance between material composition and implemented methodology progress. In general, researchers have accomplished some remarkable achievements under a greener perspective. During the past 3 years, many of MIP drawbacks concerning stability, water compatibility, higher selectivity, multifunctionality, and many others have been overcome according to recently reported works.

Conflicts of interest

The authors declare no conflict of interest.

References

[1] M.J. Whitcombe, N. Kirsch, I.A. Nicholls, Molecular imprinting science and technology: a survey of the literature for the years 2004–2011, J. Mol. Recognit. 27 (6) (2014) 297–401.

[2] G. Selvolini, G. Marrazza, Mip-based sensors: promising new tools for cancer biomarker determination, Sensors 17 (4) (2017) 718.

[3] O.S. Ahmad, T.S. Bedwell, C. Esen, A. Garcia-Cruz, S.A. Piletsky, Molecularly imprinted polymers in electrochemical and optical sensors, Trends Biotechnol. 37 (2018) 294–309.

[4] G.Z. Kyzas, D.N. Bikiaris, N.K. Lazaridis, Selective separation of basic and reactive dyes by molecularly imprinted polymers (Mips), Chem. Eng. J. 149 (1) (2009) 263–272.

[5] F.H. Dickey, The preparation of specific adsorbents, Proc. Natl. Acad. Sci. USA 35 (5) (1949) 227–229.

[6] Y. Hu, J. Pan, K. Zhang, H. Lian, G. Li, Novel applications of molecularly-imprinted polymers in sample preparation, Trac. Trends Anal. Chem. 43 (2013) 37–52.

[7] G. Vasapollo, R.Del Sole, L. Mergola, M.R. Lazzoi, S. Anna, S. Scorrano, G. Mele, Molecularly imprinted polymers: present and future prospective, Int. J. Mol. Sci. 12 (9) (2011) 5908.

[8] C. Alvarez-Lorenzo, Handbook of Molecularly Imprinted Polymers, Smithers Information Limited, 2013.

[9] B. Sellergren, Molecularly Imprinted Polymers: Man-Made Mimics of Antibodies and Their Application in Analytical Chemistry, Elsevier, U.K., 2000.

[10] L.M. Madikizela, N.T. Tavengwa, H. Tutu, L. Chimuka, Green aspects in molecular imprinting technology: from design to environmental applications, Trends Environ. Anal. Chem. 17 (2018) 14–22.

[11] E. Fischer, Einfluss der configuration auf die wirkung der enzyme, Ber. Dtsch. Chem. Ges. 27 (3) (1894) 2985–2993.

[12] E. Mazzotta, A. Turco, I. Chianella, A. Guerreiro, S.A. Piletsky, C. Malitesta, Solid-phase synthesis of electroactive nanoparticles of molecularly imprinted polymers. A novel platform for indirect electrochemical sensing applications, Sensor. Actuator. B Chem. 229 (2016) 174–180.

[13] L. Chen, T. Muhammad, B. Yakup, S.A. Piletsky, New immobilisation protocol for the template used in solid-phase synthesis of Mip nanoparticles, Appl. Surf. Sci. 406 (2017) 115–121.

[14] D. Garcia-Mutio, A. Gomez-Caballero, A. Guerreiro, S. Piletsky, M.A. Goicolea, R.J. Barrio, Solid-phase synthesis of imprinted nanoparticles grafted on gold substrates for voltammetric sensing of 4-ethylphenol, Sensor. Actuator. B Chem. 236 (2016) 839–848.

[15] F. Canfarotta, J. Czulak, A. Guerreiro, A.G. Cruz, S. Piletsky, G.E. Bergdahl, H. Martin, B. Mattiasson, A novel capacitive sensor based on molecularly imprinted nanoparticles as recognition elements, Biosens. Bioelectron. 120 (2018) 108–114.

[16] S.S. Zunngu, L.M. Madikizela, L. Chimuka, P.S. Mdluli, Synthesis and application of a molecularly imprinted polymer in the solid-phase extraction of ketoprofen from wastewater, Compt. Rendus Chem. 20 (5) (2017) 585–591.

[17] L.M. Madikizela, L. Chimuka, Determination of ibuprofen, naproxen and diclofenac in aqueous samples using a multi-template molecularly imprinted polymer as selective adsorbent for solid-phase extraction, J. Pharm. Biomed. Anal. 128 (2016) 210–215.

[18] M.A. García Mayor, G. Paniagua González, R.M. Garcinuño Martínez, P. Fernández Hernando, J.S. Durand Alegría, Synthesis and characterization of a molecularly imprinted polymer for the determination of spiramycin in sheep milk, Food Chem. 221 (2017) 721–728.

[19] H. Wang, Y. Liu, Y. Su, P. Zhu, Selective recognization of dicyandiamide in bovine milk by Mesoporous silica sba-15 supported dicyandiamide imprinted polymer based on surface molecularly imprinting technique, Food Chem. 240 (2018) 1262–1267.

[20] H. Lu, S. Xu, Hollow Mesoporous structured molecularly imprinted polymers for highly sensitive and selective detection of estrogens from food samples, J. Chromatogr. A 1501 (2017) 10–17.

[21] K. Zhi, L. Wang, Y. Zhang, Y. Jiang, L. Zhang, A. Yasin, Influence of size and shape of silica supports on the sol–gel surface molecularly imprinted polymers for selective adsorption of gossypol, Materials 11 (5) (2018) 777.
[22] R.J. Uzuriaga-Sánchez, A. Wong, S. Khan, M.I. Pividori, G. Picasso, M.D.P.T. Sotomayor, Synthesis of a new magnetic-Mip for the selective detection of 1-chloro-2,4-dinitrobenzene, a highly allergenic compound, Mater. Sci. Eng. C 74 (2017) 365–373.
[23] Y. Yuan, Y. Liu, W. Teng, J. Tan, Y. Liang, Y. Tang, Preparation of core-shell magnetic molecular imprinted polymer with binary monomer for the fast and selective extraction of bisphenol a from milk, J. Chromatogr. A 1462 (2016) 2–7.
[24] S. Hussain, S. Khan, S. Gul, M.I. Pividori, M.Del P.T. Sotomayor, A novel Core@Shell magnetic molecular imprinted nanoparticles for selective determination of folic acid in different food samples, React. Funct. Polym. 106 (2016) 51–56.
[25] F. Puoci, G. Cirillo, M. Curcio, O.I. Parisi, F. Iemma, N. Picci, Molecularly imprinted polymers in drug delivery: state of art and future perspectives, Expert Opin. Drug Deliv. 8 (10) (2011) 1379–1393.
[26] M.N. Rhyner, A.M. Smith, X. Gao, H. Mao, L. Yang, S. Nie, Quantum dots and multifunctional nanoparticles: new contrast agents for tumor imaging, Nanomedicine 1 (2) (2006) 209–217.
[27] N. Pérez, M.J. Whitcombe, E.N. Vulfson, Surface imprinting of cholesterol on submicrometer Core–Shell emulsion particles, Macromolecules 34 (4) (2001) 830–836.
[28] D.N. Clausen, I.M.R. Pires, C.R.T. Tarley, Improved selective cholesterol adsorption by molecularly imprinted poly(methacrylic acid)/silica (Pmaa–Sio2) hybrid material synthesized with different molar ratios, Mater. Sci. Eng. C 44 (2014) 99–108.
[29] A.O. Baskakov, A. Yu Solov'eva, Y.V. Ioni, S.S. Starchikov, I.S. Lyubutin, I.I. Khodos, A.S. Avilov, S.P. Gubin, Magnetic and interface properties of the core-shell Fe3o4/Au nanocomposites, Appl. Surf. Sci. 422 (2017) 638–644.
[30] S.da S. Anacleto, H.L. de Oliveira, A.T.M. da Silva, T.A. do Nascimento, K.B. Borges, Preparation of an organic–inorganic hybrid molecularly imprinted polymer for effective removal of albendazole sulfoxide enantiomers from aqueous medium, J. Environ. Chem. Eng. 5 (6) (2017) 6179–6187.
[31] T.S. Anirudhan, J. Christa, J.R. Deepa, Extraction of melamine from milk using a magnetic molecularly imprinted polymer, Food Chem. 227 (2017) 85–92.
[32] G.Z. Kyzas, S.G. Nanaki, A. Koltsakidou, M. Papageorgiou, M. Kechagia, D.N. Bikiaris, D.A. Lambropoulou, Effectively designed molecularly imprinted polymers for selective isolation of the antidiabetic drug metformin and its transformation product guanylurea from aqueous media, Anal. Chim. Acta 866 (2015) 27–40.
[33] G.Z. Kyzas, D.N. Bikiaris, Characterization of binding properties of silver ion-imprinted polymers with equilibrium and kinetic models, J. Mol. Liq. 212 (2015) 133–141.
[34] A. Poma, A. Guerreiro, M.J. Whitcombe, E.V. Piletska, A.P.F. Turner, S.A. Piletsky, Solid-phase synthesis of molecularly imprinted polymer nanoparticles with a reusable template–"plastic antibodies", Adv. Funct. Mater. 23 (22) (2013) 2821–2827.
[35] E. Yilmaz, K. Haupt, K. Mosbach, The use of immobilized templates—a new approach in molecular imprinting, Angew. Chem. Int. Ed. 39 (12) (2000) 2115–2118.
[36] M. Arabi, M. Ghaedi, A. Ostovan, Water compatible molecularly imprinted nanoparticles as a restricted access material for extraction of hippuric acid, a biological indicator of toluene exposure, from human urine, Microchimica Acta 184 (3) (2017) 879–887.

[37] I. Polyakova, L. Borovikova, A. Osipenko, E. Vlasova, B. Volchek, O. Pisarev, Surface molecularly imprinted organic-inorganic polymers having affinity sites for cholesterol, React. Funct. Polym. 109 (2016) 88–98.

[38] S. Ansari, Application of magnetic molecularly imprinted polymer as a versatile and highly selective tool in food and environmental analysis: recent developments and trends, Trac. Trends Anal. Chem. 90 (2017) 89–106.

[39] G. Ertürk, B. Mattiasson, Molecular imprinting techniques used for the preparation of biosensors, Sensors 17 (2) (2017) 288.

[40] Y. Wu, M. Yue, J. Pan, R. Gu, J. Luo, Porous and magnetic molecularly imprinted polymers via pickering high internal phase emulsions polymerization for selective adsorption of Λ-cyhalothrin, Front. Chem. 5 (2017) 18.

[41] Z. Zhang, D. Niu, Y. Li, J. Shi, Magnetic, core-shell structured and surface molecularly imprinted polymers for the rapid and selective recognition of salicylic acid from aqueous solutions, Appl. Surf. Sci. 435 (2018) 178–186.

[42] Q. Wu, M. Li, Z. Huang, Y. Shao, L. Bai, L. Zhou, Well-defined nanostructured core–shell magnetic surface imprinted polymers (Fe3o4@Sio2@Mips) for effective extraction of trace tetrabromobisphenol a from water, J. Ind. Eng. Chem. 60 (2018) 268–278.

[43] P. Anastas, N. Eghbali, Green chemistry: principles and practice, Chem. Soc. Rev. 39 (1) (2010) 301–312.

[44] H. Xiang, M. Peng, H. Li, S. Peng, S. Shi, High-capacity hollow porous dummy molecular imprinted polymers using ionic liquid as functional monomer for selective recognition of salicylic acid, J. Pharm. Biomed. Anal. 133 (2017) 75–81.

[45] Y.P. Song, L. Zhang, G.N. Wang, J.X. Liu, J. Liu, J.P. Wang, Dual-dummy-template molecularly imprinted polymer combining ultra performance liquid chromatography for determination of fluoroquinolones and sulfonamides in pork and chicken muscle, Food Control 82 (2017) 233–242.

[46] X. He, J. Wang, X. Mei, Dummy fragment template molecularly imprinted polymers for the selective solid-phase extraction of gonyautoxins from seawater, Anal. Lett. 50 (12) (2017) 1877–1886.

[47] F. Ning, T. Qiu, Q. Wang, H. Peng, Y. Li, X. Wu, Z. Zhang, L. Chen, H. Xiong, Dummy-surface molecularly imprinted polymers on magnetic graphene oxide for rapid and selective quantification of acrylamide in heat-processed (including fried) foods, Food Chem. 221 (2017) 1797–1804.

[48] K. Booker, C. Holdsworth, M. Bowyer, M.C. Adam, Ionic Liquids as Porogens in the Synthesis of Molecularly Imprinted Polymers, 2011.

[49] S. Pardeshi, R. Dhodapkar, A. Kumar, Influence of porogens on the specific recognition of molecularly imprinted poly(acrylamide-co-ethylene glycol dimethacrylate), Compos. Interfac. 21 (1) (2014) 13–30.

[50] X. Zhu, Y. Zeng, Z. Zhang, Y. Yang, Y. Zhai, H. Wang, L. Liu, J. Hu, L. Li, A new composite of graphene and molecularly imprinted polymer based on ionic liquids as functional monomer and cross-linker for electrochemical sensing 6-benzylaminopurine, Biosens. Bioelectron 108 (2018) 38–45.

[51] G. Zhu, X. Gao, X. Wang, J. Wang, J. Fan, Influence of hydrogen bond accepting ability of anions on the adsorption performance of ionic liquid surface molecularly imprinted polymers, J. Chromatogr. A 1532 (2018) 40–49.

[52] M. Akhoundian, A. Rüter, S. Shinde, Ultratrace detection of histamine using a molecularly-imprinted polymer-based voltammetric sensor, Sensors 17 (3) (2017) 645.

[53] B. Batlokwa, J. Mokgadi, T. Nyokong, N. Torto, Optimal template removal from molecularly imprinted polymers by pressurized hot water extraction, Chromatographia 73 (2011).

[54] F.A. Trikka, K. Yoshimatsu, L. Ye, D.A. Kyriakidis, Molecularly imprinted polymers for histamine recognition in aqueous environment, Amino Acids 43 (5) (2012) 2113–2124.

[55] M.S. da Silva, R. Viveiros, P.I. Morgado, A. Aguiar-Ricardo, I.J. Correia, T. Casimiro, Development of 2-(dimethylamino)ethyl methacrylate-based molecular recognition devices for controlled drug delivery using supercritical fluid technology, Int. J. Pharm. 416 (1) (2011) 61–68.

[56] S. Rebocho, C.M. Cordas, R. Viveiros, T. Casimiro, Development of a ferrocenyl-based Mip in supercritical carbon dioxide: towards an electrochemical sensor for bisphenol A, J. Supercrit. Fluids 135 (2018) 98–104.

[57] R. Lorenzo, A. Carro, C. Alvarez-Lorenzo, A. Concheiro, To remove or not to remove? The challenge of extracting the template to make the cavities available in molecularly imprinted polymers (Mips), Int. J. Mol. Sci. 12 (7) (2011) 4327.

[58] E. Roy, S. Patra, S. Saha, D. Kumar, R. Madhuri, K. Prashant, Sharma, Shape effect on the fabrication of imprinted nanoparticles: comparison between spherical-, rod-, hexagonal-, and flower-shaped nanoparticles, Chem. Eng. J. 321 (2017) 195–206.

[59] S. Patra, E. Roy, R. Parui, R. Madhuri, K. Prashant, Sharma, Anisotropic (spherical/hexagon/cube) silver nanoparticle embedded magnetic carbon nanosphere as platform for designing of tramadol imprinted polymer, Biosens. Bioelectron. 97 (2017) 208–217.

[60] B. Khodashenas, H.R. Ghorbani, Synthesis of silver nanoparticles with different shapes, Arabian J. Chem. (in press) https://doi.org/10.1016/j.arabjc.2014.12.014.

[61] N. Li, P. Zhao, D. Astruc, Anisotropic gold nanoparticles: synthesis, properties, applications, and toxicity, Angew. Chem. Int. Ed. 53 (7) (2014) 1756–1789.

Chapter 8

Graphene aerogels for oil absorption

Likourgos V. Petridis, Nikolaos C. Kokkinos, Athanasios C. Mitropoulos and George Z. Kyzas*
Hephaestus Advanced Laboratory, Department of Chemistry, International Hellenic University, Kavala, Greece
**Corresponding author*

1. Introduction

As the world becomes more industrialized, new environmental challenges arise as well. Most of the world's energy demands are being met by fossil fuels, which can potentially harm the environment either during their production and processing or through their use. One of the most devastating environmental events that fossil fuels can cause is oil spills. A major incident can possibly release millions of barrels of oil into the environment, in a short amount of time, which can prove catastrophic for the local ecosystem. For that reason special attention is given to oil spill cleanup techniques. One promising way to fight oil spills and restrict their effects is oil absorbers. Such substances can absorb the oil, be recovered, and then be reused. With technology's progress in materials science, new materials are constantly available for such roles. One such material is graphene, which shows many extraordinary properties in itself and for absorption purposes. Its aerogel form provides many advantages that can be exploited as an oil absorber.

1.1 Oil spill pollution

Oil propelled humanity into a new age of prosperity but at an environmental cost, not just from its actual use but also from accidental discharges onto the environment while transporting or producing it (Fig. 8.1).

With over 85 millions of barrels (in 2009 [1,2]) of oil used every day, it is to be expected that some eventually leaks to the environment. These leaks are usually small and can happen both by production and consumption reasons.

Advanced Low-Cost Separation Techniques in Interface Science. https://doi.org/10.1016/B978-0-12-814178-6.00008-X

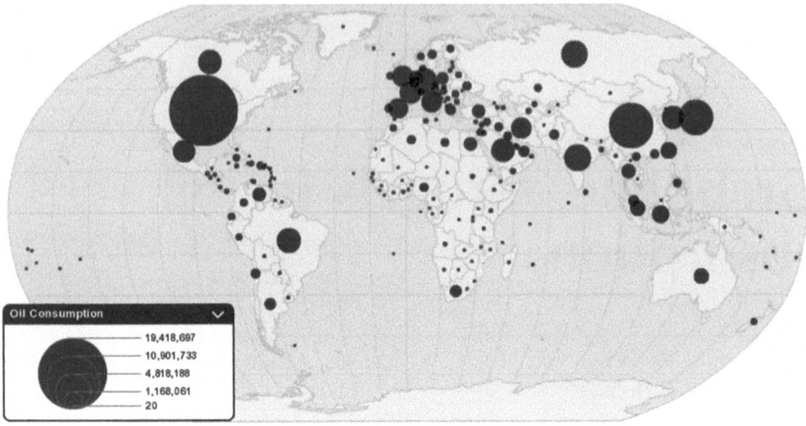

FIGURE 8.1 Daily oil consumption in barrels [1,2].

These small discharges, such as urban run-off, refinery effluents, nautical vessels, etc., along with the larger scale oil spills amount to about 1.2 million tonnes annually. It needs to be mentioned though that an oil spill can measure anywhere from a few barrels to even millions of barrels (Lakeview Gusher, Deepwater Horizon, to name a few) that has a really adverse impact.

While most man-made oil spills are very small, the majority of the oil leakage comes from major incidents. From Fig. 8.2, we can see that 93.7% of

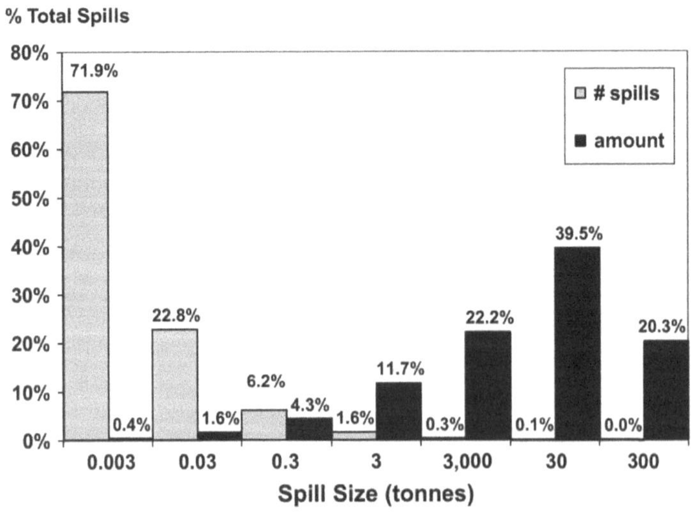

FIGURE 8.2 Size classes of US marine oil spills, 1990–99 (ERC data) [3].

spilled oil came from 0.4% of spill incidents. So it is imperative that these large spills are taken care of, as to impact the environment as little as possible.

1.2 Oil spill response methods

1.2.1 Physical responses

1.2.1.1 Booms

Containment includes the process of confining the oil so that it would not spread or to concentrate in an area so it can be further processed. The main way of containing the oil is via oil booms. A boom is a mechanical barrier, resembling a vertical curtain that extends both under and over water. They can be effective with low waves but as the wave height increases, oil can easily go over the top of the boom. Booms are usually used in a U, V, or J configuration.

Other technique's using booms include the following:

- Diversion (The boom is used to redirect oil from one location or direction of travel to a specific site for recovery.)
- Deflection (The boom is used to redirect oil away from a location to be protected or simply to change the course of the oil slick. The main difference of deflection with diversion is that diversion is not done with the purpose of recovering the oil.)
- Exclusion (The boom is in a fixed position and simply prevents oil from entering an area.) (Fig. 8.3)

1.2.1.2 Skimmers

A skimmer is a machine that removes the oil from the surface of the water. Usually skimmers perform better when the oil is more concentrated and

FIGURE 8.3 Oil boom in V configuration guiding oil to a skimmer [3].

viscous. For that reason, booms are used first to gather the oil. Skimmers are often placed in front of booms or wherever the oil has the most concentration as to remove as much oil as possible.

- A weir skimmer uses gravity to recover the oil from the water surface to a tank under the water. The most simple of these skimmers are made of the weir, the holding tank, and a pipe to some pump to extract the oil. Weir skimmers exist in many builds and models.
- Suction or vacuum skimmers use vacuum to remove oil from the water surface. These skimmers act exactly like an enlarged home vacuum and suck the oil with the addition of a floater to keep them afloat.
- Elevating skimmers make use of conveyors to move oil from the surface to another location. The conveyor belt moves the oil above the water level and into another recovery device or location.
- Submersion skimmers are like elevating skimmers but do the opposite. They have a downward incline and push the oil deeper. The oil is removed from the conveyor by scraping it. After it is removed from the conveyor, it goes up assisted by gravity into a collection well and is then pumped away from there.

Both skimmers and booms could possibly include sorbent materials to modify or enhance their scope (Fig. 8.4).

1.2.1.3 Manual recovery

Manual recovery is the arduous process of using tools such as rakes, shovels, hoses, pumps by people to clean up a beach or some other location from debris

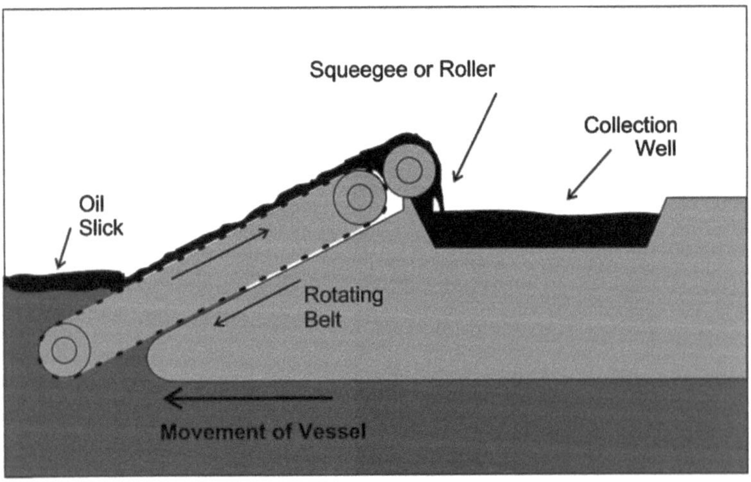

FIGURE 8.4 Elevating skimmer concept [3].

and oil. The collected materials can be sent for further processing. Manually removing the oil is hard and not very efficient but creates less waste that other cleanup methods. Heavier oils are easier to remove than lighter oils. By using this method a lot of water is collected too, requiring more storage capacity.

1.2.2 Chemical responses

1.2.2.1 Dispersants/thickeners

Dispersants are chemicals that include surfactants, emulsifiers, and solvents that cause the oil to form small droplets and help it disperse throughout the water. On the other hand there are thickeners (herding agents) that increase the surface tension between oil and water that prevents the oil spreading or facilitates easier recovery. The use of dispersants/thickeners has caused controversy over their usefulness due to their toxicity and cost [4].

1.2.2.2 Surface washing agents

Surface washing agents act in a different way than dispersants. They act like normal detergents act on stains. Surface washing agents are supposed to be used on contaminated locations to release the oil from their surface. While toxicity has been an issue, there are indeed surface washing agents that present little aquatic toxicity.

1.2.2.3 Bioremediation

The US EPA has defined bioremediation agents as microbiological cultures, enzyme additives, or nutrient additives that significantly increase the rate of biodegradation to mitigate the effects of an oil spill [5]. While bioremediation agents might perform well in laboratory studies, the same cannot be said for field studies [6–9].

1.2.2.4 Emulsion breakers and inhibitors

Emulsion breakers and inhibitors break emulsions and prevent them from forming. Water-in-oil emulsions are problematic because they can triple the volume of material that has to be processed, making storage, and disposing it harder. Water-in-oil emulsions also make it harder for pumps and skimmers to function due to increased viscosity.

1.2.2.5 Recovery enhancers

Recovery enhancers are substances that try to improve the recovery efficiency of other recovery devices such as skimmers by increasing the adhesiveness of the oil. The efficiency for sorbent surface skimmers can be increased 10-fold for nonadhesive materials like diesel. The effect on naturally adhesive substances is highly diminished.

1.2.2.6 Solidifiers

Solidifiers are used to change liquid oil to a solid state that can be easily collected by mechanical means. These are also called gelling or collecting agents. Gelling agents are the opposite of dispersants and are still in an experimental state. These agents contain cross-linkers or polymerization catalysts that cause the oil molecules to be linked together by stronger bonds and thus turning solid. They are usually in the form of powder that when used it is fused with the oil. Depending on the agent, 1/10−1/4 of the oil's mass is required for solidification.

1.2.2.7 Sinking agents

Sinking agents absorb the oil in the water and then sink. These sinking agents are a major environmental concern and most countries do not allow their use. These agents jeopardize aquatic life, and the oil can eventually be released back into the water.

2. Graphene aerogels as oil sorbents

2.1 Sorbents as an oil spill response method

Another physical response to oil spills is sorbents. Sorbents are materials that can recover oil either via adsorption or absorption. Currently, they provide a powerful tool against oil spills and have multiple uses:

- As a response for absorbing the final traces of oil spills
- As an addition to other response means, such as sorbent skimmers and booms for very small spills
- As a primary response for very small incidents
- For passive cleanup

One such example of passive cleanup is exclusion sorbent booms that absorb any remaining oil from the shoreline and stop any more oil from entering the protected site.

There are natural and synthetic sorbents:

- Natural sorbents can be made of organic materials, such as peat moss or wood products, and inorganic materials, such as vermiculite or clay.
- Synthetic sorbents are man-made materials with excellent sorbing capabilities. They are often used to clean other equipment. For cleaning equipment, they mostly come in the form of sheets or rolls.

Sorbents exist in many forms but usually in the form of granules or powder and are contained inside some kind of bag. Some other forms include pads, rolls, blankets, and pillows. A specific type of plastic sorbent be processed into flat strips or "pom-poms" that excel at absorbing heavy oil.

A significant advantage of synthetic sorbents is their potential for reusability. Synthetic sorbents can often become reusable by compressing them and recovering the oil but most sorbents show reduced performance after some cycles, and it could potentially be more expensive to recover the oil than simply using new sorbent. Additionally, sorbent filled with oil is harder to handle and can potentially leak back oil during transportation or processing.

The sorption capacity of a sorbent depends on many properties of the sorbent. One main property is the amount of surface area to which the oil can adhere to as well as the type of surface. A sorbent with complex porous structures and capillaries with high surface area is generally more fit to absorb lighter oils. Sorbents with a coarser surface are best suited for recovering heavier oils. Pom-poms that are used to recover heavy oil are made of flat plastic strips that contain no capillary structure. Another very important property is oleophilicity and hydrophobicity of the sorbent. Some sorbents that are not inherently oleophiles or hydrophobes can be treated with oleophilic and hydrophobic agents to improve the material's absorption capabilities. Natural sorbents are a prime candidate for treatment since they usually absorb large quantities of water along with the oil.

The performance of sorbents is measured in both oil recover and water intake. "Oil recovery" is the weight of the recovered oil compared with the original weight of the sorbent. It is possible for the best synthetic sorbents to potentially absorb many times their own weight in oil. It is important to also have a low water intake, with the best sorbents not absorbing any water.

Sorbents have their disadvantages though and should be used with care. Excessive use of sorbents at an oil spill, especially those that are very fine, can create problems and render skimmers inoperable at the site. Sorbents may cause plugging in pipelines or in pumps. Just like sinking agents, sorbents that sink should not be used and could possibly create similar issues. Many kinds of sorbents have problems with sinking, such as untreated peat moss, some inorganic sorbents, and many wood products. Sorbents that sink in the water are forbidden to use on most countries because both the oil and the sorbent can be harmful to the environment. For that reason, it is important to consider the recovery, transport, and final disposal of the oiled sorbent. The sorbent must be able to retain the oil inside it until the sorbent has been properly disposed (or until the oil is recovered).

Lots of promises are given by new synthetic products that show extraordinary properties that could create impressive sorbents. One such material is graphene.

3. Graphene and its synthesis

Graphene is an allotrope of carbon consisting of a single layer of carbon atoms in a hexagonal lattice, essentially a 2D material. The lateral dimension of graphene can be in nanoscale or macroscale. Graphene is considered to be the thinnest and lightest material yet. It also exhibits fantastic mechanical properties, making it stronger than diamond and 100 times stronger than steel. It even exhibits elasticity up to 20% strain [10].

While graphene was theorized even from 1947 [11], it was not separated from graphite until 2004 [12] for which Geim and Novoselov won the 2010 Nobel Prize in Physics.

These unique properties have led to graphene being used in medicine [13], mechanical design [14], electronics [15], and optics [16].

Smalls quantities of high-quality graphene can be acquired by chemical vapor deposition (CVD) and on Si or on metal substrates [17]. The "Scotch Tape" technique (mechanical exfoliation) [12] and vapor deposition provide a very small amount of graphene, though. For large-scale production, a chemical approach is usually used such as the Hummer's method [18] from which graphene oxide can be acquired [19] and subsequently graphene. The disadvantages of Hummer's method include reactions inducing high temperatures and toxic fumes, for that reason a modified version has arisen [20] which removes the need for $NaNO_3$ (Fig. 8.5).

3.1 Chemical vapor deposition

In general, graphene production via CVD follows a two step process. The first step is the dilution of carbon into the metal, in which the precursor carbon is decomposed to carbon radicals at the metal substrate surface. Gas is fed inside the reactor, and it reacts with the carbon on the surface, resulting in mono or

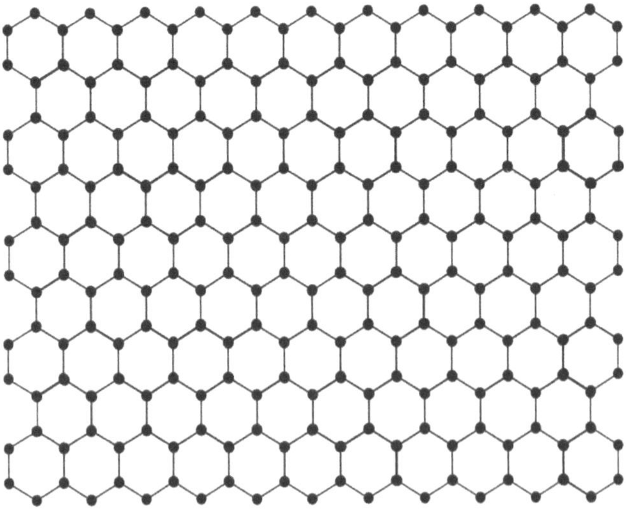

FIGURE 8.5 The honeycomb lattice of graphene [10].

few layers of graphene. The metal substrate acts as a catalyst to lower the energy barrier of the reaction and also determines the graphene deposition mechanism, which in the end affects the final quality of the graphene. Many metal substrates can be used for CVD synthesis of graphene, but Ni and Cu metal substrates are mostly used [21].

3.1.1 Carbon precursors

Many precursors can be used including solids, liquids, and gasses. Hydrocarbon gasses are a popular precursor because of its higher purity as compared with other liquids and solids. Though recently, more studies are being made into liquid precursors. Methane is one of the most common gaseous precursors to grow graphene films.

3.1.2 Graphene synthesis on Ni

First, a polycrystalline Ni film is thermally annealed (900−1000°C) in H_2 in order to increase the grain size. After that an H_2/CH_4 gas mixture is introduced. The methane is decomposed and carbon is dissolved into the Ni film forming a solid solution. The material is then cooled in Ar atmosphere. While the temperature drops, carbon atoms are diffused from the solution and precipitated on the Ni film.

Different cooling rate have a great effect on the quality and thickness of the graphene since it controls the diffusion and precipitation process, with the best results being had with medium cooling rates. Also the crystals of the Ni film seem affect the final product. It is suggested that, the graphene film on the Ni surface is not broken until it reaches some defect on the Ni structure which is on the grain's borders. If the Ni is annealed in H_2 then the grain size is increased and removes some impurities in the Ni film [22].

3.1.3 Graphene synthesis on Cu

In general, Cu foil is first annealed in hydrogen atmosphere at about 1000°C, and then a mixture of H_2/CH_4 mixture is introduced, similarly to Ni. After the graphene creation process is complete, the sample is cooled down.

Graphene grown on polycrystalline Ni exhibits multilayer flakes, while graphene grown on polycrystalline Cu is more uniformly monolayer. For graphene CVD on Ni, it has been suggested that it is a segregation process.

Cu has an extremely low carbon solubility as opposed to Ni, which means that regardless of the H_2/CH_4 concentration or reaction time, there will be a very small amount of carbon dissolved into the Cu since the carbon is originated from the catalytic decomposition of the hydrocarbon gas on the metal film.

As such, it is easier to grow monolayer graphene on low carbon solubility metals, like Cu, as opposed to fast cooled graphene growth on high carbon solubility metals like Ni since it is harder to control the growth of monolayer graphene through fast cooling processes on high-solubility metals, like Ni.

More research is put into graphene CVD, but it is still impossible to create completely uniformly monolayer graphene. It is not known how the metal crystal structure affects the segregation of carbon from the metal to form the graphene film. Also the effect the transition metal has during the connection of two graphene islands is unknown. More problems arise with the fact that graphene CVD requires very high vacuum and temperature conditions making the process more expensive. New studies show that it is possible to do graphene CVD in atmospheric pressure but results in inferior quality graphene. Also the use of valuable transition metals such as Ru, Ir, Pt, and Pd is impractical and those catalysts cannot be reused. Constant recycling of the transition metal is required to reduce the production cost and be more scalable.

3.2 Mechanical exfoliation

The idea of mechanical exfoliation is that the graphene is peeled from graphite one layer at a time. Essentially, the van der Waals forces between the flakes is overcome in a mechanical manner. The two ways to overcome these forces is via applying normal or lateral force. An example of applying normal force is through the Scotch Tape method. Due to the self-lubricating ability of graphene, it is also possible to apply lateral force to peel away layers. Another facet of mechanical exfoliation is the fragmentation of the graphite. While applying force during exfoliation, it is possible to break large graphite and graphene layers to smaller pieces. While this is unwanted for creating large area graphene, it does reduce the height of the layers and smaller graphite particles are easier to process due to smaller van der Waals forces in the smaller particles.

3.2.1 Micromechanical cleavage

With this technique one exerts normal force over the graphite to cause cleavage of layers from the main graphite bulk. Doing this repeatedly for many times, the graphite gets thinner and thinner resulting in graphene. The Scotch Tape method is an example of this. The exfoliation mechanics are dominated by a normal force. Using micromechanical cleavage, it is possible to produce high-quality and large-area graphene at the cost of intense labor and time. Hence, it is limited to producing graphene for research purposes and is not possible to be scaled for mass production.

New techniques involve feeding graphite into an oscillating lathe like diamond edge which can result in nanometers-thick graphene sheets and the usage of a three roll mill that uses Polyvinylchloride (PVC) dissolved in dioctyl phthalate (DOP) as an adhesive. While the three roll mill machine is well used in an industrial setting, the problem of residual PVC and DOP still remains. They need to be completely removed, and this is no easy task.

3.2.2 Sonication

Coleman et al. [23] were the first to achieve production of graphene using exfoliation via sonication in 2008. Graphite powder was dispersed in N,N-dimethylformamide and N-methylpyrrolidone, which was sonicated and centrifuged afterward.

While this method is an easy way to create large amounts of low-cost graphene, it does have disadvantages. When the technique first came out, the concentration of graphene was very low (0.01 mg/mL) which was far from enough for actual use. Researchers have tried to improve the concentration by changing parameters such as the sonication time, the initial graphite concentration, adding surfactants, and polymers, using solvent exchange methods, mixing solvents, etc.

Also, it has been shown that graphene synthesized through sonication is of inferior quality, exhibiting more defects than normal. This is suggested to happen due to the sonication-induced cavitation. Even though cavitations assist the exfoliation process, they can be an intense process resulting in high local temperature, pressure, and rapid heating/cooling rate.

3.2.3 Ball milling

Ball milling is a way to exfoliate graphite using lateral force, as opposed to the Scotch Tape or sonication that mainly use normal force. Ball mills, like the three roll machine, are a common occurrence in industry, for the production of fine particles. During the ball milling process, there are two factors that contribute to the exfoliation. The main factor contributing is the shear force applied by the balls. Using only shear force, one can produce large graphene flakes. The secondary factor is the collisions that occur during milling. Harsh collisions can break these large flakes and can potentially disrupt the crystal structure resulting in a more amorphous mass. So in order to create good-quality, high-area graphene, the collisions have to be minimized.

3.2.4 Fluid dynamics

Using fluid dynamics, exfoliation can be achieved while the graphite is inside a moving fluid. While sonication is an intense process, leading in cavitation, fluid dynamics can be tuned for a milder exfoliation process.

One such example of a milder fluid dynamics exfoliation process is using the vortex fluidic film method. It uses a rapidly revolving tube with the graphite inside a fluid. This method achieves exfoliation in two ways, partial lifting and slippage of the graphite layers on the walls of the tube. Parameters like the speed or orientation of the tube can be manipulated to change the fluidic film. Due to the centrifugal force, graphite flakes will start going toward the wall and because of the fluidic flow will start to get partially lifted. After the partial lift, slippage begins to occur. The graphite flakes are pushed against the tube wall by the centrifugal force and experience a shear-induced displacement along the tube, resulting in exfoliation at the tube surface.

3.3 Reduced graphite oxide

Graphite oxide (GO) is a product of using methods, such as Hummer's, to oxidize graphite. This introduces oxygen-containing groups (epoxide and hydroxyl) to the graphite crystal structure. These oxygen groups interact with the water, making the GO hydrophilic. Due to GO being hydrophilic, water is introduced between the GO sheets and causing them to disperse. It is possible to restore a big part of the graphitic network by thermally annealing or chemically reducing the GO. This method is extremely scalable and cheap. Graphite is very cheap and abundant and can easily create large amounts of reduced GO. As a final note, GO also shows promise as a material for composites.

3.4 Organic synthesis of graphene

While reducing graphene oxide, one can chemically produce micrometer-thick graphene, but there is still need for smaller molecules. While not exactly graphene, "graphene-like" polycyclic aromatic hydrocarbons (PAH) fill that niche by having just the right size between "molecule" and "macromolecule." It is possible to insert aliphatic compounds inside the aromatic structure to tune the solubility of PAHs, but they still have a really limited molecular weight range. Increasing their molecular weight reduces their solubility considerably and increases the occurrence of unwanted side reactions. Hence, tuning solubility and still keeping the wanted planar orientation is still problematic.

4. Graphene aerogels

While the exact definition of a gel might be a bit undecided, all attempts [24–26] usually follow some common ground: A gel comprises a solid and a liquid component, and it exhibits properties of a solid or a semisolid.

According to IUPAC [27], a gel is a nonfluid colloidal network or polymer network that is expanded throughout its whole volume by a fluid. An aerogel is a gel which its fluid component is a gas.

While graphene exhibits some extraordinary properties, it performs below its theoretical performance when inserted into polymers [28–36]. As such, it is hard to fully exploit graphene's capabilities in 3D structures. One structure that shows promise and has received plenty of studies are graphene aerogels (GAs) [37–47]. Aerogels in general show high porosity, extremely low density, low thermal conductivity, low refractive index, and low dielectric constant [48].

GO is used to synthesize GAs due to the reasons explained earlier, such as its scalability, low cost, and properties. GO is often produced by oxidation of graphite flakes or powder using oxidants via a modified Hummers' method [20].

4.1 Graphite oxide synthesis (Hummer's and improved Hummer's)

Graphite flakes/powder and sodium nitrate are stirred into sulfuric acid, while being stirred potassium permanganate is added slowly. An ice bath is used to keep the temperature low. After some time, water is added and the solution is finally treated with hydrogen peroxide to finish the reaction. The color of the resulting solution shows the degree of the oxidation of the graphite, with bright yellow denoting higher oxidation over green and black colors.

In improved methods there are a few modifications, the most major being that no sodium nitrate is used, which produces toxic gases. Also the acid mixture now comprises sulfuric and phosphoric acid.

After the GO is purified (rinsed, centrifuged, etc.) one can proceed in a multitude of ways to the next step, which is creating a graphene hydrogel. Ways of GA synthesis include hydrothermal reduction [45], chemical reduction [43], cross-linking [49–53], and template-directed reduction [54,55].

4.2 Graphene aerogel synthesis methods

These techniques can be used separately, or in tandem, to achieve different results.

4.2.1 Hydrothermal reduction

Hydrothermal reduction is a common way to produce a graphene gel. After the reduction, freeze drying the gel creates a GA. Hydrothermal reduction [56–58] requires high temperature and high pressure, while the solution is sealed. This provides a controllable gel rate and retains the gel integrity. During the hydrothermal process, GO self-assembles to a 3D porous structure due to π–π stacking and van der Waals forces. On the other hand, π–π stacking also causes the conjoining of layers which can reduce porosity and absorption capabilities. As such hydrothermal techniques often produce less than perfect graphene, but it still remains a popular choice.

4.2.2 Chemical reduction

This method treats the GO with some reducing agents like vitamin C or sodium ascorbate [59–62] and is considered to be better than hydrothermal reduction, which can require chemical cross-linkers and high temperatures and pressures [63]. Electrochemical reduction of GO is an easy way to create graphene [64,65].

4.2.3 Template-directed reduction

This method is favorable for creating porous GA due to the fact that it reduces the occurrences of random interconnections while also aiding to the creation of controllable and uniform macropores and tailorable microstructures [66]. Another method is ice template–directed method [55] which uses freeze drying. Porous GA in aqueous solutions [67,68] is produced this way, whose porosity can be controlled via the cool rate and temperature.

4.2.4 Cross-link

GO is hydrophilic and forms a stable solution in water [69], but this is not true in acidic solutions due to insufficient mutual repulsion [70]. In general, cross-linking aids the gelling of GO layers by increasing the bond forces between them. Cross-linkers usually used on GO are hydroxyl [55], oxygen-containing, or nitrogen functional groups [70]. Cross-linked GO sheets have uses in other areas too [71].

4.3 Freeze and supercritical drying

After the GO hydrogel has been formed, it is necessary for the water to be removed in order to be modified into an aerogel. This is usually accomplished by freeze drying. In products containing water, direct transition of water from liquid to gas could potentially damage them. While the water is being evaporated, the liquid body pulls against any solid structures the liquid might is in contact with and the surface tension can destroy sensitive structures.

4.3.1 Freeze drying

Freeze drying (sometimes also called lyophilization or cryodesiccation) is a dehydration technique which instead of facilitating high temperatures to evaporate water uses low temperatures to freeze the product. After freezing it, the pressure is lowered and the ice is sublimated into steam, essentially removing the water. By avoiding the use of high temperatures, sensitive products (mainly food or other biological matter) can be safely dehydrated with minimal loss of quality.

4.3.2 Supercritical drying

Some products though are so structurally sensitive that even just the creation of large ice crystals or the process of sublimation could potentially damage or alter their structure. Supercritical drying solves this by going through the supercritical region of the phase diagram. Each substance has a critical point with specific temperature and pressure in which there is no distinct liquid and gas phases. While at the critical point, the substance behaves both as a gas and a liquid, and both the liquid and the gas phase have the same properties. By increasing both the pressure and temperature, the water can go from liquid to gas through the supercritical area.

4.4 Graphene aerogels as oil sorbers

Depending on the production method of the GA, its properties vary and could lead to different applications, from supercapacitors or electrodes to absorbing oil and other organic wastes.

GAs provide an excellent candidate as oil sorbers due to the fact that they can be created to be both highly hydrophobic and highly oleophilic. Along with GA's durability, reusability, and high absorption per weight ratio, they show lots of promise for use as a waste sorber.

There are plenty of recent studies [72–93] showing off GA's excellent properties for oil absorption. These GAs show high absorption rate (~100–250 times their weight depending on substance absorbed and the GA at hand), low density, high porosity, and high reusability with minimal performance loss. It needs to be noted though, that ultralight graphene assemblies have a natural advantage on the absorption rate due to the way it is calculated (i.e., oil weight absorbed/GA weight).

Cheng et al. [72] in 2013 reported a way to prepare a new type of highly hydrophobic and ultralight 3D porous GA with GO as a precursor and melamine formaldehyde microspheres as a spacer and pore forming agent, followed by hydrothermal reduction, freeze drying, and calcination process. It resulted in the creation of a GA that famously won the Guinness World Record for lightest material. Its absorption capacity was 230 times its weight for diesel oil and retained 91% of its absorption capabilities after 10 cycles.

Hong et al. [73] reported creation of functionalized GA with high porosity and hydrophobicity that is prepared by surface modification of self-assembled graphene oxide aerogels. Fluorinated functional groups are introduced into the surface of three-dimensional macroporous GA through a one-step solution immersion method. The functionalized GA shows low density, high porosity, mechanical stability, and hydrophobicity. Its absorption capacity for various types of oils and organic solvents was reported for up to 112 times its weight.

Zhao et al. [74] report graphene/carbon nanotube aerogels that are prepared through hydrothermal reduction, freeze drying, and high temperature heat treatment with the blending of GO and carbon nanotubes. They choose ascorbic acid as a reduction agent to explore the best preparation process. When the carbon nanotube content is 50%, these aerogels show the best performance. Its absorption capacity was reported at 100 times its weight.

Luo et al. [75] report polyurethane sponge-reinforced GAs with low surface energy, high sorption capacity, and excellent recyclability for use as oil sorbents. They used waste sponges from plastic production plants to prepare spongy GAs via precompaction and ice-templated assembly. Its absorption is up to 29–54 times its own mass. They found that this GA has selective and high-volume absorbability (>100%) and can efficiently separate oil from water under continuous pumping action.

Xu et al. [76] report a GA that was prepared by the one-step chemical reduction–induced assembly of graphene oxide with L-phenylalanine as the reducing agent in a simple and scalable process. The resulting GA had an absorption capacity of 100–260 times its weight for organic solvents, mainly depending on the density of the solvent.

Liu et al. [77] report a highly compressible 3D with anisotropic porous structure that is fabricated by directional freezing of graphene hydrogel using anisotropically grown ice crystals as templates followed by freeze drying. This anisotropic GA exhibits high compressibility and can be recycled by burning, distilling, or squeezing. Its absorption capability was shown to be ~150 times its weight for n-heptane and n-hexane.

Xiao et al. [78] report the synthesis of a GO/nanofiber aerogel. Cellulose acetate nanofibers are introduced into the GO suspension and freeze dried. Cellulose acetate was chosen due to its abundance, hydrolytic stability, and eco-compatibility. One of their samples was modified using vapor deposition of hexadecyltrimethoxysilane. While that sample shows extreme absorption capabilities for some organic solvents (up to ~700–800 times its weight), its n-hexane capability is around ~200–240 times which is still impressive.

Huang et al. [79] report a GA that was synthesized via an ethylenediamine-ammonia double hydrothermal reduction approach, with an absorption capacity of up to 130 times its weight for oil. Most importantly though it kept the same absorption capabilities despite the density of oil.

Wang et al. [80] report the synthesis of an ultralight, hydrophobic GA in a very simple process via reducing GO using only ammonia in an autoclave, freeze drying, and finally thermal annealing. While the thermally annealed GA performed better on all absorption tests than the nonannealed sample, it requires very high temperatures (750°C) compared with the 90°C needed for just the autoclave. Their absorption capacity for oils was ~150 times its weight while the annealed sample reached 200–250 times. They found the best way to recycle this particular GA was through squeezing it, instead of burning or distilling it.

Wan et al. [81] studied the effect of the reducing agent, reaction time, and temperature have on the hydrothermal formation of GAs focused on oil absorption. Namely, the reduction agents they tested were vitamin C, ethanediamine, and ammonia (in decreasing reduction potency) at temperatures ranging from 80 to 180°C and reaction times ranging from 4 to 24 h. The samples exhibited different mechanical properties again with vitamin C having the best and ammonia the worst, which is important because they can affect their reusability. On the other hand, absorption capacity was best for ammonia and worst for vitamin C, but again ammonia was the most sensitive to temperature and reaction times. The more intense reduction caused by vitamin C to the GO's oxygen groups seem to make the resulting aerogel's walls more compact.

While not strictly only applicable to oil absorption, Han et al. [82] created a GA via ammonia reduction. They found that the concentration of ammonia solution is crucial since it differentiates the density and volume of the GA. An increase in ammonia concentration corresponds to a volume decrease while the macropore walls of the material become thicker, giving it a larger porosity.

Mi et al. [83] synthesized a highly compressible elastic anisotropic cellulose/GA through a bidirectional freeze drying process. The resulting aerogel was then grafted with long carbon chains via CVD, making it hydrophobic. It displays excellent mechanical properties, which allow up to 85% of the absorbed oil to be removed via compression and with minimal loss of absorption capacity over several cycles. Its absorption capacity for gasoline was 100 times its weight.

Shafiq et al. [84] studied the effects of oil viscosity, agitation speed, and oil–sand wettability on the adsorption efficiency of heavy crude oil from contaminated sand using GA. They contaminated sand with heavy oil and tried adsorbing it using pre-bought GA. It seems to be a two step process, where the oil rapidly adheres to the surface of the GA and then followed by a slower diffusion through the pores of the GA. This effect is more prominent on heavier oils. Increased oil viscosity negatively impacts the process, while agitation speed and oil–sand wettability both have a sweet spot for best results. Their adsorption kinetics study suggests that for spilled heavy viscous oils, the GA should be in smaller pieces to increase its total surface area in contact with crude oil for better adsorption.

Cao et al. [85] created a GA by using polyvinyl alcohol and stearic acid as cross-linking agents to construct a framework of reduced graphene oxide via a hydrothermal reduction process. Its absorption capacity for hexane is 115 times its weight, reaching 100 times after 10 cycles of mechanical compression.

Rahmani et al. [86] reported hydrothermal method using pyrrole as a doping agent to achieve a hydrophobic hydrogel to remove oil from oil/water systems. They also studied the effects of salt and temperature on the adsorption process. Benefiting from the abundant porous architectures, the prepared aerogels exhibited high adsorption capacity. Doping of graphene with heteroatoms such as N, B, S, and P is becoming more commonplace to enhance its properties. It showed excellent absorption capacity at 210 times its weight for crude oil. Due to the high selectivity of the GA, capacity did not change much after adding salt. They tested temperatures of 25, 45, and 65°C and found a very slight decrease in the adsorption capacity with increasing temperature due to the changing density and viscosity of the oil.

Chen et al. [87] created graphene/cysteamine aerogels using cysteamine at various concentrations as a covalent cross-linker under a mild heating process at 90°C and subsequent freeze drying. The GO was partly reduced by the cysteamine. Its absorption capacity was ~150 times its weight for diesel.

Li et al. [88] synthesized a peanut hull/GA, which uses an agricultural waste to create a cheaper alternative for oil absorption in a simple hydrothermal process. Its absorption capacity is 60 times its weight for diesel.

Ren et al. [89] reported on a sulfur and nitrogen codoped GA with an additional GO membrane on it, added via an electrostatic spraying method. Due to its low density and hydrophobicity, the nonsprayed GA can fly up from the water surface with strong winds. This codoped GA has extreme Janus wettability allowing it to remain on the interface despite harsh conditions. It is noted that the hydrophobic mechanism of this GA is due to the changes of surface electrostatic potentials and charge densities due to the codoping into the graphitic structure and the decrease of oxygen-containing functional groups. Its absorption capacity is ~80 times its weight for diesel.

Luo et al. [90] describe a step-by-step in situ biosynthesis process to create a highly porous, robust, and ultralight GA. It involves the self-assembly of 2D few-layered reduced graphene oxide into 3D monolithic bacterial cellulose network. Since bacterial cellulose is produced in very large quantities, this process can potentially be cheaper than pure graphene GAs or other hybrid aerogels. Its absorption capacity is 340 times its weight for diesel and 250 times for gasoline.

Dong et al. [91] created a GA by reducing GO using vitamin C. They heeded special care on specific parameters like the carbon size, GO concentration, GO/vitamin C ratio, and the PH of the solution, with optimal results for pump oil absorption at 500 mesh, 2.5 mg/mL, ≤ 2.5, and natural pH, accordingly.

Ren et al. [92] reported on a GA created using Pluronic P123 copolymer as a template and a novel drying process. P123 has been widely used to prepare mesopore core–shell structures. Both freeze and vacuum drying were attempted resulting in different pore results, with vacuum drying generally resulting in better oil absorption capabilities.

Yang et al. [93] developed a way to create a structurally intact macrostructure of GA about 1 square meter via a surfactant-foaming sol–gel method. They also achieved this using a simple fridge and air drying, further reducing down the cost for energy and special equipment. Essentially, they added a foaming agent to introduce nanobubbles as to obtain uniform large GO crystals using eco-friendly alkyl polyglucoside. This allowed the heterogeneous GO liquid crystals to reform to a more homogenous state, facilitating the creation of a macroscopic assembly with less structural damage. Its absorption capacity for gasoline is 300 times its weight and retains elasticity up to 99% strain. Also this process should be easily scalable for mass production.

5. Conclusions

Ever since graphene's isolation, the scientific community has been interested by graphene's extraordinary properties, and more effort is constantly put forth into graphene and its products. With an ever-increasing need to protect the environment, GAs could potentially provide a feasible alternative way to ameliorate oil spill treatment and restrict damage caused by massive oil spills by sorbing unwanted oil and wastes, after which it is possible to recover both the GA and large portion of the spilled oil. It is important to also notice that the same hydrophobic and oleophilic GAs can also be used for the treatment of organic wastes in general, further increasing their potential.

While GAs exhibit extremely valuable properties for sorbing oil, their use in actual spills still remains to be seen. They possess huge absorbing capacity for oils (up to 300 times their weight) and can be recycled multiple times using compression, distilling, or simply even burning. Most are also very mechanically sound and can withstand many times their weight in force and keep their elasticity for huge amounts of strain (50%–99%). Their problem lies in their expensive synthesis process and general inability to easily scale it to industrial sizes. Important steps for the creation of a GA for oil absorption are (1) creation of GO; (2) reduction of GO; (3) forming of hydrogel; (4) hydrogel drying; (5) thermal annealing. Hummer's method for creating GO has received modifications over the years to make it more eco-friendly and safer by removing $NaNO_3$ and adding phosphoric acid.

The reduction of GO is usually achieved by using ammonia, vitamin C, ethylenediamine, hydrazine, or some other reduction agents and can be assisted by a hydrothermal process. The reduction of the GO's oxygen-containing groups is very important since they are hydrophilic. Higher degrees of reduction generally correspond to better mechanical properties but reduced absorption capacity. It is very important to monitor all aspects of the reduction like agent concentration, temperature, and reaction time. It is possible to use a template to guide the creation of the GA. The template can range anywhere from cheap bacterial cellulose to polymers or nanobubbles. A sufficiently reduced GO should form a hydrogel by itself. Adding cross-linkers or acidizing the mixture will aid the gelation process. Drying the GA is a necessary step and usually requires special equipment to perform freeze or supercritical drying. New techniques are arising, that allow vacuum or air drying. Thermally annealing the GA is optional, but it enhances its properties and requires very high temperatures. In the end, improving any of these steps will improve the whole process. If GA is to be used commercially for oil spills, the economic factor has also to be considered. Every step of the synthesis has to be simplified, and expensive additional materials be avoided as much as possible. It is possible that a simple GA created by reducing GO using ammonia is the best solution instead of more expensive GAs with better properties. It all comes down to the scalability and how easy it would be to produce larger quantities of GA.

References

[1] BP Statistical Review of World Energy (Ed.), BP Statistical Review of World Energy, June 2009, BP p.l.c., London, UK, 10th September, 2009.

[2] Central Intelligence Agency, The World Factbook 2009, ISSN 1553-8133, Washington, DC, 23rd September, 2009.

[3] M. Fingas, Oil Spill Science and Technology, 2011. https://doi.org/10.1016/C2009-0-19703-9.

[4] C.A. Page, J.S. Bonner, T.J. McDonald, R.L. Autenrieth, Behavior of a chemically dispersed oil in a wetland environment, Water Res. 36 (15) (2002) 3821–3833, https://doi.org/10.1016/s0043-1354(02)00079-9.

[5] J. William, Nichols, The U.S. Environmental Protection Agency: National oil and hazardous substances pollution contingency plan, Subpart J Product Schedule (40 CFR 300.900), in: International Oil Spill Conference Proceedings: March 2001, vol. 2001, pp. 1479–1483. No. 2.

[6] A.J. Mearns, Cleaning oiled shores: putting bioremediation to the test, Spill Sci. Technol. Bull. 4 (4) (1997) 209–217, https://doi.org/10.1016/s1353-2561(98)00026-7.

[7] A.D. Venosa, M.T. Suidan, B.A. Wrenn, K.L. Strohmeier, J.R. Haines, B.L. Eberhart, E. Holder, Bioremediation of an experimental oil spill on the shoreline of Delaware bay, Environ. Sci. Technol. 30 (5) (1996) 1764–1775, https://doi.org/10.1021/es950754r.

[8] A.D. Venosa, D.W. King, G.A. Sorial, The baffled flask test for dispersant effectiveness: a round robin evaluation of reproducibility and repeatability, Spill Sci. Technol. Bull. 7 (5–6) (2002) 299–308, https://doi.org/10.1016/s1353-2561(02)00072-5.

[9] K. Lee, G.H. Tremblay, J. Gauthier, S.E. Cobanli, M. Griffin, Bioaugmentation and biostimulation: A paradox between laboratory and field results, in: International Oil Spill Conference Proceedings: April 1997, vol. 1997, pp. 697–705. No. 1.

[10] A.I. Rusanov, Thermodynamics of graphene, Surf. Sci. Rep. 69 (4) (2014) 296–324. https://doi.org/10.1016/j.surfrep.2014.08.003.

[11] P.R. Wallace, The band theory of graphite, Phys. Rev. 71 (9) (1947) 622–634. https://doi.org/10.1103/physrev.71.622.

[12] K.S. Novoselov, Electric field effect in atomically thin carbon films, Science 306 (5696) (2004) 666–669. https://doi.org/10.1126/science.1102896.

[13] D. Bitounis, H. Ali-Boucetta, B.H. Hong, D.-H. Min, K. Kostarelos, Prospects and challenges of graphene in biomedical applications, Adv. Mater. 25 (16) (2013) 2258–2268. https://doi.org/10.1002/adma.201203700.

[14] Y.C. Cheng, Z.Y. Zhu, U. Schwingenschlögl, Mechanical failure of zigzag graphene nanoribbons under tensile strain induced by edge reconstruction, J. Mater. Chem. 22 (47) (2012) 24676. https://doi.org/10.1039/c2jm34068b.

[15] H. Xiang, K. Zhang, G. Ji, J.Y. Lee, C. Zou, X. Chen, J. Wu, Graphene/nanosized silicon composites for lithium battery anodes with improved cycling stability, Carbon 49 (5) (2011) 1787–1796. https://doi.org/10.1016/j.carbon.2011.01.002.

[16] Z. Zhang, J. Du, D. Zhang, H. Sun, L. Yin, L. Ma, W. Ren, Rosin-enabled ultraclean and damage-free transfer of graphene for large-area flexible organic light-emitting diodes, Nat. Commun. 8 (2017) 14560. https://doi.org/10.1038/ncomms14560.

[17] A.N. Obraztsov, E.A. Obraztsova, A.V. Tyurnina, A.A. Zolotukhin, Chemical vapor deposition of thin graphite films of nanometer thickness, Carbon 45 (10) (2007) 2017–2021. https://doi.org/10.1016/j.carbon.2007.05.028.

[18] W.S. Hummers, R.E. Offeman, Preparation of graphitic oxide, J. Am. Chem. Soc. 80 (6) (1958). https://doi.org/10.1021/ja01539a017, 1339–1339.

[19] T. Ishikawa, T. Nagaoki, Shin Tanso Kogyo (New Carbon Industry), second ed., Kindai hensyusya, Tokyo, 1986, pp. 125–136.

[20] J. Chen, B. Yao, C. Li, G. Shi, An improved Hummers method for eco-friendly synthesis of graphene oxide, Carbon 64 (2013) 225–229. https://doi.org/10.1016/j.carbon.2013.07.055.

[21] C.M. Seah, S.-P. Chai, A. Mohamed, Mechanisms of graphene growth by chemical vapour deposition on transition metals, Carbon 70 (2014) 1–21. https://doi.org/10.1016/j.carbon.2013.12.073.

[22] Q. Yu, J. Lian, S. Siriponglert, H. Li, Y.P. Chen, S.-S. Pei, Graphene segregated on Ni surfaces and transferred to insulators, Appl. Phys. Lett. 93 (11) (2008) 113103. https://doi.org/10.1063/1.2982585.

[23] Y. Hernandez, V. Nicolosi, M. Lotya, M. Blighe, Fiona, Z. Sun, S. De, I. Mcgovern, B. Holland, M. Byrne, Y. Gun'ko, J. Boland, P.N. Nirmalraj, G. Duesberg, S. Krishnamurthy, R. Goodhue, J. Hutchison, V. Scardaci, A. Ferrari, J. Coleman, High-yield production of graphene by liquid-phase exfoliation of graphite, Nat. Nanotechnol. 3 (2008) 563–568. https://doi.org/10.1038/nnano.2008.215.

[24] T. Tanaka, Gels, in: A. Klingsberg, R. Piccininni (Eds.), Encyclopedia of Polymer Science and Engineering, vol. 7, John Wiley & Sons, New York, USA, 1987, p. 514.

[25] H.G. Bungenberg de Jong, A survey of the study objects in this volume, in: H.R. Kruyt (Ed.), Colloid Science, vol. II, Elsevier Publishing Company, Inc., Amsterdam, The Netherlands, 1949, pp. 1–18.

[26] P.H. Hermans, Gels, in: H.R. Kruyt (Ed.), Colloid Science, vol. II, Elsevier Publishing Company, Inc., Amsterdam, The Netherlands, 1949, pp. 483–651.

[27] PAC, Definitions of terms relating to the structure and processing of sols, gels, networks, and inorganic-organic hybrid materials (IUPAC Recommendations 2007), Pure Appl. Chem. 79 (2007) 1806.

[28] Y. Gu, Y. Xu, Y. Wang, Graphene-wrapped CoS nanoparticles for high-capacity lithium-ion storage, ACS Appl. Mater. Interfaces 5 (3) (2013) 801–806. https://doi.org/10.1021/am3023652.

[29] S.A. Ju, K. Kim, J.-H. Kim, S.-S. Lee, Graphene-wrapped hybrid spheres of electrical conductivity, ACS Appl. Mater. Interfaces 3 (8) (2011) 2904–2911. https://doi.org/10.1021/am200056t.

[30] H.-B. Zhang, Q. Yan, W.-G. Zheng, Z. He, Z.-Z. Yu, Tough Graphene–Polymer microcellular foams for electromagnetic interference shielding, ACS Appl. Mater. Interfaces 3 (3) (2011) 918–924. https://doi.org/10.1021/am200021v.

[31] S. Zhang, Y. Shao, J. Liu, I.A. Aksay, Y. Lin, Graphene–polypyrrole nanocomposite as a highly efficient and low cost electrically switched ion exchanger for removing ClO4− from wastewater, ACS Appl. Mater. Interfaces 3 (9) (2011) 3633–3637. https://doi.org/10.1021/am200839m.

[32] Y. Wang, R. Yang, Z. Shi, L. Zhang, D. Shi, E. Wang, G. Zhang, Super-elastic graphene ripples for flexible strain sensors, ACS Nano 5 (5) (2011) 3645–3650. https://doi.org/10.1021/nn103523t.

[33] Y. Zhu, S. Murali, W. Cai, X. Li, J.W. Suk, J.R. Potts, R.S. Ruoff, Graphene and graphene oxide: synthesis, properties, and applications, Adv. Mater. 22 (35) (2010) 3906−3924. https://doi.org/10.1002/adma.201001068.

[34] C.N.R. Rao, A.K. Sood, K.S. Subrahmanyam, A. Govindaraj, Graphene: the new two-dimensional nanomaterial, Angew. Chem. Int. Ed. 48 (42) (2009) 7752−7777. https://doi.org/10.1002/anie.200901678.

[35] J. Liang, Y. Wang, Y. Huang, Y. Ma, Z. Liu, J. Cai, Y. Chen, Electromagnetic interference shielding of graphene/epoxy composites, Carbon 47 (3) (2009) 922−925. https://doi.org/10.1016/j.carbon.2008.12.038.

[36] H. Kim, Y. Miura, C.W. Macosko, Graphene/polyurethane nanocomposites for improved gas barrier and electrical conductivity, Chem. Mater. 22 (11) (2010) 3441−3450. https://doi.org/10.1021/cm100477v.

[37] H.-P. Cong, X.-C. Ren, P. Wang, S.-H. Yu, Macroscopic multifunctional graphene-based hydrogels and aerogels by a metal ion induced self-assembly process, ACS Nano 6 (3) (2012) 2693−2703. https://doi.org/10.1021/nn300082k.

[38] C. Wu, X. Huang, X. Wu, R. Qian, P. Jiang, Mechanically flexible and multifunctional polymer-based graphene foams for elastic conductors and oil-water separators, Adv. Mater. 25 (39) (2013) 5658−5662. https://doi.org/10.1002/adma.201302406.

[39] C. Chen, Q.-H. Yang, Y. Yang, W. Lv, Y. Wen, P.-X. Hou, H.-M. Cheng, Self-assembled free-standing graphite oxide membrane, Adv. Mater. 21 (29) (2009) 3007−3011. https://doi.org/10.1002/adma.200803726.

[40] H. Chen, M.B. Müller, K.J. Gilmore, G.G. Wallace, D. Li, Mechanically strong, electrically conductive, and biocompatible graphene paper, Adv. Mater. 20 (18) (2008) 3557−3561. https://doi.org/10.1002/adma.200800757.

[41] J. Chen, K. Sheng, P. Luo, C. Li, G. Shi, Graphene hydrogels deposited in nickel foams for high-rate electrochemical capacitors, Adv. Mater. 24 (33) (2012) 4569−4573. https://doi.org/10.1002/adma.201201978.

[42] W. Chen, S. Li, C. Chen, L. Yan, Self-assembly and embedding of nanoparticles by in situ reduced graphene for preparation of a 3D graphene/nanoparticle aerogel, Adv. Mater. 23 (47) (2011) 5679−5683. https://doi.org/10.1002/adma.201102838.

[43] H. Sun, Z. Xu, C. Gao, Multifunctional, ultra-flyweight, synergistically assembled carbon aerogels, Adv. Mater. 25 (18) (2013) 2554−2560. https://doi.org/10.1002/adma.201204576.

[44] Y. Xu, W. Hong, H. Bai, C. Li, G. Shi, Strong and ductile poly(vinyl alcohol)/graphene oxide composite films with a layered structure, Carbon 47 (15) (2009) 3538−3543. https://doi.org/10.1016/j.carbon.2009.08.022.

[45] Y. Xu, K. Sheng, C. Li, G. Shi, Self-assembled graphene hydrogel via a one-step hydrothermal process, ACS Nano 4 (7) (2010) 4324−4330. https://doi.org/10.1021/nn101187z.

[46] S. Nardecchia, D. Carriazo, M.L. Ferrer, M.C. Gutiérrez, F. del Monte, Three dimensional macroporous architectures and aerogels built of carbon nanotubes and/or graphene: synthesis and applications, Chem. Soc. Rev. 42 (2) (2013) 794−830. https://doi.org/10.1039/c2cs35353a.

[47] Y. Qian, I.M. Ismail, A. Stein, Ultralight, high-surface-area, multifunctional graphene-based aerogels from self-assembly of graphene oxide and resol, Carbon 68 (2014) 221−231. https://doi.org/10.1016/j.carbon.2013.10.082.

[48] G. Gorgolis, D. Karamanis, Solar energy materials for glazing technologies, Sol. Energy Mater. Sol. Cell. 144 (2016) 559−578. https://doi.org/10.1016/j.solmat.2015.09.040.

[49] H. Bai, K. Sheng, P. Zhang, C. Li, G. Shi, Graphene oxide/conducting polymer composite hydrogels, J. Mater. Chem. 21 (46) (2011) 18653. https://doi.org/10.1039/c1jm13918e.

[50] S.-Z. Zu, B.-H. Han, Aqueous dispersion of graphene sheets stabilized by pluronic copolymers: formation of supramolecular hydrogel, J. Phys. Chem. C 113 (31) (2009) 13651–13657. https://doi.org/10.1021/jp9035887.

[51] X. Ma, Y. Li, W. Wang, Q. Ji, Y. Xia, Temperature-sensitive poly(N-isopropylacrylamide)/graphene oxide nanocomposite hydrogels by in situ polymerization with improved swelling capability and mechanical behavior, Eur. Polym. J. 49 (2) (2013) 389–396. https://doi.org/10.1016/j.eurpolymj.2012.10.034.

[52] J. Liu, G. Chen, M. Jiang, Supramolecular hybrid hydrogels from noncovalently functionalized graphene with block copolymers, Macromolecules 44 (19) (2011) 7682–7691. https://doi.org/10.1021/ma201620w.

[53] M.A. Worsley, P.J. Pauzauskie, T.Y. Olson, J. Biener, J.H. Satcher, T.F. Baumann, Synthesis of graphene aerogel with high electrical conductivity, J. Am. Chem. Soc. 132 (40) (2010) 14067–14069. https://doi.org/10.1021/ja1072299.

[54] C.-M. Chen, Q. Zhang, C.-H. Huang, X.-C. Zhao, B.-S. Zhang, Q.-Q. Kong, D. Sheng Su, Macroporous "bubble" graphene film via template-directed ordered-assembly for high rate supercapacitors, Chem. Commun. 48 (57) (2012) 7149. https://doi.org/10.1039/c2cc32189k.

[55] Z. Wang, X. Shen, N.M. Han, X. Liu, Y. Wu, W. Ye, J.-K. Kim, Ultralow electrical percolation in graphene aerogel/epoxy composites, Chem. Mater. 28 (18) (2016) 6731–6741. https://doi.org/10.1021/acs.chemmater.6b03206.

[56] Z.-S. Wu, S. Yang, Y. Sun, K. Parvez, X. Feng, K. Müllen, 3D nitrogen-doped graphene aerogel-supported Fe3O4 nanoparticles as efficient electrocatalysts for the oxygen reduction reaction, J. Am. Chem. Soc. 134 (22) (2012) 9082–9085. https://doi.org/10.1021/ja3030565.

[57] L. Zhang, W. Fan, W.W. Tjiu, T. Liu, 3D porous hybrids of defect-rich MoS2/graphene nanosheets with excellent electrochemical performance as anode materials for lithium ion batteries, RSC Adv. 5 (44) (2015) 34777–34787. https://doi.org/10.1039/c5ra04391c.

[58] M.A. Worsley, T.T. Pham, A. Yan, S.J. Shin, J.R.I. Lee, M. Bagge-Hansen, A. Zettl, Synthesis and characterization of highly crystalline graphene aerogels, ACS Nano 8 (10) (2014) 11013–11022. https://doi.org/10.1021/nn505335u.

[59] K. Sheng, Y. Xu, C. Li, G. Shi, High-performance self-assembled graphene hydrogels prepared by chemical reduction of graphene oxide, Carbon 49 (8) (2011) 2878. https://doi.org/10.1016/j.carbon.2011.02.032.

[60] S. Pei, J. Zhao, J. Du, W. Ren, H.-M. Cheng, Direct reduction of graphene oxide films into highly conductive and flexible graphene films by hydrohalic acids, Carbon 48 (15) (2010) 4466–4474. https://doi.org/10.1016/j.carbon.2010.08.006.

[61] I.K. Moon, J. Lee, R.S. Ruoff, H. Lee, Reduced graphene oxide by chemical graphitization, Nat. Commun. 1 (6) (2010) 1–6. https://doi.org/10.1038/ncomms1067.

[62] S. Park, J. An, J.R. Potts, A. Velamakanni, S. Murali, R.S. Ruoff, Hydrazine-reduction of graphite- and graphene oxide, Carbon 49 (9) (2011) 3019–3023. https://doi.org/10.1016/j.carbon.2011.02.071.

[63] W. Chen, L. Yan, In situ self-assembly of mild chemical reduction graphene for three-dimensional architectures, Nanoscale 3 (8) (2011) 3132. https://doi.org/10.1039/c1nr10355e.

[64] K. Parvez, Z.-S. Wu, R. Li, X. Liu, R. Graf, X. Feng, K. Müllen, Exfoliation of graphite into graphene in aqueous solutions of inorganic salts, J. Am. Chem. Soc. 136 (16) (2014) 6083–6091. https://doi.org/10.1021/ja5017156.

[65] V.V. Singh, G. Gupta, A. Batra, A.K. Nigam, M. Boopathi, P.K. Gutch, R. Vijayaraghavan, Greener electrochemical synthesis of high quality graphene nanosheets directly from pencil

and its SPR sensing application, Adv. Funct. Mater. 22 (11) (2012) 2352–2362. https://doi.org/10.1002/adfm.201102525.

[66] B.G. Choi, M. Yang, W.H. Hong, J.W. Choi, Y.S. Huh, 3D macroporous graphene frameworks for supercapacitors with high energy and power densities, ACS Nano 6 (5) (2012) 4020–4028. https://doi.org/10.1021/nn3003345.

[67] J.L. Vickery, A.J. Patil, S. Mann, Fabrication of graphene-polymer nanocomposites with higher-order three-dimensional architectures, Adv. Mater. 21 (21) (2009) 2180–2184. https://doi.org/10.1002/adma.200803606.

[68] L. Estevez, A. Kelarakis, Q. Gong, E.H. Da'as, E.P. Giannelis, Multifunctional graphene/platinum/nafion hybrids via ice templating, J. Am. Chem. Soc. 133 (16) (2011) 6122–6125. https://doi.org/10.1021/ja200244s.

[69] H. Bai, C. Li, X. Wang, G. Shi, On the gelation of graphene oxide, J. Phys. Chem. C 115 (13) (2011) 5545–5551. https://doi.org/10.1021/jp1120299.

[70] D.H. Everett, Basic Principles of Colloid Science, The Royal Society of Chemistry, London, 1988.

[71] S. Park, D.A. Dikin, S.T. Nguyen, R.S. Ruoff, Graphene oxide sheets chemically cross-linked by polyallylamine, J. Phys. Chem. C 113 (36) (2009) 15801–15804. https://doi.org/10.1021/jp907613s.

[72] Y. Cheng, P. Xu, W. Zeng, C. Ling, S. Zhao, K. Liao, Y. Sun, A. Zhou, Highly hydrophobic and ultralight graphene aerogel as high efficiency oil absorbent material, J. Environ. Chem. Eng. 5 (2017) 1957–1963. https://doi.org/10.1016/j.jece.2017.04.005.

[73] D. Zhao, L. Yu, D. Liu, Ultralight graphene/carbon nanotubes aerogels with compressibility and oil absorption properties, Materials 11 (4) (2018) 641. https://doi.org/10.3390/ma11040641.

[74] J.-Y. Hong, E.-H. Sohn, S. Park, H.S. Park, Highly-efficient and recyclable oil absorbing performance of functionalized graphene aerogel, Chem. Eng. J. 269 (2015) 229–235. https://doi.org/10.1016/j.cej.2015.01.066.

[75] Y. Luo, S. Jiang, Q. Xiao, C. Chen, B. Li, Highly reusable and superhydrophobic spongy graphene aerogels for efficient oil/water separation, Sci. Rep. 7 (1) (2017). https://doi.org/10.1038/s41598-017-07583-0.

[76] L. Xu, G. Xiao, C. Chen, R. Li, Y. Mai, G. Sun, D. Yan, Superhydrophobic and superoleophilic graphene aerogel prepared by facile chemical reduction, J. Mater. Chem. 3 (14) (2015) 7498–7504. https://doi.org/10.1039/c5ta00383k.

[77] T. Liu, M. Huang, X. Li, C. Wang, C.-X. Gui, Z.-Z. Yu, Highly compressible anisotropic graphene aerogels fabricated by directional freezing for efficient absorption of organic liquids, Carbon 100 (2016) 456–464. https://doi.org/10.1016/j.carbon.2016.01.038.

[78] J. Xiao, W. Lv, Y. Song, Q. Zheng, Graphene/nanofiber aerogels: performance regulation towards multiple applications in dye adsorption and oil/water separation, Chem. Eng. J. 338 (2018) 202–210. https://doi.org/10.1016/j.cej.2017.12.156.

[79] J. Huang, H. Liu, S. Chen, C. Ding, Graphene aerogel prepared through double hydrothermal reduction as high-performance oil adsorbent, Mater. Sci. Eng., B 226 (2017) 141–150. https://doi.org/10.1016/j.mseb.2017.09.014.

[80] F. Wang, Y. Wang, W. Zhan, S. Yu, W. Zhong, G. Sui, X. Yang, Facile synthesis of ultra-light graphene aerogels with super absorption capability for organic solvents and strain-sensitive electrical conductivity, Chem. Eng. J. 320 (2017) 539–548. https://doi.org/10.1016/j.cej.2017.03.082.

[81] W. Wan, F. Zhang, S. Yu, R. Zhang, Y. Zhou, Hydrothermal formation of graphene aerogel for oil sorption: the role of reducing agent, reaction time and temperature, New J. Chem. 40 (4) (2016) 3040–3046. https://doi.org/10.1039/c5nj03086b.

[82] Z. Han, Z. Tang, S. Shen, B. Zhao, G. Zheng, J. Yang, Strengthening of graphene aerogels with tunable density and high adsorption capacity towards Pb2+, Sci. Rep. 4 (1) (2014). https://doi.org/10.1038/srep05025.

[83] H.-Y. Mi, X. Jing, A.L. Politowicz, E. Chen, H.-X. Huang, L.-S. Turng, Highly compressible ultra-light anisotropic cellulose/graphene aerogel fabricated by bidirectional freeze drying for selective oil absorption, Carbon 132 (2018) 199–209. https://doi.org/10.1016/j.carbon.2018.02.033.

[84] Y. Muhammad Shafiq, W.K. Cheong, E.V. Lau, Graphene aerogel – recovery of heavy crude oil from contaminated sand, J. Environ. Chem. Eng. 5 (2) (2017) 1711–1717. https://doi.org/10.1016/j.jece.2017.03.009.

[85] J. Cao, Z. Wang, X. Yang, J. Tu, R. Wu, W. Wang, Green synthesis of amphipathic graphene aerogel constructed by using the framework of polymer-surfactant complex for water remediation, Appl. Surf. Sci. 444 (2018) 399–406. https://doi.org/10.1016/j.apsusc.2018.02.282.

[86] Z. Rahmani, A.M. Rashidi, A. kazemi, M.T. Samadi, A.R. Rahmani, N-doped reduced graphene oxide aerogel for the selective adsorption of oil pollutants from water: isotherm and kinetic study, J. Ind. Eng. Chem. 61 (2018) 416–426. https://doi.org/10.1016/j.jiec.2017.12.041.

[87] C. Chen, X. Zhu, B. Chen, Covalently cross-linked graphene oxide aerogel with stable structure for high-efficiency water purification, Chem. Eng. J. 354 (2018) 896–904. https://doi.org/10.1016/j.cej.2018.08.034.

[88] N. Li, Q. Yue, B. Gao, X. Xu, R. Su, B. yu, One-step synthesis of peanut hull/graphene aerogel for highly efficient oil-water separation, J. Clean. Prod. (2018). https://doi.org/10.1016/j.jclepro.2018.10.038.

[89] X. Ren, H. Guo, X. Ma, G. Hou, L. Chen, X. Xu, L. Ci, Improved interfacial floatability of superhydrophobic and compressive S, N co-doped graphene aerogel by electrostatic spraying for highly efficient organic pollutants recovery from water, Appl. Surf. Sci. 457 (2018) 780–788. https://doi.org/10.1016/j.apsusc.2018.06.289.

[90] H. Luo, J. Xie, J. Wang, F. Yao, Z. Yang, Y. Wan, Step-by-step self-assembly of 2D few-layer reduced graphene oxide into 3D architecture of bacterial cellulose for a robust, ultralight, and recyclable all-carbon absorbent, Carbon 139 (2018) 824–832. https://doi.org/10.1016/j.carbon.2018.07.048.

[91] S. Dong, L. Xia, T. Guo, F. Zhang, L. Cui, X. Su, J. Sun, Controlled synthesis of flexible graphene aerogels macroscopic monolith as versatile agents for wastewater treatment, Appl. Surf. Sci. 445 (2018) 30–38. https://doi.org/10.1016/j.apsusc.2018.03.132.

[92] J. Ren, X. Zhang, D. Lu, B. Chang, J. Lin, S. Han, Fabrication of controllable graphene aerogel with superior adsorption capacity for organic solvents, Res. Chem. Intermed. 44 (9) (2018) 5139–5152. https://doi.org/10.1007/s11164-018-3414-0.

[93] H. Yang, Z. Li, B. Lu, J. Gao, X. Jin, G. Sun, L. Qu, Reconstruction of inherent graphene oxide liquid crystals for large-scale fabrication of structure-intact graphene aerogel bulk toward practical applications, ACS Nano (2018). https://doi.org/10.1021/acsnano.8b06380.

Chapter 9

Nanotechnology and wood science

Antonios N. Papadopoulos[a] and George Z. Kyzas[b]
[a]*Laboratory of Wood Chemistry and Technology, Eastern Macedonia and Thrace Institute of Technology, Kavala, Greece;* [b]*Hephaestus Advanced Laboratory, Department of Chemistry, International Hellenic University, Kavala, Greece*

1. Introduction

Wood, a versatile material, has been used for centuries for many reasons, due to its fibrous nature. It varies in color and density and is considered a primary raw material in building duc to its high strength in combination with low weight and some durability. It is therefore a raw material that can be used in indoor applications and, if treated efficiently, in outdoor application as well. However, wood has two main disadvantages which restrict its wider use, namely, susceptibility to biodegradability by microorganisms and dimensional instability when subjected to varied moisture content. Most wood species deteriorate rapidly under biological factors; the most important biological decay is caused by fungi. On the other hand, when wood is subjected to fluctuating moisture, dimensional and conformational instability occur. These drawbacks are mainly due to the nature of the cell wall main polymers, and in particular due to their high abundance of hydroxyl groups (OH) [1−3]. This is solved so far, either by using imported tropical woods or by using conventional biocides which are usually based in the use of toxic chemicals. In addition, the disposal of the treated wood has caused many restrictions which have to be taken into consideration upon the utilization of conventional chemical treatments. However, both these options are nowadays under political and consumer pressure, and the seeking of alternatives has become an imperative need [1]. This need led to attention to nonbiocide treatments, such as modification of wood with nontoxic chemicals. By applying chemical modification, mainly acetylated and furfurylated wood, low-durability species can be easily transformed to new wood products with improved properties, this approach therefore can be considered as the ultimate solution to the problem [3,4].

Advanced Low-Cost Separation Techniques in Interface Science. https://doi.org/10.1016/B978-0-12-814178-6.00009-1

Another option to overcome this problem, perhaps more attractive, is to investigate the potentials that nanotechnology may offer. It is well established that the wood cell wall shows porosity of molecular scale dimensions, this is caused by the partial filling of spaces between the cellulose microfibrils mainly by polyoses and lignin [5,6]. The small-sized nanoparticles can effectively and deeply penetrate into the wood, alter its surface chemistry, improve its properties resulting in a hyper-performance product.

So, this chapter briefly reviews the research milestones in the area of wood chemical modification, focusing on acetylated and furfurylated wood which have been scaled up and exploits the solutions that nanotechnology can offer to wood protection, as an alternative green innovative approach.

2. Chemical modification

The main target of applying chemical modification onto wood is to alter the molecular structure of the main cell wall polymers, by converting the hydrophilic OH^- groups into hydrophobic groups. By this way, the cell wall of wood is in a permanently swollen situation and attracts no or very little moisture [7,8]. At the same time, the chemically modified wood is not recognized by the degrading fungi, since the lower moisture content does not promote decay. Various technologies in the area of wood modification have been developed, mainly in laboratory and semi-industrial scale since there was no economic or environmental urgency to scale up these technologies. Recently, due to environmental and legislative issues on the use of biocides, new avenues for the chemical modification have been created. Wood modification technologies may be categorized, related to the mode of action of the chemical as follows [1,3]:

- Filling the lumen with a chemical, usually resin. This decreases the sorption of water vapor but the sorption behavior over a longer period of time is not reduced.
- Bulking the cell wall and the cell lumen with a chemical. This reduces the swelling of wood and has a beneficial effect on the long-term sorption behavior.
- Modifying treatments. The chemistry of the main cell wall components is changed and covalent bonds are formed by converting the hydrophilic OH-groups into hydrophobic groups. This results to new wood products with enhanced properties.

Chemical modification involves the linking of a chemical to the main wood cell wall polymers, especially the hydroxyl groups via a covalent bond [7], without leaving toxic residues within the wood. Chemical modification of wood has been extensively reviewed elsewhere [1,3,8–10].

The remaining of the chapter focuses on the most prominent processes that have been scaled up today to industrial level, namely, acetylation and furfurylation.

2.1 Acetylation

A great amount of research in this area involves the reaction of acetic anhydride with wood. Anhydrides react with the hydroxyl groups of the wood cell wall polymers via a nucleophillic reaction pathway [11,12]. Great attention of the researchers has been attracted the reaction of hydroxyl groups with acetic anhydride, a process known as acetylation. The chemical reaction is depicted in Fig. 9.1. The wood hydroxyl groups are replaced by the acetyl groups of the acetic anhydride, and the wood remains in swollen condition due to the bulking action of the acetyl groups within the cell walls. Acetylation is a single-site reaction which means that one acetyl group is added per reacted hydroxyl group with no further polymerization. Wood acetate and acetic acid as by-products are produced.

The first attempt of acetylation has been recorded in 1865, when Schutzenberger acetylated cellulose. Fuchs [13] and Horn [14] in the late 1920s were the first researchers who acetylated wood flour. Whole wood is first acetylated by Stamm and Tarkow [15], using pyridine as a catalyst. Since then, a lot of laboratories worldwide have tried to acetylate wood with a variety of ways [16]. The first attempt, however, to commercialize the process of acetylation failed in the United States (1961), Russia (1977), and Japan (1984). In semi-industrial level, the first successful scaled-up acetylation was performed at Stichting Hout Research (SHR, The Netherlands), by Prof. Militz and coworkers [17]. Nowadays, acetylated wood is scaled-up in the Netherlands, under the commercial name Accoya, which utilizes radiate pine (*Pinus radiata*) and alder wood (*Alnus* sp.) with a 20% acetyl weight gain. Accoya wood undergoes a nontoxic acetylation process which modifies wood for permanent.

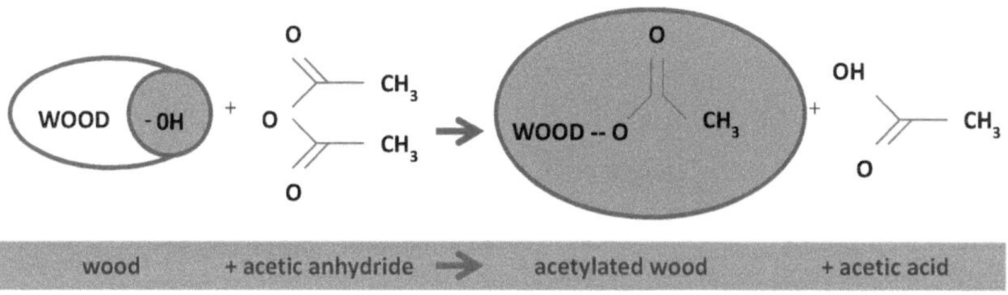

FIGURE 9.1 Reaction of wood with acetic anhydride [20].

2.1.1 Water vapor sorption of acetylated wood

Many studies have investigated the sorption of water vapor of acetylated wood and the overall stabilization was well documented [18,19]. A study by Papadopoulos and Hill examined the effect of molecular size of substituent group on sorption of water vapor, as wood was modified with a series of anhydride molecules of different size which at the same time produced similar levels of cell wall bulking at different levels of hydroxyl substitution [20,21]. The sorption data were analyzed by applying the Hailwood–Horrobin sorption model, which enables the separation of the total sorption into monomolecular and polymolecular sorption. They found that the reduction in all types of sorption is mainly a bulking effect and is not determined by the number of hydroxyl groups which have been substituted (Fig. 9.2).

2.1.2 Durability of acetylated wood

Resistance against fungi. The durability of acetylated wood against fungi and other organisms has been extensive, and a weight gain of *c.* 20% is required to achieve full protection [22]. The protection threshold levels for acetylated pine were determined by Beckers et al. [23]. They found that a weight gain of 18% was required against the brown-rot fungi *Coniophora puteana* and a value over 20% against the white-rot fungi *Postia placenta*. A value of 20% is reported to

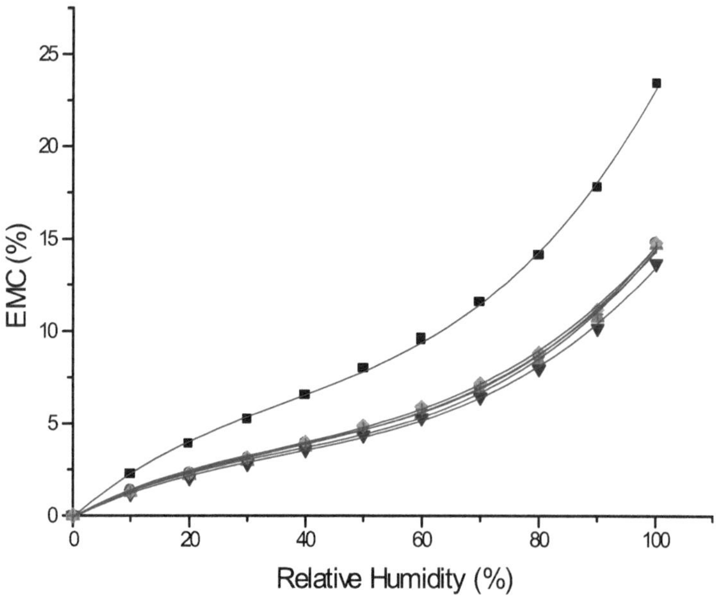

FIGURE 9.2 Adsorption isotherms for pine wood: control (■); acetic 15.8% WPG (●); propionic 15.3% WPG (▲); butyric 16% WPG (▼); valeric 15.5% WPG (♦); hexanoic 15.7% WPG (+) [20].

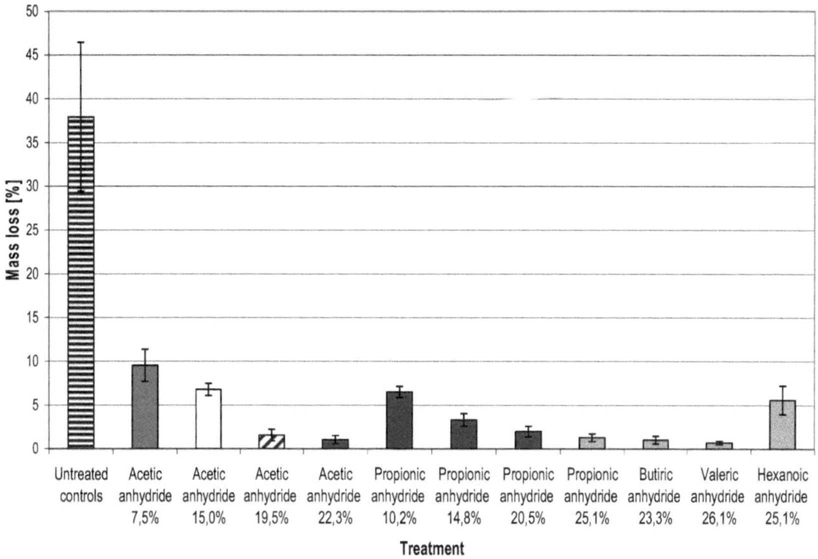

FIGURE 9.3 Mass loss (%) caused by soft-rot decay of wood modified with linear chain anhydrides [27]. *Copyright (2010) Elsevier.*

prevent attack by brown, white, and soft-rot fungi, in ground contact stake tests [24]. Suttie et al. [25] reported that the bulking effect is the only factor that determines protection from decay. Hill and Papadopoulos [26] reaffirmed this conclusion when they presented a comprehensive assessment of the effectiveness of linear chain anhydrides against the brown-rot fungi *C. puteana*. This conclusion is further reaffirmed when pine samples are exposed to soft-rot decay [27], as depicted in Fig. 9.3.

Resistance against marine wood borers and termites. Few studies have been published so far that focused on the protection against invertebrates [28] including marine wood borers. Papadopoulos et al. [29] used a rapid laboratory test developed by Borges et al. [30] and assessed if the modification of pine wood with a series of anhydrides confers resistance to the crustacean wood borer *Limnoria*. The results revealed that modification resulted in a reduction in the number of fecal pellets produced and that the type of anhydride employed had little effect on feeding.

A recent study has demonstrated that modified pine wood with a series of anhydrides afforded significant protection in feeding above 16% weight gain against the subterranean termites *Reticulitermes flavipes* [31].

2.2 Furfurylation

Stamm [32] has initiated research work relating to chemical modification with furfuryl alcohol, known as furfurylation. Furfurylation involves impregnation of wood with a mixture of furfuryl alcohol and catalysts, subsequent heating which in turns causes polymerization [33]. The final product looks like more those of a polymer-filled cell wall rather than a reacted cell wall [34].

The Norwegian company Kebony AS applies the technology and produces two distinct furfurylated wood products, namely (1) Kebony Clear, a hard furfurylated wood with typical weight gain of 35% and (2) Kebony Character, a light furfurylated wood with typical weight gain of 20%.

2.2.1 Properties of furfurylated wood

An excellent review on properties of furfurylated wood was recently performed by Mantanis [34]. According to his findings:

- Furfurylated wood exhibits very good dimensional stability and resistance to weathering.
- Furfurylated wood has comparable resistance against microorganisms with the wood treated with conventional toxic biocides. Furfurylated wood is resistant to marine borers.
- Furfurylated wood is a green product known with the trade name "Swan" in the Scandinavian market.

3. Nanomaterials

At this point, it is mandatory to explain why nanotechnology and especially nanomaterials are so necessary in the above-proposed technology (wood protection). Nanomaterials are a special topic of recent research and are a milestone of nanoscience and nanotechnology. Nanoscale materials are a series of substances/compounds, in which at least one dimension has smaller size than 100 nm. Nanomaterials have a broad area of development which day-by-day is rapidly growing. Their impact to commercial applications is huge with respective to academia and education. A high percentage of nanomaterials exist naturally, but the great interest is for the type of nanomaterials which can be synthesized. These are called "engineered nanomaterials (ENs)." These nanomaterials are formed at the molecular (nanometer) level in order to present the positive of the small/tiny size and innovative properties. Those properties are not generally observed in their common/bulk counterparts/nanoparticles. The big bang was very crucial for the nanomaterials science, given that many theories refer the existence of nanostructures after falls of early meteorites, and until now they are under examination with new nanoforms. All above clearly indicate that the "world" of nanomaterials is explosively growing in various sections.

At this point, it is mandatory to give a brief definition of nanomaterials. So, nanoscale materials are a series of substances/compounds, in which at least one dimension has smaller size than 100 nm. One nm is equal to 10^{-6} mm, meaning 10^5 times smaller than a possible size of a human hair (obviously measuring the diameter). Another question which must be replied is why nanomaterials have such big interest. The answer is not difficult. Since at this scale interesting optical, attractive, electrical, and some other different properties rise. These innovative properties have the potential for extraordinary effects in various gadgets, especially in biomedicine, biochemistry, or other relatively similar fields.

Another crucial question is where nanomaterials can be found or how they are "born"? A high percentage of nanomaterials exist naturally, but the great interest is for the type of nanomaterials which can be synthesized (ENs). This type of materials is extensively used in many processes producing even commercial products. They can be found in such things as sunscreens, cosmetics, sporting goods, stain-resistant clothing, tires, electronics. They also exist in everyday things and especially products of medicine and biomedicine (for example, drug deliver, imaging, or even diagnosis purposes).

ENs are formed at the molecular (nanometer) level in order to present the positive of the small/tiny size and innovative properties. Those properties are not generally observed in their common/bulk counterparts/nanoparticles. Two major explanations can be given to answer why these materials at nanoscale have different properties. These materials showed high (or better increasing) relative surface area and some new types of quantum effects. Also, it is noteworthy that their conventional forms present much lower surface area to volume ratio than the new forms (EN). The latter causes higher chemical reactivity, influencing also their strength. Going to nanoscale, the importance of quantum effects is great especially regarding the determination of materials properties and characteristics; this confirms some new optical, electrical, and magnetic behaviors. Nanomaterials are already used for commercial purposes in industry. The available commercial products of nowadays belong to a broad technologic area, including stain-resistant and wrinkle-free textiles, cosmetics, sunscreens, electronics, paints, and varnishes. Nanocoatings and nanocomposites are finding uses in diverse consumer products, such as windows, sports equipment, bicycles, and automobiles [35]. There are novel UV-blocking coatings on glass bottles which protect beverages from damage by sunlight, and longer-lasting tennis balls using butyl-rubber/nanoclay composites [36]. Nanoscale titanium dioxide, for instance, is finding applications in cosmetics, sun-block creams, and self-cleaning windows, and nanoscale silica is being used as filler in a range of products, including cosmetics and dental fillings [36].

Historical points. The big bang was very crucial for the nanomaterials science given that many theories refer the existence of nanostructures after falls of early meteorites. After that start, the nature physically created some other nanostructure-like formations such as seashells, skeletons, etc. However, the scientific/technological story and official start of nanomaterials can be pinned much later. Michael Faraday has synthesized colloidal gold particles synthesized in 1857. In 1930s, nanostructured catalysts were also investigated. During 1940s, the market of the United States and Germany were filled after manufacturing with precipitated and fumed silica nanoparticles as substitutes for ultrafine carbon black for rubber reinforcements. Then, from 1960 to 1970s, some metallic-like nanopowders used to produce magnetic recording tapes were also developed. In 1976, Granqvist and Buhrman published a work reporting a new term, the popular inert gas evaporation technique according to which nanocrystals were synthesized for the first time.

Nowadays, the engineering of nanophase or generally nanotechnology is rapidly developing, investigating more and more structural and functional materials (inorganic and organic). Therefore, there is possibility to change/modify or even treat some important electric, magnetic, mechanical, catalytic, optical, and electronic functions. The synthesis of materials in nanophase or the so-called cluster-assembled materials is based on the separated small clusters formation. Then, these are fused into a material (majorly in bulk form).

4. Nanotechnology and wood

Nanotechnology, an attractive science, seems to have a remarkable potential to create products of a new generation with enhanced properties [37]. The term nano is usually referred to materials whose dimensions are less than 100 nm. Nanomaterials have a high surface–volume ratio that enables them to show greater activities in surface-related phenomena compared with bulky systems with identical mass [38,39].

The change in material properties is primarily due to the large interfacial area that is developed per unit of volume, since the level of added particles is reduced to nanometers. Nanomaterials enhance the properties of the original material, show great compatibility with the traditional materials, and cause limited alteration of their original features [40,41].

Nanotechnology is being developed for a variety of applications, from medical uses to material behavior improvement, like wood and wood products. Studies on the application of nanotechnology in wood science are mainly focused on dimensional stability and resistance to attack by microorganisms. The main advantage of applying nanotechnology to wood is the unique ability of the compounds to penetrate deeply in wood substrate [41,42]. It is well established that the wood cell wall shows porosity of molecular scale dimensions, this is caused by the partial filling of spaces between the cellulose microfibrils mainly by polyoses and lignin [5,6,43]. The small size nanoparticles can effectively and deeply penetrate into the wood, alter its surface chemistry, improve its properties resulting in a hyper-performance product. In addition, complete penetration and uniform distribution can be achieved, if the size of the nanoparticles will be smaller than the diameter of the wood pits [6].

Currently, nanomaterials are applied to wood and are performed with three different and distinctive ways, namely, (1) the direct impregnation of nanosized materials in to wood (nanosized metals), (2) with controlled release of nanomaterials embedded in a nanocarrier (polymeric nanocarriers), and (3) coating treatments [44]. These, are reviewed herein.

4.1 Nanosized metals

Nanosized metals can be basically synthesized using mainly two chemical approaches, namely (1) solution-based synthesis (sol-gel, sonochemical, solvothermal) and (2) vapor-based synthesis (combustion, chemical vapor deposition) [44–48]. It is well documented that the method of synthesis determines the physicochemical characteristics of the nanosized metals [46]. Their main unique characteristics are high surface to volume ratio, uniform size distribution of the particles, and good stability [37,43,49].

Their use in wood has the objective to improve its physical and mechanical properties and its durability against microorganisms, since it is generally acceptable that nanosized metals interact with the bacterial elements, leading gradually to the cell death [40,42,49] or even to the disruption of the enzyme function [41,47]. A common practice is to disperse the nanosized metals in an organic polymer resin, in order to form a nanocomposite. A key factor in this process is the proper dispersion of nanoparticles in order to obtain maximum improvements in wood properties. It is important to examine how well the metals will interact with the wood substrate into which they are embedded, since big differences can lead to phase separation and final material failure. Accumulation of the nanosized materials due to the high surface energy that they possess, on the other hand, can result to the formation of microcomposites in the wood substrate that may behave as a weak zone for potential stresses, especially at higher loadings [50]. Nanosized materials such as metal nanoparticles (copper, silver, gold) and metal oxides (zinc oxide, aluminum oxide) are nowadays widely used to confer wood protection.

As far as copper is concerned, while traditional alkaline copper preservatives are solubilized in aqueous solution, nanocopper compounds are initially dispersed in water, and the suspension is used to treat wood. Fixation of nanocopper compounds primarily occurs through deposition on cell wall layers and in pits [51,52]. It was reported that in some cases, fungi may be not able to recognize copper nanoparticles. Once nanoparticles enter fungal cell wall, they form a reactive oxygen species with the fungus cell. In addition, nanoparticles may also undergo dissolution and thereby interfere with homeostatic processes within the fungal cell [48,49]. Akhtari and Nicholas found that nanocopper formulation can reduce the weight loss of wood due to termite attack from 46.8% to 0.2% [53]. Treating wood with nano copper oxide in presence of polystyrene improved dimensional stability [54].

The interest in silver nanoparticles for different types of applications has grown worldwide. This type of nanoparticle has also gained popularity in improving wood properties and has been the subject of numerous studies. The possibility of improving the hygroscopicity of wood by applying a nanosilver compound was studied by Mantanis and Papadopoulos [55]. The application of the compound reduced the total sorption of water vapor to a great extent. Similar results were also reported elsewhere [56,57]. Silver nanoparticles applied to improve durability in a series of tropical series [57]. In all cases, the treated woods were classified as highly resistant for white decay fungi, as opposed to untreated wood, whose weight losses were more than 20%.

Zinc-based nanocompounds have also applied to wood to improve its resistance against mold and decay fungi and the subterranean termites [58]. It was found that the mass loss was significantly inhibited by the zinc-based preparations. In another study, lime wood treated with zinc-based nanocompounds had about 56% higher resistance against the white-rot fungus *Trametes versicolor* and 40% higher resistance against the brown-rot fungus *C. puteana* [59,60].

Both Mantanis et al. [58] and Kartal et al. [61], in their comprehensive studies, applied nanozinc and nanocopper to examine termite and mold resistance. The results revealed that nanozinc is more effective in terms of termite mortality, inhibition of termite feeding, and decay by the white-rot fungus (Figs. 9.4 and 9.5). Nanozinc seems to be the most appropriate metal for wood protection [62,63]. Bak and Nemeth [64] studied the efficacy of five

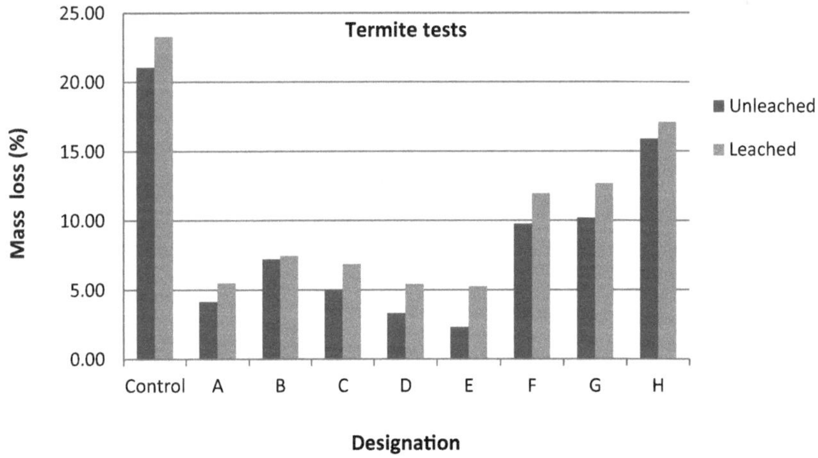

FIGURE 9.4 Mass loss (%) following termite resistance for leached and unleached pine wood: (A) nanozinc oxide; (B) nanozinc oxide plus binder A; (C) nanozinc oxide plus binder B; (D) nanozinc borate; (E) nanozinc borate plus binder A; (F) nanocopper oxide; (G) nanocopper oxide plus binder A; (H) nanocopper oxide plus binder B [58]. *Copyright (2014) Elsevier.*

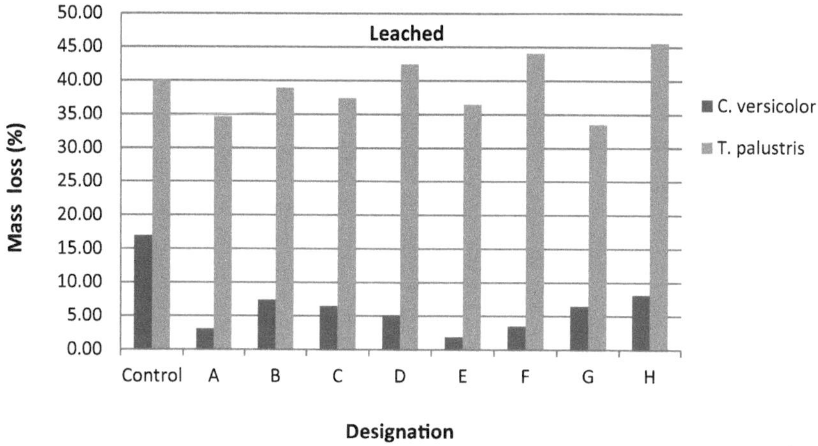

FIGURE 9.5 Mass loss (%) following decay test for leached pine wood: (A) nanozinc oxide; (B) nanozinc oxide plus binder A; (C) nanozinc oxide plus binder B; (D) nanozinc borate; (E) nanozinc borate plus binder A; (F) nanocopper oxide; (G) nanocopper oxide plus binder A; (H) nanocopper oxide plus binder B [58]. *Copyright (2014) Elsevier.*

different nanoparticles, namely (zinc-oxide, zinc-borate, copper borate, silver, copper) against fungi. They found that the most effective nanoparticle treatments were those containing borate. However, only the zinc-oxide, silver, and copper nanoparticles showed resistance to leaching, and therefore they concluded that only the zinc-oxide provided effective protection after leaching.

Titanium dioxide (TiO_2 or titania), another nanomaterial, is a photocatalyst which is activated by ultraviolet light [65]. Furthermore, titania nanoparticles are used for surface treatment which allows a thin water film to be formed on treated substrates [41,42,66], this reduces the water availability and prevents the biological growth.

4.2 Polymeric nanocarriers

Polymeric nanocarriers are also present in the wood preservation industry in improving the impregnation of wood with preservatives; however, they received by far less attention by the research studies. Encapsulation of the active ingredient into polymeric nanocarriers can be conducted through several techniques such as nanoprecipitation [67]. Fig. 9.6 depicts various types of polymeric nanocarriers that may be used for active ingredient delivery [44,67]: (a, b) Polymeric nanoparticles where active ingredients are conjugated to or are encapsulated in polymers. (c) Polymersomes composed of hydrophilic–hydrophobic block copolymers, arranged in a lipophilic bilayer vesicular system and with a hydrophilic inner core.

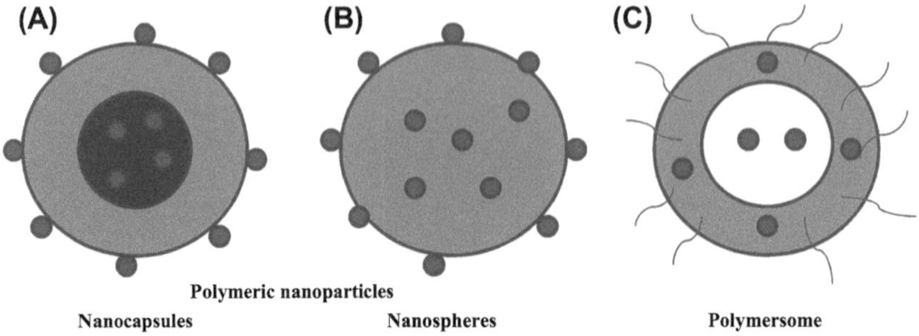

FIGURE 9.6 Types of polymeric nanocarriers. Polymeric nanoparticles: (A) nanocapsules and (B) nanospheres; (C) polymersome.

The incorporation of organic biocides into polymeric nanoparticles and the subsequent introduction of the nanoparticles into wood may be advantageous. Conventional biocides, even those with low solubility, can be easily dispersed in a solid polymeric nanoparticle that can be in turn suspended in water and then can be applied into wood with conventional water-borne treatments. With this way, biocides with particular low solubility that so far had limited portion in the market may be now found a place in wood preservation industry. In this case, the polymer wood substrate will act as a storage reservoir, which will control the release rate of the biocide and at the same time it will protect the unreleased biocide from exposure to the environment. This may cause protection of the material for a long term [68].

Fungicides, such as tebuconazole and chlorothalonil, were incorporated into polymeric nanoparticles with a great success [69]. Polyvinylpyridine and polyvinylpyridine-co-styrene were employed as the polymer matrix. Wood treated with biocide-containing nanoparticles showed great resistance against the white-rot fungi *Gloeophyllum trabeum*.

Nanotubules are other potential nanocarriers that can be a promising material for use as a carrier for biocides due to their hollow structure and high contact surface area. Carbon nanotubules are one of the best-known nanotubules, which offer several advantages that include high carrying capacity, high biocompatibility, and high surface area to volume ratio.

4.3 Coating treatments

Nanotechnology offers further improvements or even new performance properties that can be achieved in wood coatings. Fillers, such as silicate-based minerals, are commonly used in coatings in order to increase hardness and stiffness of a polymer matrix or to minimize cost. Their application in wood, when they are micron size, has two main disadvantages: reduces flexibility of the material and decreases the transparency of the coating system [50].

Polymeric coatings filled with nanoparticles such as spheres or tubes improve wood properties. Nanosized materials such as metal nanoparticles are widely used to achieve these improvements. Nanomaterials are usually added to the coating formulations by two different ways, either through solution blending (physical approach) or in situ (chemical approach) [44,50]. Solution blending means addition to the polymer in an appropriate solvent and then application of forces for dispersion. The nano-based coating is then applied on wood surface by spraying, brushing, or dipping [55,70]. In situ addition involves the addition of the nanomaterials directly to the monomers, dispersing them, and polymerizing the mixture. The applied nanomaterials are in situ synthesized on wood surface by chemical reactions like sol–gel deposition and hydrothermal methods. The chemical approach seems to be beneficial since small molecules can much more easily diffuse between nanoparticles and therefore ease their separation.

A popular coating material is the development of hydrophobic surfaces on wood; in this case the nanomaterials act as water repellents and dimensional stabilizers. For, example, beech wood treated with zinc-oxide nanoparticles provided substantial water resistance and dimensional stability [71].

Another promising nanoparticle is nanosilica. An advantage of this nanoparticle is that it can be easily modified, to further improve compatibility with the polymer. Nanosilica is found to reduce water absorption and to improve UV absorption characteristics [72]. This behavior is more pronounced when nanoclays are used [73]. Nanoclays are silicates which have platelet structure and high aspect ratio, a combination that produces a path for molecules that has to track a longer route through the coating. As a result, the water absorption of the material is reduced and the UV absorption characteristics are improved. Nanoclay is also used in intumescent coatings. These are protective coatings that are used to prolong the time that wood can remain load-bearing during exposure to fire [50].

Nanocellulose is another nanofiller which may has the shape of fibers or rodlike crystals. The elongated structure of nanocellulose offers a good potential for stress transfer and can provide high reinforcement of binders. Semicrystalline nanofibrillated cellulose is produced through mechanical shearing, while nanocellulose crystals are made by acid hydrolysis which removes the amorphous regions [74].

Furthermore, nano-based coatings with strong bacterial properties, such as silver, zinc-oxide, copper, and titanium dioxide nanoparticles, also can provide decay resistance to wood and wood products and therefore reduce the microbial activity. When these types of materials are considered for use in coatings, the slow and controlled release of the active ingredient is of great importance due to its prolonged positive effect and its lesser environmental impact [50].

5. Conclusions

This chapter briefly reviewed the research milestones in the area of wood chemical modification, focusing on acetylated and furfurylated wood which have been scaled up and exploited the solutions that nanotechnology can offer to wood protection, as an alternative green innovative approach. It is well established that the wood cell wall shows porosity of molecular scale dimensions due to the partial filling of spaces between the cellulose microfibrils mainly by polyoses and lignin. The small-sized nanoparticles can effectively and deeply penetrate into the wood, alter its surface chemistry, improve its properties resulting in a hyper-performance product. In addition, complete penetration and uniform distribution can be achieved if the size of the nanoparticles will be smaller than the diameter of the wood pits. Currently, nanomaterials are applied to wood, and it is performed in three different and distinctive ways, namely, the direct impregnation of nanosized materials in to wood (nanosized metals), with controlled release of nanomaterials embedded in a nanocarrier (polymeric nanocarriers), and coating treatments. The change in material properties is primarily due to the large interfacial area that is developed per unit of volume, since the level of added particles is reduced to nanometers. Nanomaterials enhance the properties of the original material, shows great compatibility with the traditional materials and cause limited alteration of their original features. Their use in wood has the objective to improve its physical properties, mainly its hygroscopicity and its resistance against mold, decay fungi, marine borers, and subterranean termites. Furthermore, increased UV protection may also be achieved.

References

[1] R.M. Rowell, Handbook of Wood Chemistry and Wood Composites, second ed., CRC Press, Taylor and Francis Group, Boca Raton, Florida, USA, 2012.

[2] P. Gerardin, New alternatives for wood preservation based on thermal and chemical modification of wood-A review, Ann. For. Sci. 73 (2016) 559–570.

[3] C.A.S. Hill, Wood Modification – Chemical, Thermal and Other Processes, John Wiley and Sons Ltd, West Sussex, UK, 2006.

[4] R.M. Rowell, Acetylation of wood-A review, Int. J. Lignocellul. Prod. 1 (1) (2014) 1–27.

[5] T.H. Wegner, J.E. Winandy, M.A. Ritter, Nanotechnology opportunities in residential and non-residential construction, in: 2nd International Symposium on Nanotechnology in Construction. Bilbao, Spain, 2005, pp. 23–31.

[6] C.A.S. Hill, A.N. Papadopoulos, A review of methods used to determine the size of the cell wall microvoids of wood, J. Inst. Wood Sci. 15 (2001) 337–345.

[7] R.M. Rowell, Chemical modification of wood, For. Prod. 6 (12) (1983) 366–382. Abstracts.

[8] S. Kumar, Chemical modification of wood, Wood Fiber Sci. 26 (1996) 270–280.

[9] D.N.S. Hon, Chemical Modification of Lignocellulosics, Marcel Dekker, New York, 2006.

[10] A.N. Papadopoulos, Chemical modification of solid wood and wood raw materials for composites production with linear chain carboxylic acid anhydrides: a brief Review, Bioresources 5 (1) (2010) 499–506.

[11] M.L. Bender, Mechanisms of catalysis of nucleophilic reactions of carboxylic acid derivatives, Chem. Rev. 60 (1960) 53–113.

[12] V. Gold, E.G. Jefferson, The hydrolysis of acetic anhydride. Part III. The catalytic efficiency of a series of tertiary amines, J. Chem. Soc. (1953) 1409–1415.

[13] W. Fuschs, Zur kenntnis des genuimen lignis, I: de acetylierung des Finchtenholzes, Ber. Dtsch. Chem. Ges. 61 (1928) 948–951.

[14] O. Horn, Beriche der deutsehem, Chemischem Gasellschaft 61B (1928) 2542.

[15] A.J. Stamm, H. Tarkow, Dimensional stabilisation of wood, J. Colloid Chem. 51 (1947) 493–505.

[16] R.M. Rowell, The Chemistry of Solid Wood, ACS, Washington,D.C, 1984.

[17] E.P.J. Beckers, H. Militz, Acetylation of solid wood. Initial trials on lab and semi industrial scale, in: Proceedings of the Second Pacific Rim Bio-Based Composites Symposium. Vancouver, Canada, 1994, pp. 125–135.

[18] R. Yasuda, K. Minato, M. Norimoto, Moisture adsorption thermodynamics of chemically modified wood, Holzforshung 49 (6) (1995) 548–554.

[19] A.J. Stamm, Wood and Cellulose Science, Ronald Press Co, New York, 1964.

[20] A.N. Papadopoulos, Swelling, Cell Wall Porosity and Chemical Modification of Wood, Ph.D. Thesis, University of Wales, Bangor, 2001.

[21] A.N. Papadopoulos, C.A.S. Hill, The sorption of water vapour by anhydride modified softwood, Wood Sci. Technol. 37 (2003) 221–231.

[22] I.S. Goldstein, E.,B. Jeroski, A.E. Lund, J.F. Nielson, J.W. Weaver, Acetylation of wood in lumber thickness, For. Prod. J. 11 (1961) 363–370.

[23] E.P.J. Beckers, H. Militz, M. Stevens, Acetylated solid wood. Laboratory durability test (part II) and field tests, in: International Research Group on Wood Preservation, 1995. Document no. IRG/WP/95-40048.

[24] P. Larsson-Brelid, R. Simonson, O. Bergman, T. Nilsson T, Resistance of acetylated wood to biological degradation, Holz als Roh- Werkst. 58 (2000) 331–337.

[25] E.D. Suttie, C.A.S. Hill, D. Jones, R.J. Orsler, Chemically modified solid wood. I. Resistance to fungal attack, Maerialst und Organisms 32 (1999) 159–182.

[26] A.N. Papadopoulos, C.A.S. Hill, The biological effectiveness of wood modified with linear chain carboxylic acid anhydrides against *Coniophora puteana*, Holz als Roh-und Werkstoff 60 (2002) 329–332.

[27] A.N. Papadopoulos, H. Militz, A. Pfeffer, The biological behaviours of pine wood modified with linear chain carboxylic acid anhydrides against soft rot decay, Int. Biodegrad. Biodeterior. 64 (2010) 409–412.

[28] S.N. Kartal, T. Yoshimura, Y. Imamura, Decay and termite resistance of boron treated and chemically modified wood by in situ co-polymerization of allyl glycidyl ether (AGE) with methyl methacrylate (MMA), Int. Biodegrad. Biodeterior. 53 (2004) 111–117.

[29] A.N. Papadopoulos, P. Duquesnoy, S.M. Cragg, A.J. Pitman, The resistance of wood modified with linear chain carboxylic acid anhydrides to attack by the marine wood borer *Limnoria quadripunctata* Hothius, Int. Biodeterior. Biodegrad. 61 (2008) 199–202.

[30] L.M.S. Borges, S.M. Cragg, J.R. Williams, Comparing the resistance of a number of lesser known species of tropical hardwoods to the marine borer Limnoria using a short-term laboratory assay, in: The International Research Group on Wood Preservation, 2003. Document no. IRG/WP 03-10500.

[31] A.N. Papadopoulos, D. Avtzis, N. Avtzis, The biological effectiveness of wood modified with linear chain carboxylic acid anhydrides against the subterranean termites *Reticulitermes flavipes*, Holz als Roh-und Werkstoff 66 (2003) 249–252.

[32] A.J. Stamm, Dimensional stabilization of wood with furfuryl alcohol, Wood Technol. 43 (1977) 141–149.

[33] M. Westin, Development and evaluation of new alternative wood preservation treatments, in: Final Report to the Swedish Council for Forestry and Agricultural Research, 1996, pp. 1–25.

[34] G.I. Mantanis, Chemical modification of wood by acetylation or furfurylation: a review of the present scaled-up technologies, BioResources 12 (2) (2017) 4478–4489.

[35] H.P. Boehm, R. Setton, E. Stumpp, Nomenclature and terminology of graphite intercalation compounds, Carbon 24 (2) (1986) 241–245.

[36] P.J. Borm, et al., The potential risks of nanomaterials: a review carried out for ECETOC, Part. Fibre Toxicol. 3 (1) (2006). Article ID 11.

[37] M. Roco, Nanotechnology's future, Sci. Am. (2006). August, https://www.scientificamerican.com/article/nanotechnologys-future/.

[38] G. De Filpo, F.P. Nicoletta, G. Chidichimo, Flexible nano-photoelectrochromic film, Chem. Mater. 18 (2006) 4662–4666.

[39] G. De Filpo, S. Mormile, F.P. Nicoletta, G. Chidichimo, Fast, self-supplied, all solid photoelectrochromic film, J. Power Sources 195 (2010) 4365–4369.

[40] H.R. Taghiyari, O. Schimdt, Nanotachnology in wood based composite panels, Int. J. Bioinorg. Hybrid Nanomater. 3 (2014) 65–73.

[41] G.B. Goffredo, S. Accoroni, T. Totti, T. Romagnoli, L. Valentini, P. Munafò, Titanium dioxide based nanotreatments to inhibit microalgal fouling on building stone surfaces, Build. Environ. 112 (2017) 209–222.

[42] G. De Filpo, A.M. Palermo, F. Rachiele, F.P. Nicoletta, Preventing fungal growth in wood by titanium dioxide nanoparticles, Int. Biodeterior. Biodegrad. 85 (2014) 217–222.

[43] T.H. Wegner, P. Jones, Advancing cellulose-based nanotechnology, Cellulose 13 (2014) 115–118.

[44] T. Teng, M. Arip, K. Sudesh, H. Lee, Conventional technology and nanotechnology in wood preservation: a review, BioResources (2018) in press.

[45] M.A. Polleti Papi, F.R. Caetano, M.F. Berganini, L.H. Marcolino-Junior, Facile synthesis of a silver nanoparicles/polypyrrole nanocomposite for non-enzymatic glucose determination, Mater. Sci. Eng. C75 (2017) 88–94.

[46] A. Gupta, R. Srivastana, Zinc oxide nanoleaves: a scalable disperser assisted sonochemical approach for synthesis and an antibacterial application, Ultrason. Sonochem. 41 (2018) 47–58.

[47] R. Moya, A. Zuniga, A. Berrocal, J. Vega, Effect of silver nanoparticles synthesized with NPsAg-ethylene glycol on brown decay and white decay fungi of nine tropical woods, J. Nanosci. Nanotechnol. 17 (2017) 1–8.

[48] C. Civardi, J. Kaiser, C. Hirsch, C. Mucchino, A. Wichser, P. Wick, F. Schwarze, Release of copper amended particles from micronized copper pressure treated wood during mechanical abrasion, J. Nanotechnol. 14 (2016) 1–10.

[49] C. Civardi, F. Schwarze, P. Wick, Micronized copper wood protection: an efficiency and potential health and risk assessment for copper based nanoparticles, Environ. Pollut. 200 (2015) 20–32.

[50] M. Nikolic, J.M. Lawther, A.R. Sanadi, Use of nanofilers in wood coating: a scientific review, J. Coat. Technol. Res. 12 (3) (2015) 445–461.

[51] H. Matsunaga, M. Kiguchi, P. Evans, Microdistribution of copper carbonate and iron oxide nanoparticles in treated wood, J. Nanoparticle Res. 11 (2009) 1087–1098.

[52] W. Xue, P. Kennepohl, W. Ruddick, Chemistry of copper preservative treated wood, in: 45th International Group on Wood Protection, IRG/WP, USA, 2014, pp. 1–8, 14-30651.

[53] M. Akhtari, D. Nicholas, Evaluation of particulate zinc and copper as wood preservatives for termite control, Eur. J. Wood Wood Prod. 71 (2013) 395–396.

[54] A. Jafari, A. Omidvar, D. Rasouli, The effect of nano copper oxide on physical properties and leaching resistance of wood polystyrene polymer, J. Wood Forest Sci. Technol. 25 (1) (2018) 9–25.

[55] G. Mantanis, A.N. Papadopoulos, The sorption of water vapour of wood treated with a nanotechnology compound, Wood Sci. Technol. 44 (2010) 515–522.

[56] H.T. Sahin, G.I. Mantanis, Nano-based surface treatment effects on swelling, water sorption and hardness of wood, Maderas Ciencia y technologia 13 (1) (2011) 41–48.

[57] R. Moya, A. Berrocal, A. Zuniga, J. Baudrit, S.C. Noguera, Effect of silver nanoparticles on white rot wood decay and some physical properties of three tropical wood species, Wood Fiber Sci. 16 (4) (2014) 1–12.

[58] G.I. Mantanis, E. Terzi, S.N. Kartal, A.N. Papadopoulos, Mold, decay and termite resistance of pine wood treated with zinc and copper based nanocompounds, Int. Biodeterior. Biodegrad. 90 (2014) 140–144.

[59] L. Reinprecht, Z. Vidholdova, M. Kozienka, Decay inhibition of lime wood with zinc oxide nanoparticles used in combination with acrylic resin, Acta Facultatis Xylologiae Zvolen 57 (1) (2015) 43–52.

[60] L. Reinprecht, Z. Vidholdova, Biological resistance and application properties of particleboards containing nano-zinc oxide, Adv. Mater. Sci. Eng. 2018. Article ID 2680121.

[61] S.N. Kartal, F. Green, C.A. Clausen, Do the unique properties of nanometals affect leachability or efficacy against fungi and termites? Int. Biodeterior. Biodegrad. 63 (2009) 490–495.

[62] H.H. Huang, C. Lin, K. Hsu, Comparison of resistance improvement to fungal growth on green and conventional building materials by nano-metal impregnation, Build. Environ. 93 (2015) 119–127.

[63] E. Terzi, S. Kartal, N. Yilgor, L. Rauktari, T. Yoshimura, Role of various nano-particles in preservation of fungal decay, mold growth and termite attack in wood and their effect on weathering properties and water repellency, Int. Biodeterior. Biodegrad. 107 (2016) 77–87.

[64] M. Bak, R. Nemeth, Effect of different nanoparticle treatments on the decay resistance of wood, BioResources 13 (4) (2018) 7886–7899.

[65] Z. Zhang, J. MacMullen, H.N. Dhakal, J. Radulovic, C. Herodotou, M. Totomis, N. Bennett, Biofouling resistance of titanium dioxide and zinc oxide nanoparticulate silane/siloxane exterior facade treatments, Build. Environ. 59 (2013) 47–55.

[66] G.B. Goffredo, B. Citterio, F. Biavasco, F. Stazi, S. Barkeli, P. Munafo, Nanotechnology on wood: the effect of photocatalytic nanocoatings againstAspergillus nige, J. Cult. Herit. 27 (2017) 125–136.

[67] S. Khoee, A. Haslemi, Synthesis and characterization of IUdR loaded PEG/PCL/PEG polymersome in mixed DCM/DMF solvent: experimental and molecular dynamics insights into the role of solvent composition and star architecture in drug dispersion and diffusion, Eur. J. Pharm. Sci. 114 (2018) 1–12.

[68] Y. Liu, L. Yan, P. Heiden, P. Laks, Use of nanaparticles for controlled release of biocides in solid wood, J. Appl. Polym. Sci. 79 (2001) 458–465.

[69] Y. Liu, P. Laks, P. Heiden, Controlled release of biocides in solid wood. Efficacy against brown rot wood decay fungus, J. Appl. Polym. Sci. 86 (2002) 596–607.

[70] M. Havrlik, P. Pyranova, Protection of wooden materials against biological attack by using nanotechnology, Acta Polytechnika 55 (2015) 101–108.
[71] M. Soltani, A. Najafi, E, E. Bakar, Water repellent effect and dimensional stability of beech wood impregnated with nano-zinc oxide, BioResources 8 (2013) 6280–6287.
[72] F. Bauer, R. Mehnert, UV curable acrylate nanoparticles: properties and applications, J. Polym. Res. 12 (2005) 483–491.
[73] M. Ghaemy, S. Bekhradnia, Thermal and photocuring of an acrylat-based coating resin reinforced with nanosilica, J. Coat. Technol. Res. 9 (2012) 569–578.
[74] X. Xu, F. Liu, L. Jiang, J.Y. Zhu, D. Haagenson, D.P. Wiesenborn, Cellulose nanocrystals vs. Cellulose nanofibrils: a comparative study on their microstructures and effects as polymer reinforcing agents, ACS Appl. Mater. Interfaces 5 (8) (2013) 2999–3009.

Chapter 10

Green techniques for wastewaters

Gassan Hodaifa,[a,]* Ombretta Paladino,[b] Ana Malvis,[a]
Mahdi Seyedsalehi[c] and Matteo Neviani[d]

[a]*Molecular Biology and Biochemical Engineering Department, Chemical Engineering Area, Faculty of Experimental Sciences, University of Pablo de Olavide, Seville, Spain;* [b]*DICCA, Department of Civil, Chemical and Environmental Engineering, University of Genoa, Genoa, Italy;* [c]*School of Environment, Tsinghua University, Beijing, China;* [d]*DIME, Department of Mechanical, Energy, Management and Transportation Engineering, University of Genoa, Genoa, Italy*

**Corresponding author*

1. Introduction

Water, a highly demanded natural resource, constitutes nowadays an essential element in modern anthropogenic developments, especially in the industrial sector. The total amount of water on Earth is estimated to be 14×10^8 km^3; however, only 45×10^3 km^3 is considered as freshwater, representing almost 0.003% of the total volume [1,2]. From this small portion of freshwater sources, between 9×10^3 and 14×10^3 km^3 are economically accessible for consumption [1]. According to the consumption pattern, irrigation water in the agricultural sector represents the predominant water application (over 70%) in developing countries, followed by its use in the industrial sector (Fig. 10.1).

Water is widely distributed through all different types of ecosystems; nevertheless, water sources are not uniformly dispersed, which leads to water stress and scarcity in many countries. According to Ref. [3], 4000 million global civilians undergo water scarcity nowadays, and it is assumed that around 9000 million people will suffer from water shortage in 2075. Furthermore, Africa, Latin America, and Asia represent three continents with huge risk for accessing hygienic and safe waters. On the one hand, water demand is continuously increasing every day due to the rapid growth of population. On the other hand, the industrial expansion, commercial establishments, agricultural developments, and urbanization are resulting in the production of massive amounts of wastewater. Therefore, the development of

Advanced Low-Cost Separation Techniques in Interface Science. https://doi.org/10.1016/B978-0-12-814178-6.00010-8

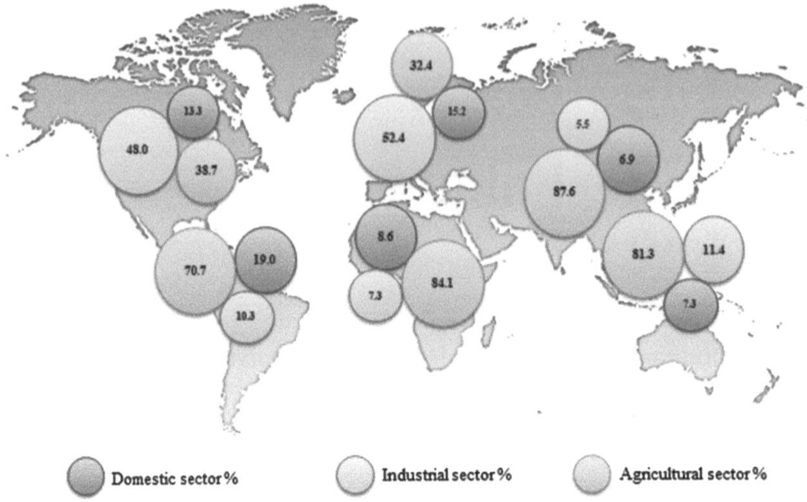

FIGURE 10.1 Percentages of water consumption distribution in different continents [1].

novel wastewater purification systems is highly relevant since wastewater generation is difficult to avoid in the contemporary world. Likewise, the preservation of high-quality freshwater is vital to ensure the human survival complying with health and hygiene standards. Wastewater recovery, reuse, and recycling are considered as vital functions of sustainable water consumption and constitute an effective solution to water growing demand. Wastewater is typically originated from both natural and human activities. However, human contamination is prioritized in the development of modern cleaning processes since its production is enormous, fast, and continuous. The direct exposition of wastewater to natural sources creates severe negative consequences to all forms of life. For instance, polluted waters are responsible for a large range of negative effects as spreading diseases and animal–human–plant death. In addition, natural scale replenishment of human wastewater is a difficult task, thus it is not a viable solution to face the increasing water demand in the current society. In order to find a proper solution, wastewater treatment technologies are currently focused on green processes that able to treat large wastewater quantities, maintaining the balance between human well-being and environmental resilience.

The unavailability of data providing information about the total volume of wastewater generated and treated by each country hampers the estimation of total wastewater volumes which are generated and treated at global level [4]. The annual generation of wastewater in North America and Middle East–North Africa regions is 85 and 22.3 km^3 whereas volumes of 61 and 11.4 km^3 are treated, respectively. According to the available data, Latin America,

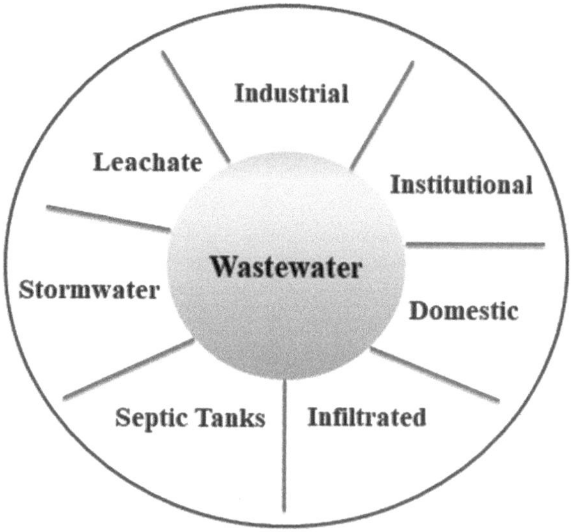

FIGURE 10.2 Classification and categories of wastewater [6].

Europe, Russian Federation, and Asia yearly treat 20%, 71%, 28%, and 32% of wastewater, respectively [5].

Wastewater origin and its proper characterization are important factors to consider selecting an appropriate treatment technique and improving its quality. Fig. 10.2 shows wastewater main origins. Wastewater composition is highly heterogeneous and variable depending on its origin and can be characterized according to three criteria: physical, chemical, and biological properties. Physical properties consist of parameters such as odor, color, turbidity, and temperature. Chemical properties are broader and can be subdivided into inorganic, organic, and gaseous constituents (Table 10.1). Inorganic constituents include parameters such as pH, alkalinity, heavy metals, chlorides, nitrogen, phosphorus, and sulfur. Organic constituents comprise parameters such as phenols, pesticides, carbohydrates, proteins, fats, oils, grease, suspended solids, filtered solids, volatile organic compounds, surfactants, and other related contaminants. Lastly, gaseous constituents include hydrogen sulfide, methane, oxygen, carbon dioxide, nitrogen, and ammonia. On the other hand, wastewater biological properties depend on the availability of macro- and microbiota such as plants, animals, eubacteria, archaebacteria, and viruses in the wastewater stream. In this sense, it is highly relevant to make an accurate determination of existing pathogens in the wastewater as some biota such as viruses and bacteria constitute a source of multiple diseases.

TABLE 10.1 Characterization of different wastewaters [6,7].

Urban/Industrial wastewater	Parameters
Urban/domestic wastewater (with minor industrial wastewater)	BOD (230–560 mg/L), COD-total amount (500–1200 mg/L), COD-suspended portion (300–720 mg/L), COD-soluble portion (200–480 mg/L), nitrogen total amount (30–100 mg/L), Ammonia-N (20–75 mg/L), phosphorus-total amount (6–25 mg/L), ortho-P (4–15 mg/L), volatile suspended solids (200–480 mg/L), total suspended solids (250–600 mg/L), volatile fatty acids (10–80 mg/L)
Textiles	BOD (6000 mg/L), organics, total solids, metals
Brine (naturally black olive oil)	BOD (3000–3830 mg/L), COD (11000–60000 mg/L), dissolved organic solids (95300–118800 mg/L), total suspended solids (100–150 mg/L), total phenolic compounds (3200–5200 mg/L), NaCl (56000–77000 mg/L), Cl^- (33300–45500 mg/L)
Dairy	BOD (3000 mg/L), organics, fats
Iron and steel	BOD (500 mg/L), COD, metals, acids, oil, phenols, cyanide, suspended solids, ammonia
Pulp and paper	BOD (4000 mg/L), COD, solids, chlorinated organic compounds
Petrochemicals and refineries	BOD (750 mg/L), COD, phenols, mineral oils, chromium
Plastics and resins	BOD (2500 mg/L), organics, oils, phenol
Vegetables and fruits	BOD (2000 mg/L), sugars, dissolved organics
Meat and poultry	BOD (2–5000 mg/L), organics, fats, manure, blood

Among different types of wastewaters, domestic and industrial are especially difficult to efficiently treat due to the large volumes which are generated and the high diversity of pollutants which are contained in these wastewaters. In this sense, the composition of wastewater must be carefully determined when designing a suitable sewer system [6]. In addition, before discharging treated industrial wastewater to natural water bodies, industries should meet country-specific standards to ensure whether treated water is in an acceptable level for human consumption. Achieving efficiently treated wastewater, alongside with the possibility to deliver an adequate supply of uncontaminated water, has become a foremost concern for many societies in today's world. In this chapter, green technologies for wastewater treatment are detailed.

2. Conventional technologies for wastewater treatment

Conventional techniques are nowadays the most widely used processes for wastewater treatment. The configuration of a conventional wastewater treatment plant (WWTP) involves four treatment stages: preliminary, primary, secondary, and tertiary. Along WWTP, each treatment is constituted by two lines: water and sludge line. In some cases, additional operations (such as ion exchange, disinfection, etc.) are implemented at the outlet of the WWTP in order to meet the specific requirements of the discharge zone or with the unique aim to improve the final quality of the treated water. In some specific cases, preliminary treatment is not required and the treatment process can be directly started with the primary treatment. Conventional treatment uses the combination of physicochemical and biological operations to improve the water quality.

2.1 Preliminary treatment

In this treatment stage, all large particles contained in the wastewater are separated, e.g., stones, carton, hygienic towel, pieces of wood, etc. This treatment mainly consists of four steps known as grit removal, coarse screening, comminution, and chemical treatment. In grit removal chamber, high-velocity water or air is applied to avoid organic solids sedimentation. Removing dense inorganic particles before entering wastewater to the primary clarifier is the main aim of grit removal. In many cases, this step is not used in small-scale WWTPs. Coarse screening mostly eliminated shreds, rods, and other big wreckages, can be problematic for downstream treatment. The objective of comminution is the decrease of particle magnitude by cutting and grinding before flowing to downstream treatment processes, used as complementary operation of the preliminary screening. Reduced size particles are discharged as sludge with consequent treatment steps. Chemical treatments are eventually used to improve the performance of downstream systems. It is essential for the use of flow measurement instruments throughout the preliminary treatment operation [8].

2.2 Primary treatment

The main target of primary treatment is the elimination of floating materials and the reduction of settleable coarse and bulky inorganic and organic solids (i.e., suspended solids) via sedimentation (normally natural sedimentation). In addition, primary treatment is intended to prevent the blockage of equipment such as sewage pipelines, pumping equipment in downstream processes, etc., and safeguard them. The accumulated garbage is released to landfill sites. Primary treatment results in the decrease of total suspended solids (TSS), settleable solids, biological oxygen demand (BOD_5), floating grease and oils presented in wastewater stream. Besides those mentioned, associated heavy metals, organic phosphorus, and some organic nitrogen are also removed with sludge; however, dissolved and colloidal components can remain in wastewater. In the primary treatment, TSS, sedimentable solids, BOD_5, grease and oils are reduced by 50%–75%, 98%–99%, 20%–50%, and 65%–99%, respectively [8,9].

The overall of primary treatment can be summarized as follows. After the preliminary treatment, wastewater is discharged to manually or mechanically operating screen chambers to remove floating and massive nondecomposable solids and garbage such as fiber, leaves, fecal matter, cloth, paper, etc. Second, wastewater is passed through fine physical barriers and membranes to fulfill filtration step. Slightly cleaned water is discharged to primary clarifiers or sedimentation tanks to settle dense solids at the bottom and float lighter particles at the surface. Floating materials like grease, oil are difficult to digest, they should be directed to skimming tanks for its removal, and other light materials such as plastics should be excluded at the tank surface. The primary sludge generated is then centrifuged to remove noncolloidal solids. It is highly relevant the development of natural settling of solids in the sedimentation tank along with pH adjustment. Chemical coagulants may be also added to improve the sedimentation process. After the sludge separation, remaining liquid is directed to secondary treatment, and the sludge obtained is dumped to a landfill or conducted to anaerobic digestion [9,10].

2.3 Secondary treatment

The moderately treated wastewater from primary treatment is then subjected to secondary treatment for further reducing the levels of BOD_5, TSS, dissolved solids, and residual organics. In this treatment, microbial biological processes are commonly applied, such as aerobic biological reactor to remove colloids and dissolved organics. Aerobic microorganisms break down complex organic substances into simple inorganic products such as carbon dioxide, water, and ammonia. Aerobic process can decrease decomposable organics concentration approximately to 90% in the wastewater stream. The treated wastewater from the aerobic reactor is conducted to settling tank to separate sludge (majority bacteria). On the other hand, the anaerobic process (negligible or absence of oxygen) is extensively used in the sludge line and can transform organics to combustible gas up to 95%. Anaerobic processes use facultative and anaerobic bacteria which constitute an energetically effective operation with the double advantage of biogas production and efficient heavy metals removal. As a result of anaerobic digestion, carbon dioxide and methane are produced. In a well-organized treatment plant, BOD_5 and TSS can be removed by 85% through secondary treatment [8–10].

Activated sludge process is the most common practice in secondary treatment. In addition, oxidation ditches, rotating biological contactors, and biofilters/trickling filters also highly apply in treatment processes. However, in function of the organics concentration presented in wastewaters, two technologies can be combined to yield an effective output. In addition, in most cases, air is pumped to stimulate aerobic degradation in the system. Thus, oxygen-dependent microorganisms are able to convert organic contaminants into simple end products.

Dense materials from the treated water settle at the bottom of the tank, and the oxygenated wastewater is released to final settling tanks after pH correction. Deposited secondary sludge with dense particles is separated and conducted to dumped in a landfill [9,10].

2.4 Tertiary treatment

After secondary treatment, for wastewater reuse and disposal in sensible places, it is necessary an additional stage known as tertiary or advanced treatment. Tertiary treatment is intended to completely remove the organic ions, nutrients, suspended solids, microorganisms, and remaining pollutants to finally obtain a wastewater which comply with the disposal regulations [11]. This stage can be performed by biological, physical, or chemical methods, alone or in combination [12]. The selected technology depends on several factors. First, it must be considered the wastewater composition and the amount to be treated, which is mainly determined by its origin: urban, industrial, agriculture, etc. Furthermore, reclaimed waters must comply with the regulations according to its use after the treatment: irrigation, groundwater recharge, drinking water supply, etc. The available technology and landscape integration, nutrient management, and the by-products generation are some conditions to consider in the selection of the treatment technology [11].

Several technologies have been developed as wastewater tertiary treatment intended to eliminate contaminants, decrease the amount of wastewaters, and improve the recovered water quality.

2.4.1 Physicochemical treatment

2.4.1.1 Coagulation–flocculation

Solid–liquid separation is the main target of this treatment method. First, coagulation consists on the process of increasing particles ionic strength to overcome repulsive energy between particles. Inorganic materials of metals such as aluminum, ferric chloride, ferrous sulfite, calcium chloride, etc., are the most extensively used substances that facilitate the neutralization of the surface charge of suspended solids and colloidal systems present in wastewaters, facilitating aggregation, and settlement [13]. Nevertheless, large amounts of these substances are required for flocculation as well as pH alteration in order to achieve the formation of complex precipitates of metal hydroxides [14].

On the other hand, flocculation is usually used as a secondary step to aggregate colloidal solids, which have not been settled during coagulation, through the addition of chemical reagents. Flocculation allows the formation of larger agglomerates and slow-settling particles. Anionic and long-chain nonionic polymers are the most commonly used flocculants including polyacrylamide, polyacrylic acid, polyamine, cationic, and anionic polyelectrolytes. Synthetic polymers can be categorized in four groups: cationic, anionic, amphoteric, and nonionic. They differ in molecular weight, charge, composition, and molecular weight. Contrary to coagulants, flocculants are highly efficient in small quantities, have a lower cost, and work in all pH values [14].

Conventional coagulation–flocculation is currently employed by most of the industries since it can be used in all types of wastewater, including inorganic and organic-based waters [14].

2.4.1.2 Direct flocculation

This separation operation consists of the neutralization of colloidal particles charges and subsequent formation of flocs. It allows a reduction in wastewater treatment costs and time with respect to coagulation–flocculation, since coagulant addition and pH adjustment are not needed given that high-molecular-weight polymers used in direct flocculation work in all pH values. The generation of less volume of sludge is another advantage of direct flocculation explained by the formation of stronger bridging forces between flocs resulting in denser and packed sludge [14].

This technique is extensively applied in certain types of organic-based wastewaters with a rich composition in suspended and colloidal solids. Olive mill effluents, aquaculture wastewater, and pulp and paper mill wastewater, among many others, have been successfully treated through direct flocculation [14].

2.4.1.3 Membrane technology

This technology is the most extensively used in wastewater treatment since it allows the separation of undesirable compounds as well as wastewater purification. Its enormous versatility allows its application in several types of industries such as textile, food, pharmaceutic, etc. [15]. Separation by membrane can be classified according to the driving force and membrane pore size in the following types:

- Microfiltration (MF). Pore size of MF membranes ranges from 0.05 to 10 μm, and driving force (ΔP) vary in the range 0.5–3 bar, which enables the reutilization of membranes and a decrease in replacement costs through the cleaning and reuse of the membrane by backflushing technology. Several industries use MF, as cleaning step, such as pharmaceutic, desalination plants, and food processing due to the reported capacity of MF membranes to remove numerous contaminants such as bacteria among many others [16].

➢ Ultrafiltration (UF). UF membranes can be characterized in terms of pore size, ranging from 0.1 to 0.001 μm (ΔP = 1−10 bar), or in terms of the solute molecular weight cut-off. Several industries employ this method for wastewater treatment since only low pressure is required during the process. UF membranes allow the elimination of proteins, polymers, or particles. In this sense, the high exposition of membranes in water treatment processes to bacteria and pathogens microorganisms require the use of UF membrane with antibacterial properties. The incorporation of silver nanoparticles into the membranes has been proved to significantly reduce the formation of bacteria biofilm in the membrane [16].

➢ Nanofiltration (NF). NF membranes (0.005−0.0005 μm and ΔP = 5−30 bar) are extensively used in the elimination of several compounds such as salt ions (as phosphorus and calcium or pesticides) [17]. In addition, NF membranes have the capacity to retain several compounds present in aquaculture wastewaters, such as ammonium. Nevertheless, its use for potable water obtaining is one of the most implemented due to the low requirements of energy and pressure in comparison with reverse osmosis [16].

➢ Reverse osmosis (RO). The characteristics of RO membrane are pore size = <0.0005 mm and ΔP = 10−50 bar. RO consists of a desalination process used in the elimination of salt ions, organic and inorganic compounds from wasters. This technology is based on the use of permeable membranes to water but impermeable to salts. In RO, inorganic compounds are rejected and form the concentrate stream. This fact presents a limitation in the process due to the high concentration of pollutants in the generated concentrate in comparison with the initial wastewater. In this sense, domestic wastewater treatment through RO allows a recovery between 70% and 75% of treated water, which means that 25%−30% of total volume is concentrated wastewater. Nevertheless, high treated water with improved quality is obtained through RO [18]. Apart from wastewater treatment, the use of RO in seawater desalination is gaining importance due to its low cost in comparison with other methods employed in water production [19].

2.4.1.4 Extensive systems with filtering materials

Sand, activated carbon, and zeolite are the most employed filtering materials in wastewater tertiary treatment. Sand filtration is a conventional wastewater treatment process characterized by its simplicity, low energy inputs, and easy maintenance. In this system, chemical reagents are not required, resulting in lower costs in comparison with other methods. In addition, the use of sand as wastewater filtering material has shown to be effective as tertiary treatment stage achieving high turbidity removal rates. Its use in combination with activated carbon is an effective alternative to the conventional method [20].

Second, adsorption of pollutants onto activated carbon is a widely used method due to its high efficacy as purification system. In this sense, granular activated carbon (GAC) and powdered activated carbon (PAC) have shown to be effective with organic compounds, dyes, surfactants, etc. GAC is used in columns, which facilitates its regeneration and reuse; nevertheless, the higher specific surface area of PAC results in faster adsorption rates [21].

Finally, zeolite is an aluminosilicate with crystalline structure whose properties such as ion exchanger and adsorbent characteristics make it an effective material for pollutants removal from wastewaters. Zeolite pore size ranges from 0.5 to 1.2 nm, it presents low density and high hydration/dehydration degree. Apart from wastewater treatment through cation exchange, zeolite is an efficient alternative for water salinity reduction with several advantages such as low cost, easy to regenerate, and high stability [22].

2.5 Microalgae as tertiary treatment

The use of microalgae as a tertiary wastewater treatment process started in the 1970s as an alternative to the chemical and physical processes. This system is based on the capacity of many microalgae species to grow in extreme conditions and remove nutrients, pollutants, heavy metals, nitrogen compounds, chemicals, etc., converting them into biomass [12]. Species belonging to the genera of *Chlorella*, *Scenedesmus*, *Neochloris*, or *Chlamydomonas* (among many others) have been applied for the treatment of different types of wastewaters as shown in Table 10.2 [31].

TABLE 10.2 Microalgae used in the treatment of different wastewaters.

Microalgae	Types of wastewater	References
Ankistrodesmus falcatus	Aquaculture wastewater	[23]
Botryococcus terribilis	Domestic wastewater	[24]
Chlorella pyrenoidosa	Soybean processing wastewater	[25]
Chlorella sorokiniana	Potato processing wastewater	[26]
Chlorella vulgaris	Brewery wastewater	[27]
Euglena sp.	Urban wastewater	[28]
Scenedesmus obliquus	Olive oil mill wastewater	[29]
Spirulina platensis	Dairy wastewater	[30]

Several beneficial characteristics make the treatment with microalgae an attractive and sustainable option. These include oxygen production through photosynthesis, disinfecting effect, efficient bioremediation of organic and inorganic compounds (such as N, P, and heavy metals [32]) and the ability to reduce chemical and biochemical oxygen demand [12]. In addition, the small size of microalgae allows the formation of large surface areas, increasing the nutrient uptake capacity [33]. Nevertheless, the main advantage of wastewater treatment with microalgae is the lower cost and technology that is required for the implementation of this process, in addition to the generation of biomass with high added value [32].

2.5.1 Microalgae cultivation in photobioreactors

Microalgae cultivation can be performed in open systems or in controlled closed systems, also known as photobioreactors (PBRs). The culture of microalgae in open systems has been extensively developed; nevertheless, the limitation in the contamination control hinders the growth of many species. On the other hand, closed systems allow the maintenance of single specie cultures and the use of a great variety of designs and configurations [34].

2.5.1.1 Open systems

This cultivation system includes natural waters, such as lakes and lagoons, and artificial ponds, such as circular ponds, raceway ponds, and inclined systems, which differ in size, shape, materials, agitation system, and inclination [35]. Among these types, inclined systems, circular ponds, and raceway ponds are the most extensively operated at large scale. The simplicity, low cost of construction, and high production rates are the main advantages of these systems in comparison with closed PBRs. Nevertheless, the sensibility of this method to weather conditions does not allow the control of culture conditions [36]. In addition, the poor light utilization by the cells, evaporation losses, and CO_2 diffusion to the atmosphere are some of the open ponds limitations [37].

- Natural ponds. This system represents the most economic and simple alternative for large-scale cultivation of microalgae. Natural ponds consist of nonstirred ponds with a depth of less than half a meter, which require suitable climatic conditions and enough nutrients [38]. The simultaneous growth of bacteria, protozoa, and viruses limit the microalgae monospecific culture [36]. In addition, low biomass yields are obtained with this method due to the dissolution of carbon dioxide from air to water and the absence of mixing. Commercial production of *Dunaliella salina* has extensively been performed in natural ponds [39].
- Circular ponds. Circular ponds are constituted by a feeding inlet, harvesting outlet, CO_2 inlet, and a rotating agitator in the central part. These ponds are characterized by a depth of about 25–30 cm and cannot exceed 50 m in diameter for an efficient agitation [38]. This system is not

employed in commercial-scale plants due to the high costs of construction and operation [40]; nevertheless, its use in Asia for *Chlorella* sp. commercial production is widely extended [39].

- Raceway ponds. Raceway ponds are the most popular open pond system for the commercial culturing of several species such as *Chlorella* sp., *Spirulina platensis, Hematococcus* sp., and *D. salina.* They are characterized by having a depth between 15 and 25 cm and can be constructed as a single channel or as a group of channels [38]. These ponds are usually made of concrete and are formed by paddlewheels which provide agitation and keep microalgae in suspension [35]. Furthermore, this agitation system creates a turbulent mixing that allows microalgae to receive sunlight and the exchange of CO_2 with the atmosphere [39]. Despite being a simple growing technique, it has some limitations such as the large area required and the lower biomass productivity in comparison with closed PBRs [40].
- Inclined systems. In these systems, agitation is generated by the continuous culture suspension flow from the top to the bottom of the surface. Despite obtaining higher cell concentrations in comparison with other systems as well as higher surface–volume ratio, these systems are limited by the cells sedimentation and higher energy costs for the continuous pumping of the culture [36].

2.5.1.2 Photobioreactors

In this type of cultivation system, the light must pass through the transparent wall of the reactor to reach the culture. With closed PBRs, some of the problems associated to open systems can be resolved: (1) Limitation in the exchange of contaminants such as dust or microorganisms between the culture and the atmosphere and (2) maximum reduction in water evaporation [36]. In addition, PBRs allow the culture of a single microalgae specie as well as control and regulation of all cultivation parameters. According to its design, PBRs can be classified in tubular and flat plate PBRs [40].

- Tubular photobioreactors (TPBRs). TPBR is one of the most employed configurations within the closed cultivation systems. TBPRs are comprised by a series of transparent tubes with small diameter ($\leq$0.1 m) which can be arranged in different orientations (vertical, horizontal, helical, alpha-shaped, etc.). In addition, TPBRs are also constituted by a harvesting unit to separate algae and suspension, a degassing column for gas exchange and a refrigeration and circulation pump [41].
- Vertical tubular reactors (VTRs). VTRs are cylinder-shaped columns made of polyethylene or glass, with a maximum ratio equal to 0.2 m and a height of up to 4 m [42]. Within this group, VTR can be found in different configurations such as bubble column and airlift reactors (internal loop

airlift or split-column PBRs), with similar structure but different components. In the first case, bubble column is constituted as column with an air injector at the bottom and a freeboard at the top to separate gas and liquid. Bubble column reactors have the advantages of achieving high mass transfer and high surface area to volume ratio [39]. Second, airlift reactors are constituted by two parts: the riser zone, where air is injected to move the liquid to the top zone, and the downcomer (irradiated) zone, where liquid falls after been degassed at the top of the column. Good mixing, low shear stress, efficient mass transfer, and the possibility to immobilize microalgae are the main advantages of this type of PBR [39]. Airlift reactors can be found in different configurations such as internal loop airlift PBR, constituted by a transparent column with a smaller internal column equipped with an air injector at the bottom. A variant of this configuration is the external loop airlift PBR, with an external circulation column. Last, split-column airlift PBR is characterized by having a flat plate which divides the column into two parts: the riser, where air is injected to move the liquid to the top, and the downcomer, where liquid falls after been degassed at the top of the column [42].

VTRs are easy to operate, compact, and low-cost. In addition, high mass transfer as well as good mixing with low shear stress is achieved, thanks to the CO_2 injection from the bottom of the column via bubbling [37].

- Horizontal tubular reactors (HTRs). HTRs consist of a series of tubes in parallel and placed horizontally with a gas exchanger through which CO_2 is injected. In this type of reactors, culture medium is pumped along the tubes to maintain a high turbulent flow regime to prevent microalgae sedimentation. Different configurations such as walls, helices or panels can be found within this group [39]. The main advantage of HTR is the high light conversion efficiency due to the possibility of orienting the reactor toward sunlight, without the necessity of artificial illumination [35]. Nevertheless, a large area of land is required due to the high surface area of HTR [39].
- Helical tubular reactors (HETRs) consist of flexible and small-diameter tubes that are arranged in a spiral forming helical-shaped PBRs. Tubes are typically made of polyethylene or PVC and have a diameter between 2.5 and 5 cm. In this configuration, several bands of tubes are disposed around a cylindrical support and connected to a pump which allows the constant flow of the suspension [36]. This kind of reactors is apt for outdoor cultures employing sunlight, which allows the reduction of production costs [38].
- Alpha-shaped reactors. They are commonly constituted by sets of transparent tubes (usually made of PVC) equipped with an airlift pump, which promotes the culture agitation, and a series of CO_2 injectors along the

tubes. Among the several advantages this system present, the flow of the culture suspension in a single direction allows the achievement of high flow rates without the requirement of high air supply, reducing the costs of operation. In addition, this configuration is apt for its use outdoor due to the efficient light harvesting achieved [43].

➢ Flat plate reactors (FPRs). FPRs consist of a series of narrow panels made of transparent materials and designed to achieve a high area/volume ratio for the efficient use of light [43]. This system can be used outdoor and indoor with sun or artificial light. The short light path and high illumination surface area enables the achievement of high photosynthetic productivities, resulting in high microalgae biomass concentrations [37]. In addition, low cost, simplicity, and the suitability of FPR with suspended and immobilized cells are some of the main advantages of FPRs. However, this system presents some limitations such as hydrodynamic stress caused by the aeration as well as the possibility of biomass adhesion to the walls of the panels [39]. Table 10.3 shows a comparison between open and closed systems.

3. Advanced oxidation technologies for wastewater treatment

Advanced oxidation systems are nonconventional technologies. In these systems of oxidation, hydroxyl radicals are generated in sufficient concentration to have a structural effect on wastewater composition [44]. AOTs as wastewater treatment methods are based on pollutants removal via oxidation reactions (performed by oxidant agents such as ozone and peroxides) in combination with catalytic and photocatalytic processes [45]. Its use for the treatment of industrial effluents has been growing in recent years due to its high effectiveness. Table 10.4 summarizes the main AOTs used in wastewater treatment [49].

3.1 Ozonation

This technology is based on the decomposition of contaminants and wastewater disinfection using ozone (O_3) produced from air by ozone generators. O_3 is a powerful oxidizing agent with high redox potential which attacks organic and inorganic compounds with aromatic rings and double bonds [49].

The reaction of ozonation in aqueous solution is known as homogeneous catalytic ozonation. In this reaction, the catalyst are transition metals (Fe(II), Mn(II), Ni(II), etc.), which determine the reaction rate and selectivity [44]. The oxidation reaction can be directly performed by molecular ozone or via ozone decomposition in hydroxyl radicals, characterized by high redox

TABLE 10.3 Advantages and disadvantages of the main microalgae culture systems.

Culture system	Advantages	Disadvantages	References
Tubular photobioreactors (PBRs)	- High illumination area - Suitable for its use indoor and outdoor - High biomass productivities - Economic - Easy to operate	- Generation of pH gradients - Biomass adhesion to the walls - Requirement of large area of land	[42]
Vertical column	- High mass transfer - Good mixing with low shear stress - Low energy inputs - Easy to clean and sterilize	- Low illumination surface - Algal cell sedimentation	[39]
Airlift column	- Simplicity - High biomass productivity - Efficient light utilization	- Difficult to scale-up	[42]
Flat plate PBRs	- High illumination area - Suitable for its use indoor and outdoor - High biomass productivity - Low cost - Easy to clean and sterilize	- Difficult control of temperature - Hydrodynamic stress - Difficult to scale-up - Biomass adhesion to the walls	[42]
Open systems	- Low cost - Easy to clean - Large scale	- Difficult control of operating conditions - Low productivity - Large area of land required - Contamination - Limited to certain species - Evaporation losses and CO_2 diffusion limitation	[38]

TABLE 10.4 Main advanced oxidation technologies (AOTs) used in wastewater treatment with their principle reactions and chemicals used.

AOT	Reactions	Chemicals used	References
Ozonation	$O_3 + OH^- \rightarrow HO_2^- + O_2$ $O_3 + HO_2^- \rightarrow HO_2^\bullet + O_3^{-\bullet}$	O_3	[46,47]
Fenton	$H_2O_2 + Fe^{2+} \rightarrow Fe^{3+} + OH^\bullet + OH^-$	Fe^{2+} H_2O_2	[46,47]
Photo-Fenton	$Fe(OH)^{2+} \xrightarrow{hv} Fe^{2+} + OH^\bullet$ ($\lambda < 580$ nm) $H_2O_2 \xrightarrow{hv} 2OH^\bullet$ ($\lambda < 310$ nm)	Fe^2 H_2O_2 UV	[46,48]
Photocatalysis	$TiO_2 \xrightarrow{hv} e^- + h^+$ $TiO_2(h^+) + H_2O_{ad} \rightarrow TiO_2 + HO^\bullet_{ad} + H^+$ $TiO_2(h^+) + HO_{ad} \rightarrow TiO_2 + HO^\bullet_{ad}$ $TiO_2(h^+) + RX_{ad} \rightarrow RX^{\bullet+}_{ad}$	TiO_2/ZnO UV	[46,47]

potential and high reactivity with the majority of the compounds in aqueous solutions [50].

On the other hand, heterogeneous catalytic ozonation is a combined system in which ozone and a solid catalyst are used. The most extensively used catalysts are metal and metal oxides (MnO_2, TiO_2, Al_2O_3) [50]. In addition, metal oxide supports are also used in combination with the catalyst ($Cu-TiO_2$, $Cu-Al_2O_3$, $Ru-CeO_2$). The efficiency of the reaction is dependent on the selected catalyst and the pH of the solution, which determine ozone decomposition in aqueous solutions [44].

3.2 Fenton

Fenton processes are based on pollutants oxidation by hydroxyl radicals generated from hydrogen peroxide (H_2O_2) as the oxidant reagent and iron ions (Fe^{2+}) as catalyst of the reaction [51]. The reaction efficiency depends on several factors such as pH, temperature, pressure, and the concentration of reagents. First, acidic pH (2.8−3.0) is required to achieve Fe^{2+} maximum catalytic activity. Furthermore, increased temperature enhances the reaction kinetics but favors H_2O_2 decomposition, so ambient temperature and pressure are commonly used. The required initial concentration of H_2O_2 and Fe^{2+} is determined by the initial wastewater pollutant load in terms of chemical oxygen demand [52].

Fenton oxidation as treatment method of numerous types of wastewaters (aquaculture, pharmaceutical, industrial, etc.) has been extensively applied in recent years due to the high-standard oxidation potential and the high reaction rates of hydroxyl radicals [52]. This results in low energy inputs, short reaction times, high effectiveness, and high diversity of target contaminants such as phenols, chlorophenols, formaldehyde, amines, etc. Nevertheless, the necessities to incorporate high-cost reagents as well as the generation of an effluent with high iron concentration are some of the limitations of the process [51].

3.3 Photo-Fenton

This technique consists of an energetically improved treatment method based on the combination of Fenton reaction with photodegradation [53]. The use of ultraviolet artificial lamps or visible light reduces the catalyst concentration required, enhances the catalytic activity, and accelerates the degradation of organic matter [48,53]. Furthermore, photoreduction of Fe^{3+} results in Fe^{2+} photochemical regeneration, which reacts with H_2O_2, generating new hydroxyl radical and ferric ion [54].

The highest efficiency during Photo-Fenton is achieved at pH equal to 3.0, explained by the higher solubility of the complex hydroxy-Fe^{3+} and higher photoactivity of $Fe(OH)^{2+}$ [54].

3.4 Photocatalysis

In this method, the exposition of a catalyst to light exhibits high oxidative and hydrophilic properties, decompose organic compounds from wastewaters into intermediates, which are finally converted in CO_2 and H_2O [55].

Among different catalysts, TiO_2 is one of the most widely used in heterogeneous photocatalysis due to its stability, resistance, and high photocatalyst activity even after several cycles. The difficult recovery of TiO_2 after water treatment is one of the major drawbacks of this system [56].

3.4.1 AOTs as full wastewater treatment system

Technical, economic, and environmental factors must be considered to implement an AOT as full system for wastewater treatment. A preliminary evaluation of the different AOTs available must be conducted [46].

i. Selection of the process and technical suitability. Wastewater characteristics are the major factor influencing the AOT selection. The selected AOT must handle wastewater physicochemical characteristics including biological oxygen demand, nitrogen and phosphorus content, total solids, etc. [46]. In addition, operating conditions must be considered in this first stage. Within operating requirements, it must be highlighted the pH value. In this sense, ozonation requires high pH value; nevertheless, Fenton and photo-Fenton processes are performed at low pH and require strict control. With regard to photocatalytic processes, the selected catalyst or oxidizing reagent will determine the pH value [46]. Added operational factors to consider are temperature and pressure as well as the required UV lamps with specific wavelength in the case of photocatalytic processes. Other equipment is also needed in some cases such as ozone, for which a reactor vessel and an ozone generator and a sparger are needed [46].

ii. Economic costs. The main factor influencing the economic costs associated to the implementation of a process is the specific energy requirement, defined in terms of kWh/kL, which is specific to each AOP. In this sense, several studies have applied the system known as the "first principle engineering cost estimation method" in which four variables are considered to estimate the treatment cost associated with an AOP: reaction rate constant (k/min), base reactor volume (L), unit treatment cost ($/1000 US galloon), and specific energy (kWh/kL), [47].

iii. Environmental impact. The environmental footprint of each AOT can be measured according to different parameters such as carbon footprint, expressed in terms of kilogram of CO_2 per kWh, and greenhouse gas emissions from the combustion of fossil fuels. Added environmental effects must be considered for each AOT, including the amounts of sludge generated, the negative impact of the chemical reagents needed, etc. [46].

For example, solar photo-Fenton requires high quantities of chemical reagents and generates effluents rich in metal ions as well as exhausted solid catalyst [57].

iv. Last, in case of reusing treated water, this must comply with the environmental discharge legislation and the limits for public health and environmental risk [46].

3.4.2 Combined process based on AOTs and biological treatment

The use of AOTs in combination with biological processes is a suitable alternative for wastewaters whose treatment is difficult or impossible to achieve through conventional physicochemical and biological processes [58]. An exhaustive physicochemical characterization of the wastewaters is especially relevant when applying combined AOTs and biological treatment for several reasons. First, the optimal chemical oxidant concentration must be selected to not result in an insufficient carbon source concentration in the biotreatment stage. In addition, the optimal pH values for AOTs must be favorable for the growth of microorganisms [51]. It is also important to take into account that some organic and inorganic compounds contained in wastewaters can compete with the target pollutant for the oxidizing compound, decreasing the efficiency of the process [58].

3.4.2.1 AOTs followed by biological treatment

Wastewater treatment performed by the combined system of Fenton followed by biotreatment can improve the efficiency of the process. This system has been used in wastewaters containing polluting compounds such as pharmaceutical and personal care products, pesticides, surfactants, etc. These compounds, which include tetracycline, alkanolamines, sulfamethoxazole, benzene ester, etc., are resistant to biodegradation, therefore, the application of Fenton prior to biotreatment has demonstrated high effectiveness [51].

Some of the microorganisms which have been used in this combined system are isolated bacterial species of *Pseudomonas* sp., reaching removal rates of 84% and 99% of pyridine and 3-cyanopyridine, respectively [59]. In addition, yeast *Candida oleophila* after Fenton process allowed 95% elimination of Reactive black 5 in dyeing and textile wastewaters [60]. *Thiobacillus ferrooxidans* culture in leather wastewaters after Fenton process allowed elimination percentages equal to 93%, 98%, and 100% of COD, BOD and chromium, respectively [61].

3.4.2.2 Biological treatment followed by AOTs

The application of biological treatment prior to Fenton and photo-Fenton processes has also been proved to be effective in the treatment of some types of wastewaters. For example, the combination of anaerobic digestion followed by photo-Fenton allowed removal percentages between 92% and 97% of azo dyes [62]. Furthermore, the treatment of pulp mill wastewater was performed through *Cryptococcus podzolicus* culture followed by solar photo-Fenton, achieving COD removal greater than 90% [63].

Nevertheless, this technology is not applicable to all types of wastewaters. If the real wastewater contains high amounts of biodegradable compounds, an oxidation posttreatment stage will not result in improved biodegradability of the wastewater, supposing an unnecessary chemicals expense. In this case, it would be reasonable a cost-effective biological pretreatment, able to remove the biodegradable fraction, followed by and AOT, intended to convert the nonbiodegradable fraction into biodegradable compounds, followed by a last biological stage. This particular system has been proved to be effective in olive oil mill wastewater and effluents containing a high initial concentration of biodegradable compounds such as sugar or proteins [58].

References

[1] FAO, Food and Agriculture Organization.Water at a Glance, 2014. http://www.fao.org/docrep/016/ap505e/ap505e.pdf.

[2] S. Tongesayi, T. Tongesayi, Water quality and public health, in: Chemistry and Water, 2017, pp. 553–596.

[3] A. Hemati, M.A. Rippy, S.B. Grant, K. Davis, D. Feldman, Deconstructing demand: the anthropogenic and climatic drivers of urban water consumption, Environ. Sci. Technol. 50 (23) (2016) 12557–12566.

[4] J. Bogner, M. Abdelrafie Ahmed, C. Diaz, A. Faaij, Q. Gao, S. Hashimoto, K. Mareckova, R. Pipatti, T. Zhang, Waste management, in: B. Metz, O.R. Davidson, P.R. Bosch, R. Dave, L.A. Meyer (Eds.), Climate Change: Mitigation. Contribution of Working Group III to the Fourth Assessment Report of the Intergovernmental Panel on Climate Change, Cambridge University Press, Cambridge, United Kingdom and New York, NY, USA, 2007.

[5] T. Sato, M. Qadir, S. Yamamoto, T. Endo, A. Zahoor, Global, regional, and country level need for data on wastewater generation, treatment, and use, Agric. Water Manag. 130 (2013) 1–13.

[6] M. Henze, in: M. Henze, M.C.M. van Loosdrecht, G.A. Ekama, D. Brdjanovic (Eds.), Biological Wastewater Treatment: Principles Modelling and Design, IWA Publishing, London, UK, 2008, ISBN 9781843391883. Published by.

[7] E. Papadaki, F.T. Mantzouridou, Current status and future challenges of table olive processing wastewater valorization, Biochem. Eng. J. 112 (2016) 103–113.

[8] M.B. Pescod, Wastewater treatment, in: Wastewater Treatment and Use in Agriculture, Food and Agriculture Organization (FAO), Rome, 1992. http://www.fao.org/docrep/T0551E/t0551e05.htm.

[9] G.K.C. Ding, Wastewater treatment and reuse—the future source of water supply, in: Encyclopedia of Sustainable Technologies, 2017, pp. 43–52.

[10] H. Patel, R.T. Vashi, Introduction, in: Characterization and Treatment of Textile Wastewater, 2015, pp. 1–20.

[11] M. Salgot, M. Folch, S.S. Unit, Wastewater treatment and water reuse, Curr.Opin. Environ. Sci. Heal. 2 (2018) 64–74. https://doi.org/10.1016/j.coesh.2018.03.005.

[12] N. Abdel-Raouf, A.A. Al-Homaidan, I.B.M. Ibraheem, Microalgae and wastewater treatment, Saudi J. Biol. Sci. 19 (2012) 257–275. https://doi.org/10.1016/j.sjbs.2012.04.005.

[13] K.E. Lee, N. Morad, T.T. Teng, B.T. Poh, Development, characterization and the application of hybrid materials in coagulation/flocculation of wastewater: a review, Chem. Eng. J. 203 (2012) 370–386. https://doi.org/10.1016/j.cej.2012.06.109.

[14] C.S. Lee, J. Robinson, M.F. Chong, A review on application of flocculants in wastewater treatment, Process Saf. Environ. Protect. 92 (2014) 489–508. https://doi.org/10.1016/j.psep.2014.04.010.

[15] Y. Zhang, S. Wei, Y. Hu, S. Sun, Membrane technology in wastewater treatment enhanced by functional nanomaterials, J. Clean. Prod. 197 (2018) 339–348. https://doi.org/10.1016/j.jclepro.2018.06.211.

[16] L.Y. Ng, C.Y. Ng, E. Mahmoudi, C.B. Ong, A.W. Mohammad, A review of the management of inflow water, wastewater and water reuse by membrane technology for a sustainable production in shrimp farming, J. Water Process Eng. 23 (2018) 27–44. https://doi.org/10.1016/j.jwpe.2018.02.020.

[17] W.P. Cathie-Lee, S.-K. Maha, C.P. Leo, T.Y. Wua, S.-P. Chai, Phosphorus removal by NF90 membrane: optimisation using central composite design, J. Taiwan Inst. Chem. Eng. 45 (4) (2014) 1260–1269.

[18] E. Sahinkaya, A. Sahin, A. Yurtsever, M. Kitis, Concentrate minimization and water recovery enhancement using pellet precipitator in a reverse osmosis process treating textile wastewater, J. Environ. Manag. 222 (2018) 420–427. https://doi.org/10.1016/j.jenvman.2018.05.057.

[19] M.A. Al-Obaidi, J.P. Li, C. Kara-Zaitri, I.M. Mujtaba, Optimisation of reverse osmosis based wastewater treatment system for the removal of chlorophenol using genetic algorithms, Chem. Eng. J. 316 (2017) 91–100. https://doi.org/10.1016/j.cej.2016.12.096.

[20] J. Li, Q. Zhou, L.C. Campos, The application of GAC sandwich slow sand filtration to remove pharmaceutical and personal care products, Sci. Total Environ. 635 (2018a) 1182–1190. https://doi.org/10.1016/j.scitotenv.2018.04.198.

[21] V. Kårelid, G. Larsson, B. Bj, Pilot-scale removal of pharmaceuticals in municipal wastewater : comparison of granular and powdered activated carbon treatment at three wastewater treatment plants, J. Environ. Manage 193 (2017). https://doi.org/10.1016/j.jenvman.2017.02.042.

[22] J. Wen, H. Dong, G. Zeng, Application of zeolite in removing salinity/sodicity from wastewater: a review of mechanisms, challenges and opportunities, J. Clean. Prod. 197 (2018) 1435–1446. https://doi.org/10.1016/j.jclepro.2018.06.270.

[23] F.A. Ansari, P. Singh, A. Guldhe, F. Bux, Microalgal cultivation using aquaculture wastewater: integrated biomass generation and nutrient remediation, Algal Res 21 (2017) 169–177. https://doi.org/10.1016/j.algal.2016.11.015.

[24] I.T.D. Cabanelas, Z. Arbib, F.A. Chinalia, C.O. Souza, J.A. Perales, P.F. Almeida, J.I. Druzian, I.A. Nascimento, From waste to energy: microalgae production in wastewater and glycerol, Appl. Energy 109 (2013) 283–290. https://doi.org/10.1016/j.apenergy.2013.04.023.

[25] S. Hongyang, Z. Yalei, Z. Chunmin, Z. Xuefei, L. Jinpeng, Cultivation of Chlorella pyrenoidosa in soybean processing wastewater, Bioresour.Technol. 102 (2011) 9884–9890. https://doi.org/10.1016/j.biortech.2011.08.016.

[26] D. Hernández, B. Riaño, M. Coca, M.C. García-González, Treatment of agro-industrial wastewater using microalgae-bacteria consortium combined with anaerobic digestion of the produced biomass, Bioresour. Technol. 135 (2013) 598–603. https://doi.org/10.1016/j.biortech.2012.09.029.

[27] W. Farooq, Y.C. Lee, B.G. Ryu, B.H. Kim, H.S. Kim, Y.E. Choi, J.W. Yang, Two-stage cultivation of two *Chlorella* sp. strains by simultaneous treatment of brewery wastewater and maximizing lipid productivity, Bioresour. Technol. 132 (2013) 230–238. https://doi.org/10.1016/j.biortech.2013.01.034.

[28] T.V. Ramachandra, M. Durga Madhab, S. Shilpi, N.V. Joshi, Algal biofuel from urban wastewater in India: scope and challenges, Renew. Sustain. Energy Rev. 21 (2013) 767–777. https://doi.org/10.1016/j.rser.2012.12.029.

[29] G. Hodaifa, M. Eugenia, R. Órpez, S. Sánchez, Inhibitory effects of industrial olive-oil mill wastewater on biomass production of Scenedesmus obliquus, Ecol. Eng. 42 (2012) 30–34. https://doi.org/10.1016/j.ecoleng.2012.01.020.

[30] S.D. Kulkarni, T. Auti, S. Saraf, Bioremediation study of dairy effluent by using *Spirulina platensis*, Res. J. Life Sci. Bioinformatics, Pharm. Chem. Sci. 1 (2016) 317–326.

[31] Y. Wang, S.H. Ho, C.L. Cheng, W.Q. Guo, D. Nagarajan, N.Q. Ren, D.J. Lee, J.S. Chang, Perspectives on the feasibility of using microalgae for industrial wastewater treatment, Bioresour. Technol. 222 (2016a) 485–497.

[32] J.K. Pittman, A.P. Dean, O. Osundeko, The potential of sustainable algal biofuel production using wastewater resources, Bioresour. Technol. 102 (2011) 17–25. https://doi.org/10.1016/j.biortech.2010.06.035.

[33] A.F. Mohd Udaiyappan, H. Abu Hasan, M.S. Takriff, S.R. Sheikh Abdullah, A review of the potentials, challenges and current status of microalgae biomass applications in industrial wastewater treatment, J. Water Process Eng. 20 (2017) 8–21. https://doi.org/10.1016/j.jwpe.2017.09.006.

[34] E. Molina, J. Fernández, F.G. Acién, Y. Chisti, Tubular photobioreactor design for algal cultures, J. Biotechnol. 92 (2001) 113–131. https://doi.org/10.1016/S0168-1656(01)00353-4.

[35] R.N. Singh, S. Sharma, Development of suitable photobioreactor for algae production – a review, Renew. Sustain. Energy Rev. 16 (2012) 2347–2353. https://doi.org/10.1016/j.rser.2012.01.026.

[36] A. Richmond, in: A. Richmond (Ed.), Handbook of Microalgal Culture: Biotechnology and Applied Phycology, 2004. Orton.Catie.Ac.Cr 472, https://doi.org/10.1002/9780470995280.

[37] C.U. Ugwu, H. Aoyagi, H. Uchiyama, Photobioreactors for mass cultivation of algae, Bioresour.Technol. 99 (2008) 4021–4028. https://doi.org/10.1016/j.biortech.2007.01.046.

[38] S.A. Razzak, M.M. Hossain, R.A. Lucky, A.S. Bassi, H. De Lasa, Integrated CO_2 capture, wastewater treatment and biofuel production by microalgae culturing – a review, Renew. Sustain. Energy Rev. 27 (2013) 622–653. https://doi.org/10.1016/j.rser.2013.05.063.

[39] K.W. Chew, S.R. Chia, P.L. Show, Y.J. Yap, T.C. Ling, J.S. Chang, Effects of water culture medium, cultivation systems and growth modes for microalgae cultivation: a review, J. Taiwan Inst. Chem. Eng. 0 (2018) 1–13. https://doi.org/10.1016/j.jtice.2018.05.039.

[40] M. Faried, M. Samer, E. Abdelsalam, R.S. Yousef, Y.A. Attia, A.S. Ali, Biodiesel production from microalgae: processes, technologies and recent advancements, Renew. Sustain. Energy Rev. 79 (2017) 893–913. https://doi.org/10.1016/j.rser.2017.05.199.

[41] P.J.M. Van Beveren, Algal Growth in Horizontal Tubular Reactors Algal Growth in Horizontal Tubular Reactors, 2011.

[42] B. Wang, C.Q. Lan, M. Horsman, Closed photobioreactors for production of microalgal biomasses, Biotechnol. Adv. 30 (2012c) 904–912. https://doi.org/10.1016/j.biotechadv.2012.01.019. https://doi.org/10.1016/j.biortech.2016.09.106.

[43] A.P. Carvalho, L.A. Meireles, Microalgae reactors: a review of enclosed systems and performances, Biotechnol.Prog. 3 (2006) 1490–1506.

[44] B. Kasprzyk-Hordern, M. Ziółek, J. Nawrocki, Catalytic ozonation and methods of enhancing molecular ozone reactions in water treatment, Appl. Catal. B Environ. 46 (2003) 639–669. https://doi.org/10.1016/S0926-3373(03)00326-6.

[45] M. Gągol, A. Przyjazny, G. Boczkaj, Wastewater treatment by means of advanced oxidation processes based on cavitation –a review, Chem. Eng. J. 338 (2018) 599–627. https://doi.org/10.1016/j.cej.2018.01.049.

[46] M.N. Chong, A.K. Sharma, S. Burn, C.P. Saint, Feasibility study on the application of advanced oxidation technologies for decentralised wastewater treatment, J. Clean. Prod. 35 (2012) 230–238. https://doi.org/10.1016/j.jclepro.2012.06.003.

[47] N.N. Mahamuni, Y.G. Adewuyi, Advanced oxidation processes (AOPs) involving ultrasound for waste water treatment: a review with emphasis on cost estimation, Ultrason.Sonochem. 17 (2010) 990–1003. https://doi.org/10.1016/j.ultsonch.2009.09.005.

[48] N. Wang, T. Zheng, G. Zhang, P. Wang, A review on Fenton-like processes for organic wastewater treatment, J. Environ. Chem. Eng. 4 (2016b) 762–787. https://doi.org/10.1016/j.jece.2015.12.016.

[49] Y. Bar Oz, H. Mamane, O. Menashe, V. Cohen-Yaniv, R. Kumar, L. Iasur Kruh, E. Kurzbaum, Treatment of olive mill wastewater using ozonation followed by an encapsulated acclimated biomass, J. Environ. Chem. Eng. 6 (2018) 5014–5023. https://doi.org/10.1016/j.jece.2018.07.003.

[50] X. Li, W. Chen, L. Ma, H. Wang, J. Fan, Industrial wastewater advanced treatment via catalytic ozonation with anFe-based catalyst, Chemosphere 195 (2018b) 336–343. https://doi.org/10.1016/j.chemosphere.2017.12.080.

[51] D. Huang, C. Hu, G. Zeng, M. Cheng, P. Xu, X. Gong, R. Wang, W. Xue, Combination of Fenton processes and biotreatment for wastewater treatment and soil remediation, Sci. Total Environ. 574 (2017) 1599–1610. https://doi.org/10.1016/j.scitotenv.2016.08.199.

[52] P. Bautista, A.F. Mohedano, J.A. Casas, J.A. Zazo, J.J. Rodríguez, An overview of the application of Fenton oxidation to industrial wastewaters treatment, J. Chem. Technol. Biotechnol. 83 (2008) 1323–1338. https://doi.org/10.1002/jctb.

[53] C.A. García, G. Hodaifa, Real olive oil mill wastewater treatment by photo-Fenton system using artificial ultraviolet light lamps, J. Clean. Prod. 162 (2017) 743–753. https://doi.org/10.1016/j.jclepro.2017.06.088.

[54] A. Babuponnusami, K. Muthukumar, A review on Fenton and improvements to the Fenton process for wastewater treatment, J. Environ. Chem. Eng. 2 (2014) 557–572. https://doi.org/10.1016/j.jece.2013.10.011.

[55] T. Threrujirapapong, W. Khanitchaidecha, A. Nakaruk, Treatment of high organic carbon industrial wastewater using photocatalysis process, Environ. Nanotechnol. Monit. Manag. 8 (2017) 163–168. https://doi.org/10.1016/j.enmm.2017.07.006.

[56] M.N. Chong, B. Jin, C.W.K. Chow, C. Saint, Recent developments in photocatalytic water treatment technology: a review, Water Res. 44 (2010) 2997–3027. https://doi.org/10.1016/j.watres.2010.02.039.

[57] S. Foteinis, A.G.L. Borthwick, Z. Frontistis, D. Mantzavinos, E. Chatzisymeon, Environmental sustainability of light-driven processes for wastewater treatment applications, J. Clean. Prod. 182 (2018) 8–15. https://doi.org/10.1016/j.jclepro.2018.02.038.

[58] I. Oller, S. Malato, J.A. Sánchez-Pérez, Combination of advanced oxidation processes and biological treatments for wastewater decontamination-A review, Sci. Total Environ. 409 (2011) 4141–4166. https://doi.org/10.1016/j.scitotenv.2010.08.061.

[59] K.V. Padoley, S.N. Mudliar, S.K. Banerjee, S.C. Deshmukh, R.A. Pandey, Fenton oxidation: a pretreatment option for improved biological treatment of pyridine and 3-cyanopyridine plant wastewater, Chem. Eng. J. 166 (2011) 1–9. https://doi.org/10.1016/j.cej.2010.06.041.

[60] M.S. Lucas, A.A. Dias, A. Sampaio, C. Amaral, J.A. Peres, Degradation of a textile reactive Azo dye by a combined chemical-biological process: Fenton's reagent-yeast, Water Res. 41 (2007) 1103–1109. https://doi.org/10.1016/j.watres.2006.12.013.

[61] T. Mandal, D. Dasgupta, S. Mandal, S. Datta, Treatment of leather industry wastewater by aerobic biological and Fenton oxidation process, J. Hazard Mater. 180 (2010) 204–211. https://doi.org/10.1016/j.jhazmat.2010.04.014.

[62] J. García-Montaño, X. Domènech, J.A. García-Hortal, F. Torrades, J. Peral, The testing of several biological and chemical coupled treatments for Cibacron Red FN-R azo dye removal, J. Hazard Mater. 154 (2008) 484–490. https://doi.org/10.1016/j.jhazmat.2007.10.050.

[63] L. Fernandes, M.S. Lucas, M.I. Maldonado, I. Oller, A. Sampaio, Treatment of pulp mill wastewater by *Cryptococcus podzolicus* and solar photo-Fenton: a case study, Chem. Eng. J. 245 (2014) 158–165. https://doi.org/10.1016/j.cej.2014.02.043.

Chapter 11

Modeling in environmental interfaces

Ombretta Paladino,[a,*] Gassan Hodaifa,[b] Matteo Neviani,[c] Mahdi Seyedsalehi[d] and Ana Malvis[b]
[a]*DICCA, Department of Civil, Chemical and Environmental Engineering, University of Genoa Genoa, Italy;* [b]*Molecular Biology and Biochemical Engineering Department, Chemical Engineering Area, Faculty of Experimental Sciences, University of Pablo de Olavide Seville, Spain;* [c]*DIME, Department of Mechanical, Energy, Management and Transportation Engineering, University of Genoa Genoa, Italy;* [d]*School of Environment, Tsinghua University, Beijing, China*
**Corresponding author*

1. Environmental interfaces

Environmental interfaces (EIs) are defined as the regions along which different environmental compartments, either abiotic or biotic, interact each other by exchanging mass, momentum, and energy.

Physical, chemical, and biochemical processes were usually studied using each single compartment as a system and by looking at EIs as boundaries between the compartments; recently, attention is being focused in modeling these processes right at interfaces.

EIs are complex systems, they usually show nonlinear and complex dynamic behavior, their modeling often needs a multiscale approach.

EIs are encountered in saturated and vadose zone in aquifers, in surface and marine water; they are the region along which bubbles, drops, soil, ice, snow interact with air, water, and chemicals; they can be observed between enzymes, living cells, organs, human bodies and their surrounding environment; they play an important role in the development of biosensors and green separation techniques.

Modeling in EIs requires, as modeling of classical environmental systems, the definition of the scale of representation and of the control volume of the system to be described. In fact, while at large length scales, EIs can be represented as two-dimensional boundaries between the environmental

Advanced Low-Cost Separation Techniques in Interface Science. https://doi.org/10.1016/B978-0-12-814178-6.00011-X

compartments; at a smaller representation scale, they act as three-dimensional interphase volumes with which each different environmental compartment interacts; this also means that some transfer mechanisms and possible accumulation of mass, energy momentum can exist inside them.

If we consider the approach adopted in studying transport phenomena, these mechanisms are all similar in their behavior: they move a conservative "physical quantity" (mass, momentum, energy) from a place where there is a lot of it to a place where there is a smaller amount of it, in order to reach thermodynamic equilibrium. There is a driving force causing the transfer that becomes smaller as time progresses, and eventually becomes zero when no more transfer of physical quantity takes place. In modeling these transport phenomena a multiscale approach [31,75,76] with different levels of representation can be used, where each level describes the phenomenon over a specific scale of length and time. The most adopted representation scales are the following:

Macroscopic scale: description of momentum, energy, and mass exchanges due to the introduction (and/or extraction) of different streams (inputs/outputs) into the system from surroundings. The control volume, continuum, and with uniform properties, is often equivalent to the volume of the entire object to be modeled (e.g., the environmental compartment). The exchange with the EI is supposed to be concentrated in some points of its surface.

Microscopic scale: description of what is happening in a small region (local control volume) of the system to be modeled, supposed to be a continuum, homogeneous medium (e.g., water with soluble pollutants, polluted air). This approach can be used also for heterogeneous media as groundwater, emulsion, foams, where a representative elemental volume (REV) is defined. Local equations of change are used to describe mass, momentum, and energy exchanges within this small region.

Grain scale: description of what is happening in a small region with uniform properties within the environmental compartment and separated by it due to an interface. If this level is used in reaction engineering, the region usually corresponds to well individuated "macroscopic particles" inside the system (drops, pellets, grains, bubbles). In general, in multiscale modeling a *coarse-grained* scale is adopted to model individual "particles," molecules or groups of them, inside a medium [43].

Molecular scale: fundamental understanding of the mechanisms of mass, momentum, and energy transport in terms of molecular structure and intermolecular forces. Information about atoms and molecules is required.

At a quantum mechanics scale, no transport phenomena are analyzed and information about electrons is also included into the model.

1.1 Why modeling in environmental interfaces?

The need of modeling EIs arises from the fact that EIs are not simply two-dimensional *separation systems* between the different environmental compartments, but they can be complex systems (even if their thickness could be only few nanometers) strongly modifying transport phenomena between the different environmental compartments.

Let us consider a simple problem of heat conduction through two different materials having different physical properties: the interface between the two materials usually does not need to be modeled to solve engineering problems, since some simple thermodynamic equilibrium assumptions can be reasonably made at the interface. In this case, the assumption that the temperature on the left and on the right side of the interface is the same (the interface reaches immediately the thermal equilibrium) is a reasonable hypothesis for the system study if the compartments (the two materials) have a thickness much greater than that of the interface.

If mass transfer is the studied phenomenon, as, for example, in the case of one polluted liquid, immiscible with water, and interacting with seawater, the interface between the two does not need to be modeled if, in addition to temperature for thermal equilibrium and pressure for mechanical equilibrium, chemical potential is used to define the phase equilibrium at the interface. In this case, even if the driving force of mass transfer is expressed in terms of concentration of the pollutant, the assumption of the same chemical potential on the left and on the right side of the interface (hypothesis that the interface immediately reaches the thermodynamic equilibrium) is a reasonable approach for the study of the system. Otherwise, if concentration is used as the variable to model the system, some partition coefficients can be adopted to describe what happens at the interface, i.e., concentration of the pollutant is not the same on the two sides of the interface due to the difference in solubility of the pollutant in the two liquids.

If the problem is only slightly complicated, by considering again a simple but moving G–L interface, as, for example, rising bubbles of gas into a liquid, the two phases interact by exchanging both mass and momentum. Moreover, bubble interface changes its shape. In this case, we have to define models for the mass transfer at the interface (we can still adopt some equilibrium assumption, if we want) plus the models for describing both the movement of the bubbles (Navier–Stokes (NS) equations) and surface tension–pressure relationships (some constitutive equations) affecting the bubble shape changing.

In the study of groundwater pollution and related remediation criteria, the vadose zone of unconfined aquifers is a typical interesting and complex system constituted by three or more environmental compartments (soil, water, air), plus other liquid mixtures as in the case of pollution by DNAPL (dense nonaqueous phase liquid), plus the various particulate and biotic compartments which may be present in dispersed form within these phases, as when degrading bacteria are present in the sediments. EIs exist between all the phases, some of them are physical interfaces (G—L, L—L, S—L), some others are "living" interfaces, as between the cellular membrane of the bacteria and the single pollutant.

A simple modeling approach for the heterogeneous environmental compartment constituted by the aquifer can be made at the microscopic or local scale if an REV is defined where all the main chemical—physical properties are uniform in the medium, considered as homogeneous. In this case, the transport of a reacting pollutant, miscible with water, can be described by the advection—dispersion—reaction equation. Pools formation, accumulation of vapors of the pollutant in soil interspaces, pollutant partitioning, local pollutant dilution due to water pools or rain percolation, different degradation rates due to local bacteria growth cannot be described with this approach. Some deepening of the model can be made by considering separate equations of change, one for each phase (usually under the assumption of a moving liquid compartment and steady soil, steady air, and steady pollutant pools) and thermodynamic equilibrium assumptions at the EIs, for example, described by fugacity-based models. These are distribution-based models incorporating all the environmental compartments, based on steady-state fluxes of pollutants across the EIs [40] and can forecast the concentration of pollutants in the environment with respect to different space and time scales. The dynamics of the EIs, also seen as boundaries, is not modeled with this approach. To partially investigate EIs dynamics, models at grain scale taking into account the adsorption kinetics (e.g., active sites on soil grains) or biochemical kinetics (e.g., parameters of the microbial film) at the interface should be used. Finally, if particular remediation criteria are adopted, e.g., using zero valent iron nanoparticles, dual-scale models can be proposed [69,70] or, going down in the detail, the behavior of the nanoparticles at the EIs can be investigated: in this case, the L—L interface between water and DNAPL could be modeled as an interphase volume in order to describe the nanoparticle aggregation and self-assembly at the interface [37] with the goal to design proper stabilization techniques.

Doing research on membranes [5,8], cellular membranes [30,34,77], nanomaterials, and composite materials [84] often requires the modeling of transport phenomena inside the interface, seen as an interphase volume. Generally, when EIs are complex and chemically or biologically active and when they strongly interact with the transport mechanism, it becomes mandatory to investigate at least if they could represent the limiting step in the whole interaction mechanism between the environmental compartments.

Therefore, the need of this chapter starts from the assumption that some EIs are complex enough to require at least the development of simple dynamic models for describing their behavior at some level; the novelty of this chapter concerns the approach adopted for the solution of the two proposed case studies with some of the most recent modeling techniques.

2. Dynamics of environmental interfaces and modeling approaches

Dynamic models can be developed at different scales as well; regardless of the scale of representation the following definitions are valid:

A *system* is a set of elements suitably connected, organized, and interacting with each other and the external environment in such a way as to constitute a single entity responsible for obtaining a predetermined objective.
A *system* responds to an action, exerted through one or more inputs, with a reaction, called an output.
A *system* that evolves over time, indicating with this term that both the input and the output may vary over time, is called a *dynamic system.*

In general, the output at time t of a dynamic system does not depend only on the input of the system at the same time, but also on a quantity, called *state*, which represents the past history of the system inputs.

The main variables of a dynamic system include the following:

Inputs or input variables $\boldsymbol{u}(t) \in \mathbb{R}^m$: variables acting on the trajectory of the state (state as a function of time) of the system. They can be manipulated variables, that is to say, on which we can act to modify the response of the system, or disturbances/noises, on which we cannot act in any way;
Outputs or output variables $\mathbf{y}(t) \in \mathbb{R}^r$: variables from which it is possible to estimate the trajectory of the state (state as a function of time) of the system. They can be measured, depending on the observability characteristics of the system, or unmeasured or not-measurable variables.
Internal or state variables $\mathbf{x}(t) \in \mathbb{R}^n$: variables describing the internal evolution of the system. They contain information on its past history and allow to determine its future state when initial conditions and external stresses applied to it are known.
Parameters $\boldsymbol{p}(t) \in \mathbb{R}^k$: characteristic properties of the system, also varying in time.

Excluding the approaches in which EIs are considered boundaries in steady state, in all the other situations EIs are dynamic systems, hence they can be represented with the following scheme (Fig. 11.1):

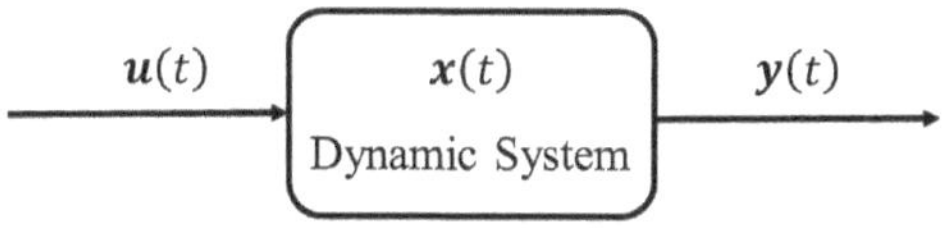

FIGURE 11.1 Representation of dynamic system.

We consider them as *linear dynamic systems* when the link between the input and state variables with the output variables is linear (this link can be expressed through matrices), i.e., $\boldsymbol{y}(t) = \boldsymbol{A}\boldsymbol{x}(t) + \boldsymbol{B}\boldsymbol{u}(t)$.

On the contrary, they are considered *nonlinear dynamic systems* when the link between the input and state variables with the output is expressed by nonlinear functions, i.e., $\boldsymbol{y}(t) = \boldsymbol{f}(\boldsymbol{u}(t), \boldsymbol{x}(t), \boldsymbol{p})$.

EIs are handled as *time-invariant dynamic systems* when their parameters do not depend on time. In case of parameters $\boldsymbol{p}$ changing in time, they can exhibit complex dynamic behavior and are classified as *time-variant dynamic systems*. The output can be written as $\boldsymbol{y}(t) = \boldsymbol{f}(\boldsymbol{u}(t), \boldsymbol{x}(t), \boldsymbol{p}(t))$.

Mathematical models of dynamic systems can be divided into two large classes according to the adopted descriptive approach.

Physically based models (or mechanistic models): they are derived from a thorough knowledge of the physics of the process and based on first principles, typically conservation laws and constitutive equations. They use a hypothetical deductive approach and can be written at each representation scale.

Phenomenological models: they are derived from empirical knowledge obtained through quantitative observations in experiments in which measurements are carried out.

Mathematical models are called *stochastic models* when the uncertainty is taken into account (random variables and random parameters are considered); on the contrary, they are called *deterministic models*.

The structure and the characteristic variables of a physically based model are assumed starting from the a priori knowledge of the system. The parameters inside the model have physical meaning and, if they are not known, they can be directly or indirectly measured with appropriate experiments. Experimental data are not needed to determine the shape of the model. A physically based model provides predictions of the output variables but also information on the state variables, so it is considered a *white-box model* and its physical meaning is perceptible. Unfortunately, the high number of variables and parameters required to set up a physically based model affects computational complexity and sometimes influences the identification procedure, requiring quite "rich" time series of all the considered variables in order to correctly estimate the high number of model parameters.

Phenomenological models do not consider a priori information about the system, and the characteristic variables are identified during the model construction. Only a class of functions suitable for linking the identified variables in cause–effect relations is chosen. The parameters inside the model have no physical meaning at first and are those for which maximum compliance with the historical series of measures is made. Experimental data (time series for dynamic models) determine the shape of the model, i.e., both parameters and order of polynomials are estimated from them [74]. The structure of a phenomenological model is relatively simple, and therefore it is useable for simulation of EIs seen as interphase volumes and when particular cause–effect relations have to be investigated: only the interesting variables are introduced into the model and risk of parameters overestimation is highly reduced. Main disadvantages of these models are related to the fact that the parameters have no physical meaning and that the model itself cannot consider explicitly the spatial variability of the system. This means that for heterogeneous systems, an error can be inserted into the model, associated with the estimation of global parameters used instead of discrete parameters.

An intermediate class of models includes the so called *data-based mechanistic (DBM) models*. An inductive approach is used for the identification of the model structure, so they are similar to phenomenological models, but some effort is made to provide at least a partial description of the physical behavior of the system (hence the term mechanistic). The physical interpretations deduced by a DBM model are not said to be in conformity with the scientific paradigms, but the analysis of time series with this models can provide some useful information on the validity of some assumptions.

Formally, they are input–output models and can be used to describe nonlinear dynamic systems. Moreover, they are stochastic models consisting of two parts of which the former is a deterministic input–output link and the latter is a stochastic disorder representing the model error. Some of the first applications of these models were presented for modeling environmental systems [78,79,82,83], and their use in modeling EIs is promising.

Regardless of the adopted modeling approach, if EIs are modeled as interphase volumes, they can exhibit complex dynamic behavior as hysteresis, multiple steady states, limit cycles, onset to deterministic chaos. This complexity usually appears at greater scales at changing parameter values when some kind of nonlinearity is present into the mathematical model: a lot of biochemical systems [18,46] such as Belousov–Zhabotinsky reaction inside continuous reactors [27], simple controlled separation or reacting process units can show this behavior [26,54,56].

The insights on these issues are beyond the scope of this chapter, which has here the sole purpose of providing some food for thought on the open problems and challenges in the field of modeling in EIs. Some theoretical and technical in-depth analysis on these issues can be found in Mihailovic et al. [45].

S1. Suggested insights: the mixing volume at macroscopic scale

Let us consider one of the simplest representative systems that can be adopted to describe a portion of environmental matrix: regardless to the representation scale adopted, this is a quiescent mixing volume, where mixing is considered ideal, exchanging mass and energy with the external only by punctual input and

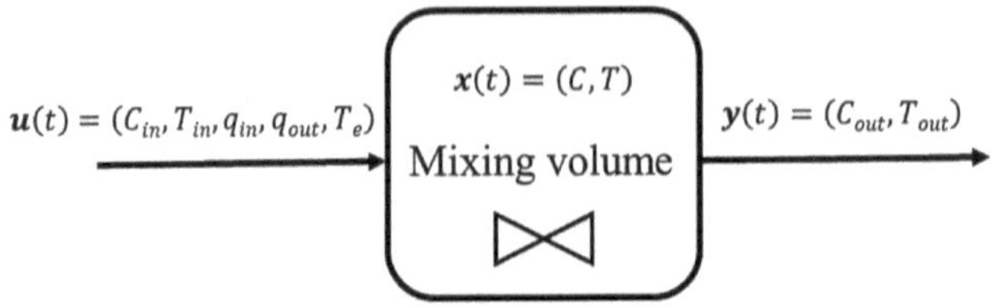

FIGURE 11.2 Mixing unit and its dynamic variables.

punctual output. At macroscopic scale, it can be a continuous stirred tank reactor used for the abatement of an organic pollutant in aqueous wastewater. State variables, input and output variables are uniform inside the mixing volume.

This system is represented in Fig. 11.2.

If a chemical reaction takes place inside the mixing volume, e.g., a biological degradation modeled with a pseudo-first order decay, under some conditions, this system can exhibit very complex dynamic behavior.

For this system, concentration C (mol m^{-3}) and temperature T (K) inside the volume are the state variables; C_{in} (mol m^{-3}), T_{in} (K), T_e (K), and flow rates q_{in} (m^3 s^{-1}) and q_{out} (m^3 s^{-1}) are the input variables (q_{out} is a dynamic input variable for the system since it can be measured and manipulated to change the system response); C_{out} (mol m^{-3}) and T_{out} (K) are the output variables. The simplest assumptions about the chemical–physical parameters is to consider reaction enthalpy ΔH_r, preexponential factor K_0, activation energy E_a, heat capacities C_p, $C_{P,in}$, $C_{P,out}$, heat transport coefficient U, and surface S as constant parameters, so the reacting mixing volume is a nonlinear time-invariant dynamic system. Its physically based, simplified, dynamic model results in:

i) mass and energy balances expressed by two ordinary differential equations (ODEs) where the state variables are the independent variables;

$$V\frac{dC}{dt} = q_{in}C_{in} - q_{out}C_{out} + Vv \tag{11.1}$$

$$V\rho C_p\frac{dT}{dt} = q_{in}C_{P,in}C_{in}T_{in} - q_{out}C_{P,out}C_{out}T_{out} - V\Delta H_r v + US(T - T_e) \tag{11.2}$$

ii) two algebraic equations linking output variables and state variables and expressing the hypothesis of ideal mixing;

$$C_{out} = C \tag{11.3}$$

$$T_{out} = T \tag{11.4}$$

iii) one constitutive equation describing the mechanism of decaying rate, depending on the state variables and on two chemical–physical parameters in case an Arrhenius-like kinetic law is assumed.

$$v = -KC \text{ where } K = K_0 e^{-\frac{E_a}{R_u T}} \tag{11.5}$$

The system is not linear, and classical methods to study roots and stability show that it can show instability and roots multiplicity, this means that the number of possible steady states changes at different values of some input variables (for example, the input temperature). Varying the parameter values has generally the effect of varying only quantitatively the properties of the orbits, while the overall picture of the dynamics in phase space remains unchanged [55].

If mixing is not ideal the system can also show input multiplicity, this means that different values of some input variables or parameters give the same output values [6,38].

Finally, if a feedback control applies on the system, e.g., a man-made controller as on a wastewater treatment unit [51], it can show all the features of the classical chaotic transitions as period doubling and limit cycles, intermittence, positive maximal Lyapunov exponents, and fractal structure of the limit attractor [53,59].

3. Physically based models for simulation of environmental interfaces

Physically based models establish their roots in first principles, namely conservation laws. These are employed, in synergy with constitutive relations characterizing the specific nature of materials or substances, as keystones to structure models that delve deeper on a priori information about the studied system or process, contrary to what happens for empirical approaches.

Confronting with the simulation of EIs means considering transport phenomena. Therefore, the balance equations on which to build the mathematical models usually include those of mass, energy, and momentum. EIs frequently involve multiple phases (liquid characterized by diverse densities and solids presenting dissimilar crystalline systems are commonly treated as separate phases) of which some are not quiescent fluids; fluid flows comprising *free moving interfaces* are thus of the utmost importance. They encompass bubbles and drops, jets, waves, and films. These, in particular, are, in essence, biphasic immiscible fluid flows [65]. The architecture of a physically based simulation model for said moving EIs rests on the conjunction between conservation of mass and energy with conservation of momentum, under constitutive equations prescribing the characteristics of superficial stresses that, in the case of Newtonian fluids, lead to the well-known NS equations.

To analyze a dynamic system subject to the action of a flow field, fluid dynamics teaches how it is necessary to choose a reference frame. The specification can fall on a point of view integral with a fluid particle, known as the Lagrangian approach, or on one that focuses on a portion of space in which the fluid flows over time, called the Eulerian approach. In the Eulerian treatment, the properties of the flow field are defined as a function of time t and position $\boldsymbol{x}$ in space with respect to an inertial reference system. The variables $(\boldsymbol{x},t)$ are called Eulerian variables and characterize a single position in the space where, as time changes, different fluid particles are found. The Lagrangian approach is instead based on the concept of a material element, that is a volume which moves with the fluid and consequently is always made up of the same liquid particles. In other words, the material element is a volume in which the distinct particles at the initial instant remain such even in the successive instants. The flow field properties can then be defined as a function of the dynamic behavior of the material element (i.e., fluid particle) considered as identified by the position $\boldsymbol{X}$ of its center of mass at a fixed initial moment t_0. Inter alia, this representation can be straightforwardly extended to a cloud of particles by considering its centroid. With the passage of time this material element will move and deform; its position at the generic instant t, calculated starting from t_0, is a function $\boldsymbol{X}(\boldsymbol{x}(t_0), t)$ linked to the Eulerian position through the trajectory. The $(\boldsymbol{X}, t)$ pairs are denoted as material coordinates. As a result of what has just been explained, it is then clear how the choice of one of the two perspectives affects the formulation of the equations of the model.

Conservation laws are, in general, nonlinear hyperbolic partial differential equations (PDEs). As such, global classical solutions oftentimes are not discernible. Instead of looking for sufficiently smooth solutions, one can either resort to weak formulations or to numerical methods. Again, an indicative example is given by the NS equations, of which the existence of a global solution is one of the most intricate and ambitious open problems taken on by the scientific community. The NS equations constitute the foundations of almost the totality of the mathematical models nowadays used in computational fluid dynamics (CFD), a now computer-aided branch of fluid mechanics that exploit numerical analysis methods to simulate the behavior of various physical properties for fluids flow. With computational power becoming less and less a limiting factor, CFD has emerged as the main tool to inspect solutions for fluids flow at various degrees of sophistication: they may consider solid interaction, phase changes, chemical reactions, heat and mass transfer.

CFD is at the forefront of multiphase flow studies. While attacking single-phase flow problems with CFD, one must choose between a Eulerian or Lagrangian specification of the field to determine a numerical solution that approximates the analytical one, for the multiphase cases the options are distinguished between Eulerian–Eulerian (E–E) and Eulerian–Lagrangian (E–L). In a nutshell, in the E–L paradigm the flow is described as a continuous fluid matrix in a Eulerian perspective while the dispersed phase, consisting of discrete elements, is followed in a Lagrangian manner. On the contrary, in E–E simulations both the carrier and dispersed phases are treated following the Eulerian point of view. In other terms, the different classifications of these numerical methods for the study of multiphase flows arise from the difference in the treatment of the characteristics of the dispersed phase. The adoption of a Lagrangian or Eulerian specification of the dispersed phase field, having marked repercussions on the mathematical formulation, determines differences also in the numerical constraints and techniques to be adopted.

The Eulerian description, sometimes referred to as multifluid method, envisions the different phases as interpenetrating continua that occupy the same volume simultaneously, treating their representative properties without distinction. Nevertheless, physically, the volume occupied by each phase cannot be shared with other ones, reason for which the concept of phasic void (or volume) fraction is introduced. These volume fractions, the sum of which equals one, are determined through averaging procedures, and those relative to the dispersed phases should yield high values as it would signify that the particles are too clustered to be treated as isolated [1]. The assumption of viewing the dispersed phases as continua holds if the smallest dimension of the boundaries of a boundary value problem proves to be larger than the average diameter of a particle. However, this argument can be legitimized by the means of an averaging perspective [21]. In this sense, adopting a jargon more akin to the field of CFD, one can assert that whenever alterations in meshgrid cell sizes or position do not substantially affect the predicted distribution of the particle concentration, the hypothesis is satisfied [39]. This means that the phases manifest mixing phenomena at length scales smaller than those intended for the simulation. In E–E models, the governing transport equations generate a set of coupled PDEs associated with each considered phase, subjected to the averaging process, resulting in a homogenization. It is possible to perform this latter step in different ways, all of which may bear effects on the equations describing the interaction mechanisms: by carrying out volume averaging, time averaging, and ensemble averaging [28]. As these procedures do not truncate the connection of some undetermined terms (stress tensors and interaction force exchanged by the phases) with details of the microscopic scale motion by directly linking them to the average variables, closure problems emerge. In the majority of cases, it is not feasible to solve them analytically, rather calling for the use of empirical expressions [36]. For that matter, phenomenological correlations, in the form of interphase exchange coefficients, represent a pivotal element in the modeling of physical quantities transfer.

The aftereffect of the discretization of the conservation equations in Eulerian form is that each node of the meshgrid contains cell-averaged particle-phase variables, opening the door to the use of the same grid for both carrier and dispersed phases. This leads to one of the main advantages of E−E methods: average interphase coupling terms are handled alike, boosting efficiency and accuracy of the method when dealing with two-way coupling effects [39].

Conversely, availing oneself of an E−L method involves, for the treatment of dispersed phases, the resolution of ODEs along the trajectories of the particles, for whose integration the interpolation of the properties of the carrier phase to the centroids of the parcels is demanded. Moreover, this passage is crucial since discontinuous phases can exchange momentum, mass, and energy with the carrier phase, with this latter described in the usual Eulerian way. As it is the case for E−E approach, E−L necessitates an averaging step, which pertains to all the volumes of the particles whose centroids are inside a discrete cell volume. The aim is to move from the distribution of the localizations of the particles to discrete volume averages, thus allowing particle concentration prediction [39]. A contraindication manifests itself in the removal of small-scale fluctuations from the model, which, if they are to be taken into account, must be restored by the means of appropriate stochastic models (e.g., Ref. [49]). Besides, collisions of particles with walls or between themselves, clustering, coalescence, and breakups are all phenomena that can be included in E−L models without excessive difficulties.

Wanting to deepen the merits and demerits of the E−E and E−L strategies, it is possible in the first instance to note the greater computational sustainability of the purely Eulerian approaches, markedly less demanding than mixed E−L ones, especially with respect to systems with higher dispersed volume fractions [20]. The inability to know the position of individual particles represents a deterrent to the field of application of the E−E specification for what regards interfaces, as it is less suitable to manage different size distributions since it implies that the speed of all the particles inside a cell is the same. Nevertheless, this is currently probably the only realistic alternative when dealing with a very large population of particles. E−E methods may suffer from numerical diffusion problems, which are nevertheless manageable through the adoption of adequately high-order discretization schemes through which the same order of accuracy of E−L models can be achieved [66]. Albeit it allows to assess the interactions between particles in a deterministic manner [7], the E−L approach not only carries a higher computational load but it is not well-suited for parallel computing as the need to couple diverse solvers dramatically heighten the complexity of the implementation when checked against single-phase code [24]. By the same token, this limit worsens with the increase in the number of particles tracked.

Where both multifluid and mixed E−L methods fail, unless adequate and specific farther modeling is introduced, are the accurate descriptions of shapes and behavior of the elements of the dispersed phases. This may represent a gleaning limitation when the focus falls on deformable interface dynamics, e.g., for the study of the behavior of bubbles. For this task, two classes of strategies may prove to be especially interesting: front-tracking (FT) and front-capturing (FC) methods. In FT methods, the interface (front) is tracked explicitly through marker particles and the mesh has to be updated as the flow evolves; FC ones hinge on the advection of implicit marker functions on fixed spatial domains. Therefore, the former can be considered as belonging to the Lagrangian class, whereas the latter exploits the Eulerian framework. In brief, the possibility to handle implicitly the changes of topologies, without the need of detecting and reconstructing the merging or splitting of fronts, represents the main advantage of FC methods, while FT techniques perform better in terms of accuracy [25]. This latter can be further improved by increasing the number of elements representing the fronts, obviously hindering the computational speed. All things considered, the main drawback ascribable to the FT class is its complexity. The FT approach, which is used in the context of fluid fronts dynamics dates back to 1969 [17] and whose development and diffusion is mainly due to research groups of Glimm and Tryggvason, performs the localization of the fronts along the trajectories of the fluid particles in a Lagrangian way by representing each interface as a set of points whose connective relationship is provided by a parametric description. These concatenated marker points of the various phase boundaries are advected by the flow velocity, interpolated from a Eulerian stationary grid with which the governing equations are handled, and the parametric fit is iteratively reevaluated and reestablished, with this algorithm being performed for each separate front [62]. Earlier versions of the FT method presented further intricacies as the moving boundaries of the phases were traced by altering the fixed mesh near the fronts so that the grid lines followed the interfaces. Adding to that, new computational points have to be incorporated, for those instances, in the regions where the grid stretches while some are removed where there is compression [73]. Also, automatic coalescence is not a characteristic of FT: for two fronts to merge the marker points have to be reordered and the parametcrization reconstructed by the new and expanded set of points.

From the use of a Lagrangian framework follows that numerical smearing can be avoided. It is an unwanted product of the advection step possibly affecting properties like density and viscosity that on the whole computational domain are assimilated to discontinuous fields (the so-called one-fluid formulation). For further information on the FT topic, it is recommended to refer the works authored by Tryggvason and collaborators [58,71,72].

The one-fluid formulation pertains not only to the FT methods but also to the FC ones. In fact, the general structure that essentially distinguishes all the FC and FT algorithms relies primarily on a flow solver that uses some projection technique to solve the NS equations and usually a Poisson pressure equation on a Eulerian mesh. On the other hand, it is necessary to take into account the different material characteristics of the phases involved, essentially contained in the density and viscosity scalar fields, as well as interfacial phenomena, such as the onset of surface forces that translate into surface tension. By crossing the front that separates two phases, the change of the values of the aforesaid scalar fields will ideally manifest itself with jump discontinuities, due to the theoretically negligible thickness of the interface. The one-fluid formulation allows to keep the notation compact, ideally combining the various multiphase equations in a single equation including the jump conditions at the interface. Practically, the modeling of the interfacial terms concentrated on the front is carried out with Dirac delta distribution δ, while the discontinuity in material properties is conveyed through the Heaviside step function H [58].

The one-fluid formulation is an essential ingredient also for the two main representatives of the FC class: the volume of fluid (VOF) method [29] and the level set method (LSM) [50]. They both diversify the characteristics of the subdomains (e.g., two fluids making up a gas–liquid system) through the marker function and thus solving the conservation equations for the entire flow field. Relying on a phase indicator field means a more refined and accurate strategy for the exact identification of fronts with respect to a multiple fluids formulation, avoiding the empiricism typical of the averaging models to predict the physics of the fronts driving the flow [35]. These fixed-grid methods handle the surface tension–induced force numerically through the adoption of the continuum surface force (CSF) method [9], which deals with the relative source term necessarily introduced in the momentum equations.

In VOF method, the scalar indicator function F is discontinuous as it is a step function that maps each point of the computational domain either to 0 (interior phase) or 1 (exterior phase). The values of the marker function between these extrema identify the presence of the front. F is then transported by the means of solving a time-dependent advection equation. Anyway, the necessity of front reconstruction arises due to the representation of the moving boundaries as a volume fraction field ($0 < F < 1$); the discontinuities across the interface cause issues in the transport step if adequately precise numerical schemes are not employed and have repercussions on the volume conservation as well. While conventional convective schemes like upwinding may not be viable due to numerical diffusion that distorts the front profile [22], various alternatives are available for the purpose of sharpening and reconstructing the fronts, the most popular of which is probably the piecewise linear interface calculation scheme [60]. This method is based on the approximation obtained by using oblique or piecewise linear segments, highly accurate, extendable to three-dimensional cases (plane segments are used instead of straight lines), and applicable even to complex flows.

The continuous counterpart of VOF is LSM, where F is substituted by the so-called level set function ϕ, a higher dimension map which zero-level set represents the front. ϕ is a time- and space-dependent signed distance field that assumes positive values on one side of the front, negative on the other, and zero at the interface. The initial condition is set so that the zero iso-contour of ϕ at starting time $t = 0$, coincides with the initial location of the front and can

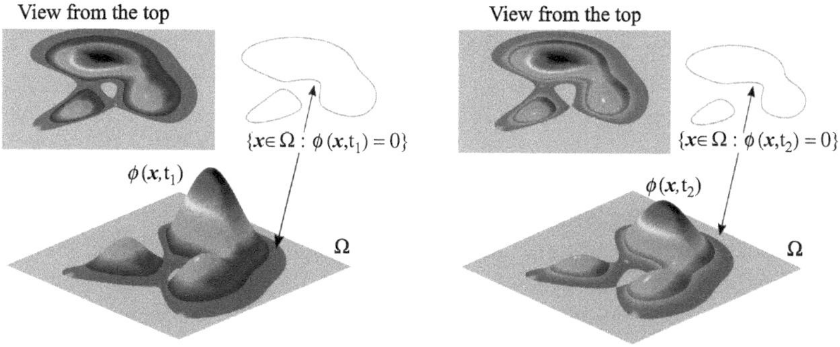

FIGURE 11.3 Illustration of the level set method.

therefore be set to the signed distance to it, meaning that the domain Ω is subdivided into two distinct regions Ω^+ and Ω^- $(\Omega^+ \cap \Omega^- = \varnothing)$ by their boundary, i.e., $\phi_0 = \phi(\boldsymbol{x}, t = 0)$. LSM presupposes the computation of the zero level set of ϕ at each time step, in order to obtain the front propagation (see Fig. 11.3).

An anticipated side effect of the interface propagation consists in the need of reinitializing the distance profile, as a loss in smoothness of field ϕ is to be expected. In fact, significant velocity gradients can determine substantial steepness or flatness of the level set function, unhitching ϕ from its distance function nature. Sussman et al. [67] proposed an iterative reinitialization of ϕ that, albeit subject to Courant−Friedrichs−Lewy limitations, bears satisfying results and has been widely employed.

Among the advantages deriving from the use of LSM in place of VOF, the streamlined extensibility of two-dimensional models to three dimensions, the easier and more precise computation of normal vector and curvature that leads to a simpler treatment of the advection procedure, is cited. On the downside, since interphase is smeared out instead of sharp, volume conservation is not achieved, fact that can jeopardize the simulations. In this regard, variation of the standard LSM was proposed over the years to cope with this critical issue, see, for example, the conservative level set method (CLSM) [47,48] or the accurate conservative level set method (ACLSM) [19]. In CLSM, the phase indicator function ϕ is substituted with Ψ, a piecewise function equitable with a hyperbolic tangent that maps the interface as its 0.5 level set while assuming 0 value for one phase and 1 for the other. Artificial compression is introduced by the means of a reinitialization equation apt to contain numerical diffusion in the solution of the advection PDE, which is written in conservative form. A trade-off between volume conservation (markedly influenced by the mesh step size) and overall accuracy characterizes CSLM. In fact, the creators of this method called for the use of low-order total variation diminishing (TVD) schemes to deal with numerical issues in interface normal vector computation, critical for a precise calculation of the surface tension force, diminishing the accuracy of the method. ACLSM proposes to increase the robustness of the numerical method by resorting to the fast marching method to reconstruct a signed distance function from which the normal vector field is then evaluated.

Other notable members of the collection of FC methods include the phase field method [32], that hinges on the definition of a density function that varies continuously at the front, and the combined level set volume of fluid method [68], that focuses on the strengths of VOF and LSM, exploiting their synergy to overcome the inaccurate evaluation of normal vectors at the interface typical of VOF and the poor volume conservation capabilities of LSM.

4. Phenomenological models for simulation of environmental interfaces

Phenomenological models are not derived from first principles and simply describe the empirical relationship among the system variables. Some authors characterize these models as the complete opposite of physically based models, in the sense that they must be totally independent from theory [44]. On the contrary, we prefer only to say that they are not derived from first principles, and in this sense they can incorporate some physics laws [23].

With this definition, many constitutive equations are grouped within the class of phenomenological models and can contain parameters related to the properties of a chemical substance, as the Fick law or the Arrhenius law. The reason is that in their first formulation many constitutive equations were mainly based on observations, and only later a physically based approach, adopted to check their consistency, allowed to attribute a physical meaning to the parameters contained in the model.

The structure of the phenomenological models is generally chosen among families of mathematical functions, so they are often linear or nonlinear algebraic equations; but nothing prevents them from being differential equations.

The term phenomenological model is often used as a synonymous of black-box model, conversely we prefer to include into the class of phenomenological models also empirical models describing what happens inside the system under study. Therefore, if the model adopted to simulate the dynamic system uses the state variables into the empirical relationship, we have a white-box phenomenological model, otherwise, if the model equations link only the input and output variables, the model is a black-box type. Intermediate empirical gray-box models can be considered.

White-box models of dynamic systems are (or can be reduced to) a state-space representation of the dynamic system, where a set of input, state, and output variables are related by first-order differential equations in the state variables and by algebraic equations in the output variables. A deterministic continuous formulation of state-space models can be written as:

$$\frac{d\boldsymbol{x}}{dt} = \boldsymbol{G}(\boldsymbol{x}(t), \boldsymbol{u}(t)) \tag{11.6}$$

$$\boldsymbol{y}(t) = \boldsymbol{H}(\boldsymbol{x}(t), \boldsymbol{u}(t)) \tag{11.7}$$

where $\boldsymbol{x}(t)$ is the vector of the state variables, and $\boldsymbol{u}(t)$ and $\boldsymbol{y}(t)$ are respectively the vectors of input and output variables.

Since state variables are the internal variables of the system and a state-space model can describe their evolution in time, the model is said "white." The behavior of the output variables is directly related to the values of the state variables and the state space is the Euclidean space in which state variables are represented on the Cartesian axes. This representation provides a compact way to develop models for multiple input and multiple output systems.

S2. Suggested insights: state-space model of the mixing volume

Taking again as an example the reacting mixing volume of Fig. 11.2, a phenomenological model of this system using Eqs. (11.6) and (11.7) can be written as:

$$\frac{dC}{dt} = G1(q_{in}, C_{in}, T_{in}, q_{out}, C, T, V); \quad \frac{dT}{dt} = G2(q_{in}, C_{in}, T_{in}, q_{out}, C, T, V) \tag{11.8}$$

$$C_{out} = H1(q_{in}, C_{in}, T_{in}, q_{out}, C, T, V); \quad T_{out} = H2(q_{in}, C_{in}, T_{in}, q_{out}, C, T, V) \tag{11.9}$$

It is straightforward to notice that the physically based model formulated using mass and energy conservation laws and described by Eqs. (11.1–11.5) is a particular case of the state-space phenomenological model described by Eqs. (11.8) and (11.9).

It has been said that the need of modeling EIs arises from the fact that EIs are not simply *separation systems* between the different environmental compartments, but they can be complex *interphase* systems inside which transport phenomena, biological and chemical transformations take place. Since sometimes the physically based interpretation of the phenomena inside the EIs is very difficult, and the phenomenological state-space models require the measurement of state variables inside the interface, the use of a data-driven, bottom-up approach based on black-box phenomenological models becomes a must.

This approach represents the intermediate choice between the simplest, steady-state algebraic model of the EI that uses global transport coefficients, and the complete representation of the EI as a dynamic interphase volume.

A black-box phenomenological model is in fact a simplified representation of the dynamics of transport and transformation processes: the knowledge of what happens inside the EIs, i.e., the dynamic behavior of the state-variables, is not taken into consideration, and the evaluation of the interface dynamics is related only to its outputs. A block-box phenomenological model (sometimes called surrogate model, even if this term is of recent use and is main related to some particular bottom-up computer simulation models of complex systems) is developed when the only available information about the system is the values of the input and output variables. This is one of the current approaches in modeling multiphase environmental systems such as contaminated groundwater [4] or composite materials [61].

A basic dynamic black-box model computes the actual output as a function of all the past history of both output and input variables; if some *additional* parameters are introduced, the model is called *adaptive black-box model* and the additional parameters can be used as independent variables (degrees of freedom) to better adapt the model to reality (by a model identification procedure and parameters optimization). In a black-box model, the errors previously made in calculating (some of) the output variables can also be introduced as predictors. The role played by the error term is to internally consider all the variables and disturbances that influence the process to be modeled, but which are not explicitly taken into account in the model.

In order to set up a phenomenological black-box model, the following steps are required:

- Selection of the set of inputs variables $\boldsymbol{u}=(u_1, u_2, \ldots, u_m)$, output variables $\boldsymbol{y}=(y_1, y_2, \ldots, y_r)$, and errors $\boldsymbol{e}=(e_1, e_2, \ldots, e_q)$, $q \leq r$, if any;
- Selection of the total sampling time, of the amplitude of sampling intervals for each variable and of delays (if any);
- Selection of the class of functions that can describe the empirical relationship between input and output variables;
- Regression of the model with respect to the observations using model parameters (whose number depends on both the chosen sampling intervals and class of functions) as independent variables.

Named $n_u=(n_{u1}, n_{u2}, \ldots, n_{um})$, $n_y=(n_{y1}, n_{y2}, \ldots, n_{yr})$, and $n_e=(n_{e1}, n_{e2}, \ldots, n_{eq})$, the number of time sampling intervals, respectively, for the input, output, and the error variables, a black-box model with error can be written as:

$$\begin{aligned}\boldsymbol{y}(t) = \boldsymbol{f}[&y_1(t-1), y_1(t-2), \ldots, y_1(t-n_{y1}), y_2(t-1), \ldots, y_2(t-n_{y2}), \ldots,\\ &y_r(t-1), \ldots, y_r(t-n_{yr}), u_1(t-1), u_1(t-2), \ldots, u_1(t-n_{u1}), \ldots,\\ &u_m(t-1), \ldots, u_m(t-n_{um}), e_1(t-1), e_1(t-2), \ldots, e_1(t-n_{e1}), \ldots,\\ &e_q(t-1), \ldots, e_q(t-n_{eq})]\end{aligned} \tag{11.10}$$

S3. Suggested insights: black-box model of the mixing volume

Taking again as an example the reacting mixing volume of Fig. 11.2, a phenomenological black-box model able to describe how the inlet temperature acts on the concentration in exit from the volume does not need the knowledge of all the involved variables and can be simply constructed as follows:

1) Selection of one input variable $\boldsymbol{u}=(u_1)=(T_e)$ and two output variables $y=(y_1,y_2)=(C_{out}, T_{out})$;
2) Selection of a total sampling time of 100 h; temperature sampling intervals (e.g., measured by a thermocouple) of 100 ms and concentration sampling intervals (e.g., by online HPLC) of 30 s.

Eq. (11.10) becomes:

$$\begin{aligned} C_{out} = f_1 \Bigg\{ & C_{out}(t-1), C_{out}(t-2), \ldots, C_{out}\left(t - 100 * \frac{3600}{30}\right), \\ & T_{out}(t-1), \ldots, T_{out}(t-2), \ldots, T_{out}\left(t - 100 * \frac{3600}{0.1}\right), \ldots, \\ & T_e(t-1), T_e(t-2), \ldots, T_e\left(t - 100 * \frac{3600}{0.1}\right), \\ & \left[C_{out}(t-1)_{sper} - C_{out}(t-1)_{model}\right], \left[C_{out}(t-2)_{sper} - C_{out}(t-2)_{model}\right], \\ & \left[C_{out}\left(t - 100 * \frac{3600}{30}\right)_{sper} - C_{out}\left(t - 100 * \frac{3600}{30}\right)_{model}\right], \\ & \left[T_{out}(t-1)_{sper} - T_{out}(t-1)_{model}\right], \left[T_{out}(t-2)_{sper} - T_{out}(t-2)_{model}\right], \\ & \left[T_{out}\left(t - 100 * \frac{3600}{0.1}\right)_{sper} - T_{out}\left(t - 100 * \frac{3600}{0.1}\right)_{model}\right] \Bigg\} \end{aligned} \tag{11.11}$$

If phenomenology suggests that the input variables have a delayed effect on the system outputs, some delay time intervals can be introduced, whose number is equal to the dimension of the input vector. The model in this case becomes:

$$\begin{aligned} \boldsymbol{y}(t) = \boldsymbol{f}[& y_1(t-1), y_1(t-2), \ldots, y_1(t-n_{y1}), y_2(t-1), \ldots, y_2(t-n_{y2}), \ldots, \\ & y_r(t-1), \ldots, y_r(t-n_{yr}), u_1(t-n_{k1}-1), u_1(t-n_{k1}-2), \ldots, \\ & u_1(t-n_{k1}-n_{u1}), \ldots, u_m(t-n_{km}-1), \ldots, u_m(t-n_{km}-n_{um}), \\ & e_1(t-1), e_1(t-2), \ldots, e_1(t-n_{e1}), \ldots, e_q(t-1), \ldots, e_q(t-n_{eq})] \end{aligned} \tag{11.12}$$

By defining the *regression vector* $\boldsymbol{\Psi}(t)$, as:

$$\begin{aligned}\boldsymbol{\Psi}(t) = [&y_1(t-1), y_1(t-2), \ldots, y_1(t-n_{y1}), y_2(t-1), \ldots, y_2(t-n_{y2}), \ldots,\\ &y_r(t-1), \ldots, y_r(t-n_{yr}), u_1(t-1), u_1(t-2), \ldots, u_1(t-n_{u1}), \ldots,\\ &u_m(t-1), \ldots, u_m(t-n_{um}), e_1(t-1), e_1(t-2), \ldots, e_1(t-n_{e1}), \ldots,\\ &e_q(t-1), \ldots, e_q(t-n_{eq})]^{\mathrm{T}}\end{aligned} \quad (11.13)$$

where its components are called the *regressors*, the choice of a nonlinear mapping $\mathbf{y}(t) = \mathbf{f}(\boldsymbol{\Psi}(t), \boldsymbol{p})$ consists in how to choose the regression vector $\boldsymbol{\Psi}(t)$ and how to choose the nonlinear mapping $\mathbf{f}(\boldsymbol{\Psi}(t), \boldsymbol{p})$.

The nonlinear mapping $\mathbf{f}(\boldsymbol{\Psi}(t), \boldsymbol{p})$ is usually expressed by the expansion of some *basis function* f_k (polynomial, sigmoid, wavelet), as:

$$\mathbf{f}(\boldsymbol{\Psi}(t), \boldsymbol{p}) = \sum a_k f_k(\boldsymbol{\Psi}(t), \boldsymbol{p}) \quad (11.14)$$

and f_k can be obtained by parameterizing a single "mother function." One of the most used expansion is the linear polynomial. A review on this subject can be found in Ref. [64].

As regards the choice of the regression vector, linear black-box structures can be summarized by the general family:

$$A(z)y(t) = \frac{B(z)}{F(z)}u(t-n) + \frac{C(z)}{D(z)}e(t) \quad (11.15)$$

where z is the shift operator, so $A(z)$ is a polynomial in z^{-1}.

When $A(z) = 1$, Eq. (11.15) reduces to the Box–Jenkins models and when $A(z) = 1$, $C(z) = 1$, and $D(z) = 1$ to the output-error model.

When both $F(z) = 1$ and $D(z) = 1$, Eq. (11.15) reduces to the linear autoregressive moving-average with exogenous (ARMAX) model written as:

$$A(z)y(t) = B(z)u(t-n) + C(z)e(t) \quad (11.16)$$

The special cases of Eq. (11.16) are known as ARX model when $C(z) = 1$, ARMA model when $B(z) = 1$, AR model when both $B(z) = 1$ and $C(z) = 1$. Finally, if both $A(z) = 1$ and $C(z) = 1$, the ARMAX model reduces to the finite impulse response model FIR. ARMAX and derived linear black-box models cannot describe multiplicity of steady states.

By following the nomenclature adopted for the linear models, we can define the corresponding NARMAX model that uses the complete *regression vector* $\boldsymbol{\Psi}(t)$, i.e., containing all the variables $\boldsymbol{y}^{t-1}$, $\boldsymbol{u}^{t-1}$, and $\boldsymbol{e}^{t-1}$ as regressors, NARX that uses $\boldsymbol{y}^{t-1}$ and $\boldsymbol{u}^{t-1}$ and NFIR that only uses regressors $\boldsymbol{u}^{t-1}$.

NARMAX models can describe complex dynamics. It is interesting to notice that a NARX model (polynomial nonlinear ARX), being nonlinear in the variables can describe complex dynamics, but since it is linear in the parameters, its identification procedure (parameters estimation) can be carried out with a least square algorithm [13].

Before concluding, it is interesting to observe that a discrete linear state-space time invariant model in predictor form can be written as:

$$\begin{aligned} x(t+1) &= Ax(t) + Bu(t) + K[y(t) - Cx(t) - Du(t)] \\ y(t) &= Cx(t) + Du(t) + e(t) \end{aligned} \tag{11.17}$$

where $y_{mod}(t) = Cx(t)+Du(t)$ is the predictor, and all the states x(t) are the regressors. If compared with black-box models, it is easy to observe that this "white" model is efficient with a smaller number of regressors [64].

S4. Suggested insights: NARX model for biosensors feasibility

A great number of research projects have been developed in the last 20 years about the design of biosensors aimed at detecting contaminants in water, soil, and food, and discriminating their effects. Biosensors production at reasonable costs could be promising in view of their massive adoption for large-scale monitoring campaigns devoted to assess ecological and human health risk.

This kind of biosensors is based on cells which are the natural target of the pollutant under study. The use of spheroids of coelomocytes of sea urchin as an in vivo model, immobilized on some nanoporous support seems to be a promising technique to detect pesticides in seawater. Sea urchin has in fact metabolic and physiological pathways similar to mammals and has notable genomic homology with high vertebrates, so it could be a biological model for human health risk assessment. The inhibition of acetylcholinesterase (AChE) by organophosphates and carbamates is the output mechanism that can be detected on these cells as a response to water contaminated by pesticides [2,57]. In standard living conditions if acetylcholine is added to cells a temporary depolarization on the cellular membrane appears, on the contrary when pesticides are present they inhibit AChE and the residual acetylcholine produces a continuous stimulation of the cells leading to membrane depolarization above a threshold. While a lot of effects can be detected offline on these cells by laboratory experiments, their use in living biosensors needs a previous feasibility study on their response dynamics to changing input conditions (pollution levels).

A detailed dynamic model of the EI represented by the cellular membrane is very difficult, but as a first step in the biosensor feasibility study a simple single input single output (SISO) NARX model can be developed, using pesticide concentration in seawater as input variable and potentiometric measurements as the output detectable variable. The choice of this model is due to the need to describe any possible nonlinear and complex dynamic behavior while maintaining a fairly simple model identification procedure. Since the AChE inhibition mechanism acts as a threshold-linked response, only few past time intervals can be taken into account inside the model.

As an example, by using two input elements $u(t-1)$, $u(t-2)$ and two output elements $y(t-1)$, $y(t-2)$, the model becomes:

$$\begin{aligned} y(t) = [&a_1 y(t-1) + a_2 y(t-2) + a_3 u(t-1) + a_4 u(t-2) + a_5 y(t-1) y(t-1) \\ &+ a_6 y(t-1) y(t-2) + a_7 y(t-1) u(t-1) + a_8 y(t-1) u(t-2) \\ &+ a_9 y(t-2) y(t-2) + a_{10} y(t-2) u(t-1) + a_{11} y(t-2) u(t-2) \\ &+ a_{12} u(t-1) u(t-1) + a_{13} u(t-1) u(t-2) + a_{14} u(t-2) u(t-2)] \end{aligned} \tag{11.18}$$

It is interesting to notice that despite the reduced number of regressors, this NARX model has 14 parameters (while using a cubic polynomial the parameters would have been 83) and parameters overestimation could occur.

DBM models are some different phenomenological models, sometimes classified within the black-box models. They too are based on a transfer function approach developed exclusively from experimental data and without any a priori mechanistic assumption, but they are then *interpreted in a physical way by inferring from the transfer function the main characteristic of the physical process.*

The term *data-based mechanistic modeling* was first adopted by Young and Lees [80], and this approach was successfully used in predicting a bunch of problems related to environmental, biological, ecological, and economic complex systems, trying to overcome any subjective belief in interpreting the modeled phenomenon [79]. Since DBM modeling requires adequate data sets to be completely successful, its development stages also include considerations about the representation scale and data availability at the chosen scale, particularly as they relate to the defined modeling objectives. However, the prior assumptions about the form and structure of the DBM model are kept at a minimum in order to avoid the prejudicial imposition of untested perceptions about the nature and complexity of the model, needed to meet the defined objectives.

Therefore, the appropriate *model structures* are identified by a process of relatively objective statistical inference applied directly to time-series data and based initially on a given generic class of linear transfer function models, whose parameters are allowed to vary over time, if it seems necessary to satisfactorily explain the data. Regardless of whether the model is identified in linear, piecewise linear or nonlinear form, the parameters characterizing the identified model structures are estimated using methods of statistical estimation for dynamic systems [83]. Furthermore, the estimated model is only accepted as a credible representation of the system if, in addition to well explaining the data, it also provides a description that has direct relevance to the physical reality of the system under study at the scale of interest. This is the most important aspect of DBM modeling and differentiates it from more classical black-box modeling techniques, such as those associated with standard Box–Jenkins transfer functions, NARMAX representations, neural networks, and neuro-fuzzy models. On the contrary, the aspect that differentiates DBM modeling from deterministic, physically based top-down methods is its inductive approach to model synthesis: rather than trusting in an a priori conceptual model form, the DBM structure is directly inferred from the observed data in relation to a more general class of models.

As a last step in dynamic modeling, common to both physically based and phenomenological modeling approaches, the estimated DBM model is tested to ensure that it is conditionally valid [81]. This step also involves predictive validation, as well as model applications under stochastic uncertainty and sensitivity analysis of model output using Monte Carlo simulation methods [63].

Since a DBM model is formally a black-box model, it is only dominated by input–output relationships. A DBM model with linear transfer functions, where nonlinearity only resides eventually in the definition of the input variables through nonlinear functions $f(u(t))$, can be written as follows:

$$y(t) = \frac{B(z)}{A(z)} f[u(t-n)] + \varepsilon(t) \tag{11.19}$$

where $\varepsilon(t)$ represents a general stochastic noise that accounts for the combined effects of different factors other than the input variables $u(t)$ affecting the output $y(t)$. Generally, $\varepsilon(t)$ is the uncertainty in the relationship between the input and output variables due to the measurement noise, to the effects of other unmeasured inputs and to the modeling error. It is usually modeled as an AR or an ARMA stochastic process [81].

In order to estimate the parameters inside the transfer function the model can be written in a vector form as:

$$\boldsymbol{y}(t) = \boldsymbol{\Psi}(t)^{\mathrm{T}} \boldsymbol{p} + \boldsymbol{A}(z) \boldsymbol{\varepsilon}(k) \tag{11.20}$$

In this case, when linearity dominates the dynamic relationship between the nonlinear functions of the input variables, the instrumental variable (IV) estimation algorithms can provide a robust approach to model identification, as they always produce consistent estimates of model parameters with minimum bias and exhibit close to optimal performance.

5. Modeling the dynamic behavior of G–L interfaces by LSM—a case study

An example of the usefulness of physically based models is offered by the tracking of the moving gas–liquid interfaces in a bioreactor. Specifically, the case study is based on the level set method adopted to model the behavior of gas bubbles, crucial aspect for the fine-tuning of the operation of reactors such as bubble columns and airlift reactors (ALRs).

Bioreactors constitute the fulcrum of any bioprocess so their optimization is the way to go for improving the efficiency of the whole process. ALRs are a class of pneumatic photobioreactors that achieve high mixing of the contained phases through the triggering of a swirling motion of natural circulation due to density gradients, all without high energy demand. Their field of application covers microorganism growth, cultured plant cells in particular. More recently, as the use of microalgae for environmental engineering purposes gained steam, their implementation for microalgae nurture aimed at wastewater treatment, CO_2 sequestration, and (third generation) biofuels production become more investigated. Interest in ALRs for microalgae growth is especially spiked by high yields net of energy consumption for medium/low mixing.

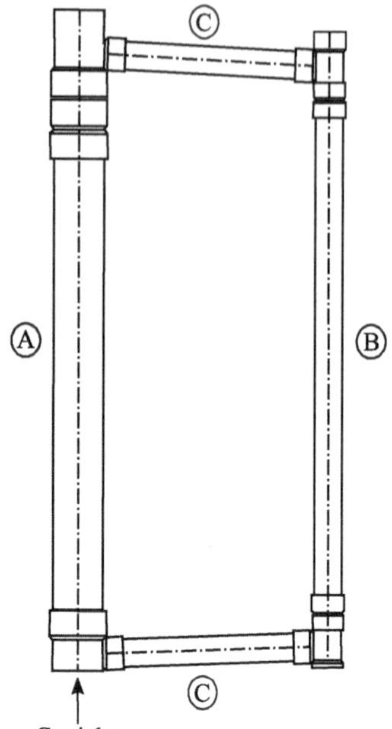

FIGURE 11.4 External loop airlift reactor: (A) riser; (B) downcomer; (C) horizontal collectors.

In the following, LSM is employed to model bubble dynamics in the riser of a pilot-scale external loop airlift photobioreactor in which microalgae are used for carbon dioxide capture from flue gas and for wastewater treatment. External loop ALRs are a particular subclass of ALR whose structural morphology articulates in two transparent, vertical tubes, possibly of different diameters, called riser and downcomer because of the fluid flow main direction, connected by as many horizontal collectors (see Fig. 11.4).

Bubble size is a pivotal aspect as it reflects itself on mass transfer. Their trajectories influence permanence time. The optimization of the size and shape of the bubbles ensures a good mass transfer between gas and liquid and guarantees the recirculation of microalgae. Knowledge of the flow regime is thus a crucial information to optimize processes inside ALRs. When biological reactants such as microalgae are used and excessive turbulence in the flow could damage the cells, the possibility of manipulating the flow regime of the bubbles during microalgae growth becomes a major goal. All this motivates the study and simulation of the phenomenon in detail.

Recalling what LSM entails (see section 11.3), indicating with $\Omega \subset R^n$ the spatial domain and with T the final time, the level set scalar function $\phi:\Omega \times [0,T] \rightarrow R$ is the solution of the Hamilton–Jacobi (HJ) equation:

$$\begin{cases} \dfrac{\partial \phi(\boldsymbol{x},t)}{\partial t} = \boldsymbol{u}(\boldsymbol{x},t)\|\nabla \phi(\boldsymbol{x},t)\| \\ \phi(\boldsymbol{x},0) = \phi_0(\boldsymbol{x}) \end{cases} \tag{11.21}$$

where $||\cdot||$ is the Euclidean norm, $\nabla=(\partial/\partial x,\partial/\partial y,\partial/\partial z)$ is the vector of spatial partial derivatives so that $\nabla\phi$ is the gradient of ϕ, $\phi_0:\Omega\rightarrow R$ a given initial condition, and $\boldsymbol{u}:\Omega\times[0,T]\rightarrow R$ is the velocity vector field. Moreover,

$$\begin{cases} \phi(\boldsymbol{x},t) > 0 & x\in\Omega^+ \\ \phi(\boldsymbol{x},t) = 0 & x\notin\{\Omega^+,\Omega^-\} \\ \phi(\boldsymbol{x},t) < 0 & x\in\Omega^- \end{cases} \tag{11.22}$$

being Ω^+ the subdomain occupied by the carrier phase, which in this case is constituted by water (and microalgae), Ω^- the dispersed one, i.e., gas. The points $\boldsymbol{x}\notin\{\Omega^+,\Omega^-\}$ form the boundary of these two sets, i.e., the G–L interfaces.

In order to solve Eq. (11.21) and obtain the propagation of the fronts, provided by the zero level set of ϕ, the knowledge of $\boldsymbol{u}$ is required. To determine it, it is necessary to solve the NS equations. The local formulation of the NS equations, obtained from momentum balance by resorting to the divergence theorem in conjunction with Reynolds transport theorem, states that:

$$\rho\frac{\mathrm{D}\boldsymbol{u}}{\mathrm{D}t} = -\nabla p + \left(\zeta + \frac{1}{3}\mu\right)\nabla(\nabla\cdot\boldsymbol{u}) + \mu\Delta\boldsymbol{u} + \boldsymbol{F}_b \tag{11.23}$$

The ratio $\mathrm{D}\boldsymbol{u}/\mathrm{D}t$ denotes the material derivative of the velocity $\boldsymbol{u}$, ρ is the density, p is the pressure of the fluid, μ is the dynamic viscosity while ζ represents the volume viscosity (also known as second viscosity coefficient), $\Delta = \nabla^2$ is the Laplacian, and $\boldsymbol{F}_b$ indicates other body forces (per unit volume) such as gravitational force per unit volume $\rho\boldsymbol{g}$.

As it is well known, for Newtonian, incompressible fluids, the continuity equation applies. It describes the velocity field $\boldsymbol{u}$ as a solenoidal vector field:

$$\nabla\cdot\boldsymbol{u} = 0 \tag{11.24}$$

It follows that, if one considers $\boldsymbol{F}_b = \rho\boldsymbol{g}$ as it is often the case,

$$\frac{\mathrm{D}\boldsymbol{u}}{\mathrm{D}t} = -\frac{1}{\rho}\nabla p + \boldsymbol{g} + \frac{\mu}{\rho}\Delta\boldsymbol{u} = -\frac{1}{\rho}\nabla p + \boldsymbol{g} + \frac{\mu}{\rho}\Delta\boldsymbol{u} \tag{11.25}$$

Eq. (11.25) holds for the usual CFD applications, not for multiphase flow ones. Nevertheless, by resorting to the one-fluid formulation according to LSM, the expression does not change sensitively, despite implicitly taking into account jump discontinuities across the fronts. In fact, in this way, material properties such as ρ and μ can be written as

$$\rho = \rho_G + (\rho_L - \rho_G)H(\phi) \tag{11.26}$$

$$\mu = \mu_G + (\mu_L - \mu_G)H(\phi) \tag{11.27}$$

where subscripts G and L indicate the gas and liquid phases, respectively, and H is the Heaviside function. From a computational viewpoint, as jump discontinuities are problematic to cope with and can lead to stiff equations and numerical issues in general, sharp changes in material properties are represented by a smeared-out version H_ε of the step function instead of H, so that:

$$H_\varepsilon(\phi) = \begin{cases} 0 & \phi < -\varepsilon \\ \frac{1}{2} + \frac{\phi}{2\varepsilon} + \frac{1}{2}\sin\left(\frac{\pi\phi}{\varepsilon}\right) & -\varepsilon \le \phi \le \varepsilon \\ 1 & \varepsilon < \phi \end{cases} \tag{11.28}$$

ε is a parameter influencing the numerical smearing and usually of the same order of magnitude as the interface thickness.

A final feature has to be considered in multiphase flows: the surface tension force $\boldsymbol{F}_{ST}$. Since it acts only at the location of the phase interface, one way to model it is by using a Dirac delta function (that, for the LSM, has ϕ as its argument), accordingly to the CSF method [9]:

$$\boldsymbol{F}_{ST} = \sigma\kappa\widehat{\boldsymbol{n}}\delta(\phi) \tag{11.29}$$

σ is the coefficient of surface tension, κ is the curvature, $\widehat{\boldsymbol{n}}$ is the normal vector, and δ is the Dirac distribution. In the LSM framework, both κ and $\widehat{\boldsymbol{n}}$ are easily computed:

$$\widehat{\boldsymbol{n}} = \frac{\nabla\phi}{||\nabla\phi||} \tag{11.30}$$

$$\kappa = \nabla\cdot\widehat{\boldsymbol{n}} \tag{11.31}$$

Similarly to what happens for H, δ_ε is used in place of δ:

$$\delta_\varepsilon(\phi) = \begin{cases} 0 & \phi < -\varepsilon \\ \frac{1}{2\varepsilon} + \frac{1}{2\varepsilon}\cos\left(\frac{\pi\phi}{\varepsilon}\right) & -\varepsilon \le \phi \le \varepsilon \\ 0 & \varepsilon < \phi \end{cases} \tag{11.32}$$

To recapitulate, the NS equations for two-phase incompressible flows are:

$$\frac{\mathrm{D}\boldsymbol{u}}{\mathrm{D}t} = -\frac{1}{\rho}\nabla p + \boldsymbol{g} + \frac{\mu}{\rho}\Delta\boldsymbol{u} = -\frac{1}{\rho}\nabla p + \boldsymbol{g} + \frac{\mu}{\rho}\Delta\boldsymbol{u} + \frac{\sigma\kappa\widehat{\boldsymbol{n}}\delta(\phi)}{\rho} \tag{11.33}$$

It may be useful to rewrite Eq. (11.33) in nondimensional form. To do so, one can introduce suitable nondimensional variables (hereinafter marked by the ★ superscript) by specifying a characteristic length L and velocity U so that:

$$\mathbf{x}^\star\mathrm{L} = \boldsymbol{x}, \frac{t^\star\mathrm{L}}{\mathrm{U}} = t, \boldsymbol{u}^\star\mathrm{U} = \boldsymbol{u}, \boldsymbol{g}^\star g = \boldsymbol{g}, p^\star\rho\mathrm{U}^2 = p \tag{11.34}$$

Then, after introducing three famous dimensionless groups, Reynolds number Re and Froude number Fr,

$$\mathrm{Re} = \frac{\rho \mathrm{LU}}{\mu} \tag{11.35}$$

$$\mathrm{Fr} = \frac{\mathrm{U}}{\sqrt{g\mathrm{L}}} \tag{11.36}$$

While expanding the material derivative in its extended form, it follows that

$$\frac{\partial \boldsymbol{u}^{\star}}{\partial t^{\star}} + \boldsymbol{u}^{\star} \cdot \nabla^{\star} \boldsymbol{u}^{\star} = \nabla^{\star} p^{\star} + \frac{1}{\mathrm{Fr}^2} \boldsymbol{g}^{\star} + \frac{\sigma \kappa \widehat{\boldsymbol{n}} \delta_{\varepsilon}}{\rho} \frac{\mathrm{L}}{\mathrm{U}^2} + \frac{1}{\mathrm{Re}} \Delta^{\star} \boldsymbol{u}^{\star} \tag{11.37}$$

However, in some instances, Eq. (11.37) is reported in a slightly different form (Eq. 11.40) that can be inferred through the use of the Weber number We

$$\mathrm{We} = \frac{\rho \mathrm{LU}^2}{\sigma} \tag{11.38}$$

and by defining two additional dimensionless variables, i.e.,

$$\frac{\kappa^{\star}}{\mathrm{L}} = \kappa, \quad \frac{\delta^{\star}}{L} = \delta \tag{11.39}$$

Thus,

$$\frac{\partial \boldsymbol{u}^{\star}}{\partial t^{\star}} + \boldsymbol{u}^{\star} \cdot \nabla^{\star} \boldsymbol{u}^{\star} = -\nabla^{\star} p^{\star} + \frac{1}{\mathrm{Fr}^2} \boldsymbol{g}^{\star} + \frac{\kappa^{\star} \widehat{\boldsymbol{n}} \delta^{\star}}{\mathrm{We}} + \frac{1}{\mathrm{Re}} \Delta^{\star} \boldsymbol{u}^{\star} \tag{11.40}$$

Once all the ingredients have been established, all that remains is to build the computational model. The simulator rests on a cascade model in which the flow field is obtained by the means of the NS PDEs system, after which the HJ equation is solved to infer the position and shape of the fronts. Accordingly, the material properties are updated and the algorithm starts with the subsequent iteration.

Considering for simplicity the two-dimensional case, the employed flow solver treats the nondimensional incompressible NS equations relying on the Chorin's projection method [14] on a Marker-and-Cell staggered grid. Chorin's projection method is a fractional step technique built on Helmholtz decomposition which states that any vector field can be split in the sum of a curl-free (irrotational) field and a divergence-free (solenoidal) one. This operator splitting method is articulated in three steps: the nonlinear convection term $\boldsymbol{u}^{\star} \cdot \nabla^{\star} \boldsymbol{u}^{\star}$ is processed with an explicit scheme while the diffusion term $\Delta^{\star} \boldsymbol{u}^{\star}$ with an implicit one, finding an intermediate velocity field that do not abide by the incompressibility constraint. The final passage consists in projecting this tentative velocity onto the space of solenoidal fields, thus imposing continuity and requiring the resolution of a Poisson's equation for pressure [41]. In other terms, by adding the pressure gradient, the corrected velocity is found.

The spatial discretization for the portion of the model that concerns the resolution of the HJ equation for computing the zero-level set is performed by an upwind second-order essentially nonoscillatory scheme while time discretization is carried out with a three-step, second-order TVD Runge-Kutta scheme. The choice of upwind approximations is motivated by the hyperbolic nature of the PDE to avoid numerical instabilities. Albeit LSM can be used with finite volume schemes, here all the equations were treated by relying on finite difference approximations.

Overall, the equation systems are numerically solved in domain $\Omega = \left[-\frac{d_r}{2}, \frac{d_r}{2}\right] \times \left[-\frac{l_r}{2}, \frac{l_r}{2}\right]$, where l_r is the length of the ALR riser, equal to 0.78 m, and d_r is its diameter, equal to 0.11 m. The meshgrid is constituted by 300 × 300 rectangular cells.

The boundary conditions employed to evaluate the velocity field are of the Dirichlet type. No-slip conditions are imposed on three sides of the domain, both in the x and y direction. The right-side wall, on the contrary, is modeled as to consider the inflow and outflow determined by the presence of the horizontal collectors. Hence, in correspondence of the openings, a noise-imposed, fairly flat profile velocity distribution typical of the turbulent fluid flow is adopted, while no-slip conditions are used elsewhere.

Before presenting the results, it has to be specified that the dynamic parameters used for the simulation (Table 11.1) come from a previous experimental campaign [52] while the remaining data, aside for the ALR dimensions, are retrieved from literature (Table 11.2).

The simulation here reported (Fig. 11.5) concerned a puff of three initially spherical bubbles. To cope with the well-known volume loss problem of level

TABLE 11.1 Data gathered in the experimental campaigns.

Geometric data	**Value**
Length of the riser	0.780 m
Diameter of the riser	0.110 m
Length of the downcomer	0.780 m
Diameter of the downcomer	0.050 m
Length of the horizontal collectors	0.385 m
Diameter of the horizontal collectors	0.050 m
Operating variables	**Value**
Temperature	293 K
Pressure	3 ÷ 4 bar
NaCl pulse injection	1 ÷ 10 g L^{-1}

TABLE 11.1 Data gathered in the experimental campaigns.—cont'd

Measured variables	Value
Gas velocity	0.57 ÷ 0.64 m s^{-1}
Liquid velocity (in the riser)	0.06 ÷ 0.14 m s^{-1}
Liquid velocity (in the downcomer)	0.17 ÷ 0.39 m s^{-1}
(Min, average, max) diameter Of the bubbles	(0.003, 0.007, 0.020) m
Average number of bubbles	2200 ÷ 2800

TABLE 11.2 Physicochemical data from literature and employed in simulation.

Physicochemical parameters	Value
Specific gas constant (CO_2)	188.9 J kg^{-1} K^{-1}
Surface tension coefficient (CO_2)	72.86×10^{-3} N m^{-1}
Liquid density (H_2O)	1000 kg m^{-3}
Liquid dynamic viscosity (H_2O)	1.002×10^{-3} Pa s
Gas dynamic viscosity (CO_2)	1.47×10^{-5} Pa s

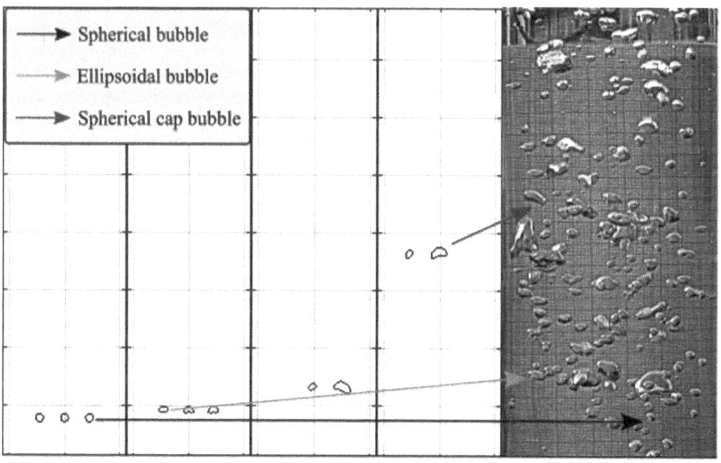

FIGURE 11.5 Comparison between simulated bubbles and experimental ones.

set method, a corrective method [65] that relocates ϕ raising or lowering it according to the need is implemented, bearing satisfactory results. The shapes of bubbles thus obtained all along the simulation time showed accordance with those seen in the experimental campaign (see Fig. 11.5) as well as the zigzagging trajectory of their ascending paths. Coalescence and breaking phenomena were successfully modeled as well, as evidenced by the formation of spherical caps as result of coalescence.

6. Modeling fate and transport of contaminants in porous media by a DBM approach—a case study

With the aim of characterizing complex solute transport in porous media some laboratory scale flow tank experiments using noninvasive experimental physical equipment can be adopted. We designed and constructed a two-dimensional apparatus that uses glass beads as porous medium and a transmitted light imaging technique in conjunction with a fluorescent dye tracer to simulate solute fate and transport into groundwater. The apparatus is made by a Perspex box of internal dimension 200 × 280 × 10 mm (its 10 mm thickness is sufficient to well approximate a two-dimensional system). Different types of injection can be obtained by a controlled syringe pump at low seepage flux to negligibly affect the flow velocity distribution. The effluent from the model is removed by a low-pulse peristaltic pump placed downstream which permits a quasi-constant flow rate in the chamber. At time zero, a selected volume of tracer at known concentration is introduced into the system, and image acquisition starts by recording the pollutant plume at regular time steps of 5 s. The acquired set of images, taken by a CCD camera, is filtered and then processed to estimate the 2D distribution of tracer concentrations inside the model by pixels intensity. A proper methodology to calibrate the physical model is necessary, and it takes into account changes in glass beads distribution by determining pixel-by-pixel calibration curves. Details about the experimental procedure can be found in Catania et al. [12]. The apparatus is shown in Fig. 11.6.

In order to describe the behavior of this experimental apparatus, different kinds of models can be used. Over the last decades a great research effort was conducted in order to comprehend the mechanisms of infiltration and redistribution of pollutants in water-saturated porous media, and most of the studies demonstrated the importance of subsurface heterogeneity on the migration and final distribution of contaminants. Indeed, heterogeneities cause mechanisms like the unstable behavior of downward pollutant flow, no uniform distribution of entrapped pollutants if not miscible with water and formation of pools on top of fine-grained layers. Stochastic physically based modeling is usually invoked in this case; this approach recognizes that hydrological properties of the porous formations are affected by uncertainty and considers them as a spatial random function. Also at a laboratory scale, solute transport in

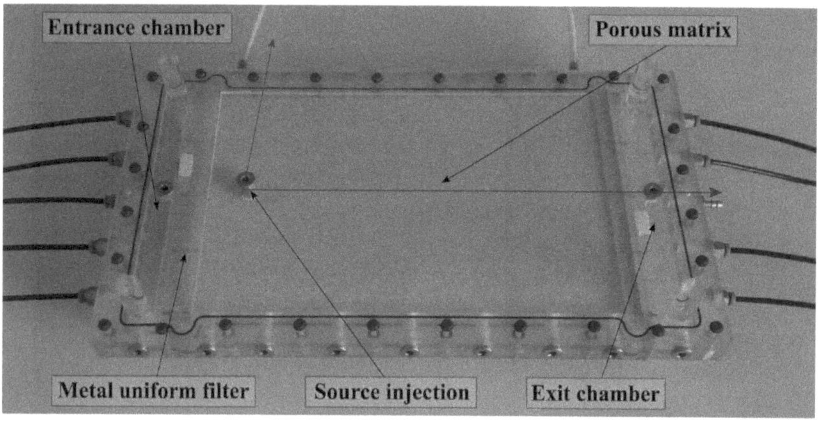

FIGURE 11.6 Flow-tank experimental apparatus for the study of solute transport in porous media.

undisturbed soil columns sometimes does not conform to the deterministic predictions of the advection–dispersion model [33]; in these cases the physically based advection–dispersion equation (ADE) must be modified, e.g., by introducing distinct components of the pore space, generally referred to as the mobile and the immobile fractions.

If a physically based deterministic model is adopted with reference to the saturated porous medium contained in the experimental apparatus, the following, first assumption can be made about the environmental compartment: the porous medium can be considered as constituted only by water, tracer, and glass beads. This assumption is valid only if a perfect pretreatment of the apparatus is performed, e.g., first by vacuum, then by fluxing carbon dioxide into the porous medium in order to eliminate residual air. Subsequently, the representation scale has to be decided. Since this is a heterogeneous porous medium, EIs exist between liquid and glass beads, supposing the tracer being completely miscible with water; otherwise L–L–S EIs must be considered. Therefore, an approach at grain scale should be adopted, and mass and momentum conservation laws should be applied at local scale to the liquid phase (or both the liquid phases) moving between the glass beads. Moreover, if photobleaching or chemical degradation occurs, a chemical decay law must be inserted into the mass balance equations of the liquid phase, and if adsorption occurs on the glass beads it must be considered at the L–S interface.

The classical ADE can only be adopted if subsurface heterogeneities are not important for the process under study: all the porous medium is considered a continuous homogeneous environmental compartment and what happens at the L–S interfaces is incorporated into some global parameters (e.g., the use of the retardation factor to model adsorption onto the solid phase). In this case, the model can be identified by estimating a limited number of parameters (pore scale velocity, dispersion coefficients, decay constant, retardation factor) from the experimental data.

Despite this simplified model of the experimental apparatus reduces to only one PDE, its analytical solution in closed form is complex enough [42] to require some particular efforts in order to estimate model parameters using a maximum likelihood procedure like in Catania et al. [11]. So, a simpler approach for a first identification of the physical model can be based on the direct calculation of the spatial moments from the acquired concentration data.

Finally, since the experimental data are taken at the pixel (CCD camera) scale, they must be converted at the REV scale in order to be used to estimate the parameters of the proposed model.

An alternative approach, based on DBM modeling [10] is here illustrated as a case study, where only a limited number of a priori assumptions are made about the physical processes occurring into the environmental compartment contained in the apparatus [78].

The main steps to define a DBM model:

i) *Considerations about the representation scale and data availability at the chosen scale*: before proceeding with the model setup it is necessary to consider the concentration as a quantity changing in a regular manner through space, especially if some physical interpretation of the DBM model is required. To do this, the REV scale is chosen, and since output concentration data can be available at pixel scale (equal to 0.77 mm), they are averaged over volumes containing many pores, so that the space average is independent of the size of the averaging domain. The REV ensuring this condition can be obtained through spatial or statistical averaging; these two approaches can be considered as equivalent if ergodicity conditions are satisfied. Being the glass beads' diameter equal to 1 mm, box thickness equal to 10 mm, and porosity of the medium equal to 0.36, by following the approach of Dagan [15,16], a completely random medium is found with characteristic REV scale dimension of 8 mm. Smoothing due to REV averaging is shown in Fig. 11.7.

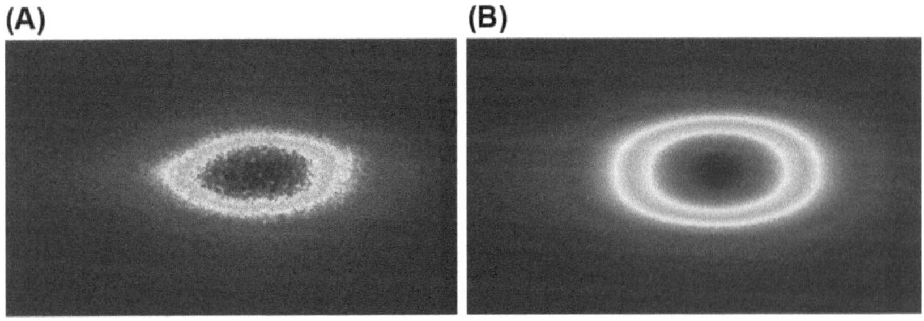

FIGURE 11.7 Imaged concentration at 450 s; (A) pixel scale (B) representative elemental volume scale.

For each frame, 25 × 35 average concentration data are available, for a total of 120 frames.

ii) *Prior assumptions about the form and structure of the DBM model kept at a minimum*: the only assumption made about the form and structure of the model was to consider polynomial functions, so Eq. (11.17) can be written as:

$$y(t) = \frac{B(z)}{A(z)}[u(t-n)] + \varepsilon(t) \tag{11.41}$$

The system is SISO and the model is estimated in discrete-time transfer function form; then the continuous time parameters are deduced from the estimated parameters of this discrete-time transfer function for all the spatial points. By using image data from experimental time duration of 600 s, corresponding to 120 frames at time steps of 5 s, DBM modeling strongly suggested that the dynamic relationship between the pulselike input concentration $\left\{\begin{matrix} u(t) = C_0 = 50\ \text{mg/L} & \text{for } t = 0 \\ u(t) = 0 & \text{for } t > 0 \end{matrix}\right\}$ and the output concentration $y(t) = C(t)$ is linear and third order with the continuous time transfer function and can be written as:

$$C(t) = \frac{b_0}{s^3 + a_1 s^2 + a_2 s + a_3}[u(t-d)] + \varepsilon(t) \tag{11.42}$$

where d is the time elapsing from the beginning of the experiment (solute injection) to the instant the input affects first the output. The physical meaning of the delay parameter d is almost clear, representing de facto the effects of the advection through the model.

An important aspect of DBM modeling is the identification of parametrically efficient (parsimonious) models, that means to find the lowest dynamic order that is consistent with the information content provided by the experimental data and whose parameters are statistically significant. A useful model order statistics is the Young information criterion defined as:

$$YIC = ln\frac{\sigma^2}{\sigma_y^2} + ln\frac{1}{np}\sum_{i=1}^{np}\frac{\sigma_{pii}^2}{a_i^2} \tag{11.43}$$

where σ^2 is the variance of the model residuals, σ_y^2 is the variance of the experimental data, $np = r + m + 1$ is the number of the estimated parameters in the transfer function model (in this case, in Eq. 11.42), σ_{pii}^2 is an estimate of the variance of the estimated uncertainty on the ith parameter estimate, and a_i is the ith element of the vector of the transfer function parameters. Eq. (11.42) satisfies this criterion.

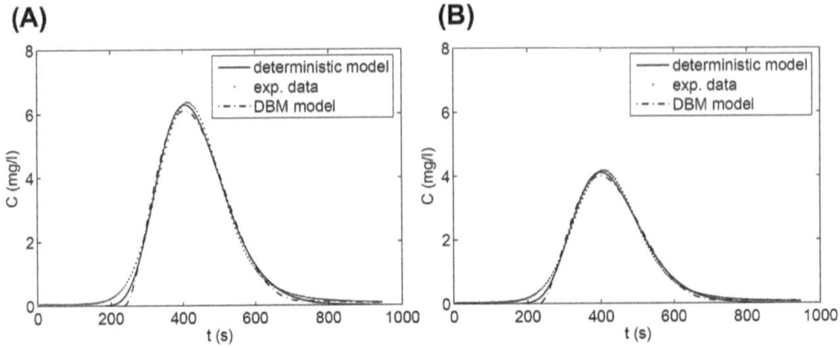

FIGURE 11.8 Comparison between physically based deterministic model, data-based mechanistic (DBM) model and experimental data. (A) point A; (B) point B.

For all the experimental data analyzed the estimated residuals are very low, with coefficients of determination between 0.9885 and 0.9982. The comparison between the experimental data and the DBM model for two points inside the experimental apparatus is reported in Fig. 11.8. Point A is located on the x-axis at 10 cm from the injection point while point B is located at the same abscissa but at 2.3 cm off the y-axis. The estimated delay d is 240 s for both the two points indicating that the estimation procedure captures well the advection process.

The last step of the DBM modeling procedure is to find some physical meaning of the DBM model. A sort of parallelism between the deterministic, physically based, model of the experimental apparatus and the DBM one can be done by considering the ADE expressing mass conservation of the generic solute at REV scale in a two-dimensional Cartesian geometry, under steady-state flow and constant dispersion parameters along the spatial domain (without solute decay or adsorption).

$$\frac{\partial C(x,y,t)}{\partial t} = -v\frac{\partial C(x,y,t)}{\partial x} + D_L\frac{\partial^2 C(x,y,t)}{\partial x^2} + D_T\frac{\partial^2 C(x,y,t)}{\partial y^2} \tag{11.44}$$

where v (m s^{-1}) is the average pore scale velocity and D_L (m^2 s^{-1}) and D_T (m^2 s^{-1}) are the longitudinal and the transverse dispersion coefficients, respectively.

For an instantaneous point source of solute, the following initial conditions apply:

$$C(x,y,0) = \frac{M}{nb}\delta(x)\delta(y-q) \tag{11.45}$$

where b is the thickness of the apparatus, n is the porosity of the medium, M is the mass really inserted in the experimental apparatus, and q is the distance of the source from the inlet chamber.

If the edge effects on y-axis are negligible, i.e., $\sqrt{2D_T t} \ll Y$, where Y is the transverse dimension of the apparatus, the system can be considered an infinite porous medium and boundary conditions are written as:

$$\begin{aligned} C(\pm\infty, y, t) &= 0 \\ C(x, \pm\infty, t) &= 0 \end{aligned} \tag{11.46}$$

Otherwise, boundary conditions must be written as:

$$\begin{aligned} &C(\pm\infty, y, t) = 0 \\ &\frac{\partial C}{\partial y}\Big|_{y=\pm Y/2} = 0 \end{aligned} \tag{11.47}$$

The solution of Eqs. (11.44–11.46) is

$$C(x, y, t) = \frac{M}{4nb\pi t\sqrt{D_L D_T}} \exp\left[-\frac{(x-q-vt)^2}{4D_L t} - \frac{(y)^2}{4D_T t}\right] \tag{11.48}$$

while using boundary condition Eq. (11.47) the solution is expressed in terms of Jacobi function as:

$$\begin{aligned} C(x, y, t) = \frac{M}{2Y}\theta_3\left[\frac{\pi y}{Y}, \exp\left(-\frac{4D_T t\pi^2}{Y^2}\right)\right]&\left\{\frac{1}{\sqrt{D_L \pi t}}\left[\exp\left(-\frac{(x-q-vt)^2}{4D_L t}\right)\right.\right. \\ &+ \left.\exp\left(\frac{(x+q-vt)^2}{4D_L t} - \frac{vq}{D_L}\right)\right] \\ &\left. - \frac{v}{D_L}\exp\left(\frac{vx}{D_L}\right)\operatorname{erfc}\left(\frac{(x+q+vt)}{\sqrt{4D_L t}}\right)\right\} \end{aligned} \tag{11.49}$$

$$\theta_3(z, r) = 1 + 2\sum_{n=1}^{\infty} r^{n^2}\cos(2nz) \tag{11.50}$$

Eq. (11.48) is here used for model comparison. The chemical–physical parameters contained in Eq. (11.48) could be estimated by a maximum likelihood procedure, since the model is expressed in closed form. For our purpose an estimation procedure based on direct calculation of the spatial moments is enough. By defining the Laplace transform $\widehat{C}_\delta(x, y, s)$ of the solute breakthrough curve at a general location (x,y) and resulting from a Dirac input at location (0,0) as:

$$\widehat{C}_\delta(x, y, s) = \frac{M\exp\left(\frac{vx}{2D_L}\right)}{2nb\pi\sqrt{D_L D_T}} K_0\left[\sqrt{v^2 + 4sD_L}\,\sqrt{y^2 D_L + x^2 D_T}\,\frac{1}{2D_L\sqrt{D_T}}\right] \tag{11.51}$$

where $K_0[x]$ is the zero-order Bessel function of x, the temporal moments of $C(x,y,t)$ can be found from the general relationship [3]:

$$m_n = (-1)^n \frac{d^n \widehat{C}_\delta(x,y,s)}{ds^n}\Big|_{s=0} \tag{11.52}$$

The total mass M_0 inserted into the experimental apparatus is estimated by the zero-order moment, the solute pore scale velocity v is estimated by the first-order moment, and the longitudinal and transverse dispersion coefficients D_L and D_T are obtained using the second-order moments.

The temporal moments obtained from the deterministic physically based ADE can be compared with the temporal moments computed from the DBM model by Taylor expanding around $s = 0$ the transfer function expressed by Eq. (11.42) that holds:

$$\begin{aligned} y(t)_{\text{Taylor}} &= \frac{b_0}{a_3} - \frac{b_0 a_2}{a_3^2} s + \left(\frac{b_0 a_2^2}{a_3^3} - \frac{b_0 a_1}{a_3^2}\right) s^2 \\ &+ \left(\frac{-b_0}{a_3^2} + \frac{b_0 a_2 a_1}{a_3^3} + \frac{b_0 a_2 \left(a_1 a_3 - a_2^2\right)}{a_3^4}\right) s^3 + O\left(s^4\right) \end{aligned} \tag{11.53}$$

As the Taylor expansion can be considered as the impulse response in Laplace domain derived exclusively from the experimental data (the DBM model is data-driven), Eq. (11.52) can be applied to Eq. (11.53) establishing in this way a direct connection among the parameters in the transfer function and the chemical–physical parameters of the system as the dispersion coefficients.

Moreover, we can factorize the DBM model approximately by partial fraction expansion, and assuming it is composed by three first-order elements arranged as one "slow" element in parallel with two "fast" elements in series, the following expression is obtained:

$$C(t) = \left(\frac{B}{s - a_{\text{slow}}} + \frac{D}{\left(s - a_{\text{fast}}\right)^2}\right) u(t - \delta) \tag{11.54}$$

The possible physical interpretation of this parallel flow decomposition is a form of "dual process flow," with the slow pathway predominantly dominated by transverse dispersion and the fast flow associated with the longitudinal dispersion.

It is possible to calculate the mean residence time inside the apparatus by considering the slow and the fast elements. Again with reference to points A and B, the computed values of parameters a_{slow} and a_{fast} and the mean residence times are reported in Table 11.3.

Finally, it can be observed that the DBM and the physically based deterministic model are comparable, as testified by the values of the coefficients of determination reported in Table 11.4.

TABLE 11.3 Mean residence time of the elements for points A and B, factorized data-based mechanistic model.

	a_{slow}	a_{fast}	Residence time (s) slow element	Residence time (s) fast element
Point A	0.0062	0.0068	160.27	69.21
Point B	0.0063	0.0069	159.91	67.68

TABLE 11.4 Coefficient of determination for point A and point B with different models.

	Physically based deterministic model	Data-based mechanistic model
Point A	0.0062	0.0068
Point B	0.0063	0.0069

Fig. 11.8 shows the results of simulation with the two models for points A and B.

7. Conclusions

EIs, defined as the regions along which the different environmental compartments interact with each other, are recently assuming great attention. They are complex systems, usually showing nonlinear dynamic behavior, and whose modeling often requires a multiscale approach. This chapter discusses the possible modeling techniques for the study of EIs, on the basis of both the physically based hypothetical deductive paradigm and the empirical data-driven phenomenological approach. As regards the physically based dynamic models, as CFD is at the forefront of multiphase flows studies, some insights are given on advantages and drawbacks of Euler–Euler and Euler–Lagrangian schemes and about FT and FC methods. In order to illustrate some of the presented simulation schemes an interesting case study about the dynamic modeling of bubbles rising into an airlift photobioreactor is illustrated. A brief review on phenomenological models is also given, where different black-box modeling schemes are analyzed. Finally, attention is devoted to some intermediate approaches such as data-based mechanistic modeling and a case study, where laboratory-scale contaminated groundwater is the environmental compartment and liquid phase–soil grains is the EI, is solved with this approach.

References

[1] E.A. Afolabi, J.G.M. Lee, An eulerian-eulerian CFD simulation of air-water flow in a pipe separator, J. Comput. Multiph. Flows 6 (2014) 133–149. https://doi.org/10.1260/1757-482X.6.2.133.

[2] C. Angelini, A. Amaroli, C. Falugi, G. Di Bella, V. Matranga, Acetylcholinesterase activity is affected by stress conditions in *Paracentrotus lividus* coelomocytes, Mar. Biol. 143 (2003) 623–628. https://doi.org/10.1007/s00227-003-1120-x.

[3] R. Aris, On the dispersion of linear kinematic waves, Proc. Roy. Soc. Lond. A 245 (1958) 268–277. https://doi.org/10.1098/rspa.1958.0082.

[4] M.J. Asher, B.F.W. Croke, A.J. Jakeman, L.J.M. Peeters, A review of surrogate models and their application to groundwater modeling, Water Resour. Res. 51 (2015) 5957–5973. https://doi.org/10.1002/2015WR016967.

[5] S. Bader, H. Kooi, Modelling of solute and water transport in semi-permeable clay membranes: comparison with experiments, Adv. Water Resour. 28 (2005) 203–214. https://doi.org/10.1016/j.advwatres.2004.11.001.

[6] V. Balakotaiah, D. Luss, Input-multiplicity in lumped-parameter systems, Chem. Eng. Commun. 39 (1985) 309–322. https://doi.org/10.1080/00986448508911679.

[7] S. Besbes, M. El Hajem, H. Ben Aissia, J.Y. Champagne, J. Jay, PIV measurements and Eulerian–Lagrangian simulations of the unsteady gas–liquid flow in a needle sparger rectangular bubble column, Chem. Eng. Sci. 126 (2015) 560–572. https://doi.org/10.1016/j.ces.2014.12.046.

[8] W.R. Bowen, J.S. Welfoot, Modelling the performance of membrane nanofiltration—critical assessment and model development, Chem. Eng. Sci. 57 (2002) 1121–1137. https://doi.org/10.1016/S0009-2509(01)00413-4.

[9] J.U. Brackbill, D.B. Kothe, C. Zemach, A continuum method for modeling surface tension, J. Comput. Phys. 100 (1992) 335–354. https://doi.org/10.1016/0021-9991(92)90240-Y.

[10] F. Catania, Solute Transport in Porous Media – Inverse Modelling at Laboratory Scale (Ph.D. thesis), University of Genoa, Genoa, Italy, 2007.

[11] F. Catania, M. Massabò, O. Paladino, Estimation of transport and kinetic parameters using analytical solutions of the 2D advection-dispersion-reaction model, Environmetrics 17 (2006) 199–216. https://doi.org/10.1002/env.770.

[12] F. Catania, M. Massabò, M. Valle, G. Bracco, O. Paladino, Assessment of quantitative imaging of contaminant distributions in porous media, Exp. Fluid 44 (2008) 167–177. https://doi.org/10.1007/s00348-007-0388-x.

[13] S. Chen, S.A. Billings, W. Luo, Orthogonal least squares methods and their application to non-linear system identification, Int. J. Control 50 (1989) 1873–1896. https://doi.org/10.1080/00207178908953472.

[14] A.J. Chorin, Numerical solution of the Navier-Stokes equations, Math. Comput. 22 (1968) 745–762. https://doi.org/10.2307/2004575.

[15] G. Dagan, Transport in heterogeneous porous formations: spatial moments, ergodicity, and effective dispersion, Water Resour. Res. 26 (1990) 1281–1290. https://doi.org/10.1029/WR026i006p01281.

[16] G. Dagan, Flow and Transport in Porous Formations, Springer-Verlag, Berlin Heidelberg, 1989.

[17] B.J. Daly, Numerical study of the effect of surface tension on interface instability, Phys. Fluids 12 (1969) 1340–1354. https://doi.org/10.1063/1.1692673.

[18] O. Decroly, A. Goldbeter, Birhythmicity, chaos, and other patterns of temporal self-organization in a multiply regulated biochemical system, Proc. Natl. Acad. Sci. Unit. States Am. 79 (1982) 6917. https://doi.org/10.1073/pnas.79.22.6917.

[19] O. Desjardins, V. Moureau, H. Pitsch, An accurate conservative level set/ghost fluid method for simulating turbulent atomization, J. Comput. Phys. 227 (2008) 8395–8416. https://doi.org/10.1016/j.jcp.2008.05.027.

[20] M.T. Dhotre, N.G. Deen, B. Niceno, Z. Khan, J.B. Joshi, Large eddy simulation for dispersed bubbly flows: a review, Int. J. Chem. Eng. 22 (2013), 2013, https://doi.org/10.1155/2013/343276.

[21] D.A. Drew, S.L. Passman, Theory of multicomponent fluids, in: Applied Mathematical Sciences, Springer-Verlag, New York, USA, 1999.

[22] B. Duz, M.J.A. Borsboom, A.E.P. Veldman, P.R. Wellens, R.H.M. Huijsmans, Efficient and accurate PLIC-VOF techniques for numerical simulations of free surface water waves, in: ICCFD9 Proceedings. Presented at the ICCFD9: 9th International Conference on Computational Fluid Dynamics, Istanbul, 2016.

[23] R. Frigg, S. Hartmann, Models in science, in: E.N. Zalta (Ed.), The Stanford Encyclopedia of Philosophy, Metaphysics Research Lab, Stanford University, 2018.

[24] M. Garcìa, Y. Sommerer, T. Schönfeld, T. Poinsot, Evaluation of Euler/Euler and Euler/Lagrange strategies for large eddy simulations of turbulent reacting flows, in: ECCOMAS Thematic Conference on Computational Combustion, 2005.

[25] F. Gibou, R. Fedkiw, S. Osher, A review of level-set methods and some recent applications, J. Comput. Phys. 353 (2018) 82–109. https://doi.org/10.1016/j.jcp.2017.10.006.

[26] M. Giona, O. Paladino, Bifurcation analysis and stability of a controlled CSTR, Comput. Chem. Eng. 18 (1994) 877–887. https://doi.org/10.1016/0098-1354(94)E0005-8.

[27] L. Györgyi, R.J. Field, A three-variable model of deterministic chaos in the Belousov–Zhabotinsky reaction, Nature 355 (1992) 808–810. https://doi.org/10.1038/355808a0.

[28] K. Hiltunen, A. Jäsberg, S. Kallio, H. Karema, M. Kataja, A. Koponen, M. Manninen, V. Taivassalo, Multiphase Flow Dynamics. Theory and Numerics, VTT Publications, Espoo, Finland, 2009.

[29] C.W. Hirt, B.D. Nichols, Volume of fluid (VOF) method for the dynamics of free boundaries, J. Comput. Phys. 39 (1981) 201–225. https://doi.org/10.1016/0021-9991(81)90145-5.

[30] H.I. Ingólfsson, C. Arnarez, X. Periole, S.J. Marrink, Computational 'microscopy' of cellular membranes, J. Cell Sci. 129 (2016) 257–268. https://doi.org/10.1242/jcs.176040.

[31] G.D. Ingram, I.T. Cameron, K.M. Hangos, Classification and analysis of integrating frameworks in multiscale modelling, Chem. Eng. Sci. 59 (2004) 2171–2187. https://doi.org/10.1016/j.ces.2004.02.010.

[32] D. Jacqmin, Calculation of two-phase Navier–Stokes flows using phase-field modeling, J. Comput. Phys. 155 (1999) 96–127. https://doi.org/10.1006/jcph.1999.6332.

[33] M. Javaux, M. Vanclooster, Scale-dependency of the hydraulic properties of a variably saturated heterogeneous sandy subsoil, J. Hydrol. 327 (2006) 376–388. https://doi.org/10.1016/j.jhydrol.2005.11.040.

[34] B.N. Kholodenko, Cell-signalling dynamics in time and space, Nat. Rev. Mol. Cell Biol. 7 (2006) 165–176. https://doi.org/10.1038/nrm1838.

[35] D. Lakehal, M. Meier, M. Fulgosi, Interface tracking towards the direct simulation of heat and mass transfer in multiphase flows, Int. J. Heat Fluid Flow 23 (2002) 242–257. https://doi.org/10.1016/S0142-727X(02)00172-8.

[36] P. Lettieri, L. Mazzei, Challenges and issues on the CFD modeling of fluidized beds: a review, J. Comput. Multiph. Flows 1 (2009) 83–131. https://doi.org/10.1260/175748209789563937.

[37] Y. Lin, H. Skaff, T. Emrick, A.D. Dinsmore, T.P. Russell, Nanoparticle assembly and transport at liquid-liquid interfaces, Science 299 (2003) 226–229. https://doi.org/10.1126/science.1078616.

[38] C.-T. Liou, C. Yu-Shu, The effect of nonideal mixing on input multiplicity in a CSTR, Chem. Eng. Sci. 46 (1991) 2113–2116. https://doi.org/10.1016/0009-2509(91)80168-X.

[39] E. Loth, Overview of numerical approaches, in: C.T. Crowe (Ed.), Multiphase Flow Handbook, CRC Press, New York, USA, 2016, pp. 79–95.

[40] D. Mackay, Multimedia Environmental Models : The Fugacity Approach, second ed., CRC Press, Boca Raton, 2001. https://doi.org/10.1201/9781420032543.

[41] S. MacNamara, G. Strang, Operator splitting, in: R. Glowinski, S.J. Osher, W. Yin (Eds.), Splitting Methods in Communication, Imaging, Science, and Engineering, Scientific Computation, Springer International Publishing, Cham, 2016, pp. 95–114. https://doi.org/10.1007/978-3-319-41589-5_3.

[42] M. Massabó, R. Cianci, O. Paladino, An analytical solution of the advection dispersion equation in a bounded domain and its application to laboratory experiments, J. Appl. Math. (2011). WWW Document, https://doi.org/10.1155/2011/493014.

[43] S. Matera, K. Reuter, Transport limitations and bistability for in situ CO oxidation at RuO2 (110): first-principles based multiscale modeling, Phys. Rev. B 82 (2010) 085446. https://doi.org/10.1103/PhysRevB.82.085446.

[44] E. Mcmullin, What do physical models tell us? in: B. Van Rootselaar, J.F. Staal (Eds.), Studies in Logic and the Foundations of Mathematics, Logic, Methodology and Philosophy of Science III Elsevier, 1968, pp. 385–396. https://doi.org/10.1016/S0049-237X(08)71206-0.

[45] D.T. Mihailovic, I. Bala, D. Kapor, Time and Methods in Environmental Interfaces Modelling, in: Developments in Environmental Modelling, first ed., Elsevier Science Publishers B. V., Amsterdam, The Netherlands, The Netherlands, 2016.

[46] B. Novák, J.J. Tyson, Design principles of biochemical oscillators, Nat. Rev. Mol. Cell Biol. 9 (2008) 981–991. https://doi.org/10.1038/nrm2530.

[47] E. Olsson, G. Kreiss, A conservative level set method for two phase flow, J. Comput. Phys. 210 (2005) 225–246. https://doi.org/10.1016/j.jcp.2005.04.007.

[48] E. Olsson, G. Kreiss, S. Zahedi, A conservative level set method for two phase flow II, J. Comput. Phys. 225 (2007) 785–807. https://doi.org/10.1016/j.jcp.2006.12.027.

[49] P.J. O'Rourke, Statistical properties and numerical implementation of a model for droplet dispersion in a turbulent gas, J. Comput. Phys. 83 (1989) 345–360. https://doi.org/10.1016/0021-9991(89)90123-X.

[50] S. Osher, J.A. Sethian, Fronts propagating with curvature-dependent speed: algorithms based on Hamilton-Jacobi formulations, J. Comput. Phys. 79 (1988) 12–49. https://doi.org/10.1016/0021-9991(88)90002-2.

[51] O. Paladino, Transition to chaos in continuous processes: applications to wastewater treatment reactors, Environmetrics 5 (1994) 57–70. https://doi.org/10.1002/env.3170050106.

[52] O. Paladino, M. Neviani, A closed loop biowaste to biofuel integrated process fed with waste frying oil, organic waste and algal biomass: feasibility at pilot scale, Renew. Energy, SI: Waste Biomass to Biofuel 124 (2018) 61–74. https://doi.org/10.1016/j.renene.2017.08.027.

[53] O. Paladino, M. Ratto, Robust stability and sensitivity of real controlled CSTRs, Chem. Eng. Sci. 55 (2000) 321–330. https://doi.org/10.1016/S0009-2509(99)00327-9.

[54] O. Paladino, M. Ratto, Stability and bifurcation analysis of systems described by nonexact models, Math. Comput. Model. 24 (1996) 43–55. https://doi.org/10.1016/0895-7177(96)00106-9.

[55] O. Paladino, M. Ratto, P. Costa, Chaos and chemical reactor models: sensitivity of dynamics on parameters uncertainty, Chem. Eng. Sci. 50 (1995) 3829–3833. https://doi.org/10.1016/0009-2509(95)00174-4.

[56] L. Pellegrini, G. Biardi, Chaotic behaviour of a controlled CSTR, Comput. Chem. Eng. 14 (1990) 1237–1247. https://doi.org/10.1016/0098-1354(90)80004-U.

[57] D. Pesando, P. Huitorel, V. Dolcini, C. Angelini, P. Guidetti, C. Falugi, Biological targets of neurotoxic pesticides analysed by alteration of developmental events in the Mediterranean sea urchin, *Paracentrotus lividus*, Mar. Environ. Res. 55 (2003) 39–57. https://doi.org/10.1016/S0141-1136(02)00215-5.

[58] A. Prosperetti, G. Tryggvason, Computational Methods for Multiphase Flow, Cambridge University Press, 2009.

[59] M. Ratto, O. Paladino, Analysis of controlled CSTR models with fluctuating parameters and uncertain parameters, Chem. Eng. J. 79 (2000) 13–21. https://doi.org/10.1016/S1385-8947(00)00139-X.

[60] W.J. Rider, D.B. Kothe, Reconstructing volume tracking, J. Comput. Phys. 141 (1998) 112–152. https://doi.org/10.1006/jcph.1998.5906.

[61] R. Rikards, H. Abramovich, J. Auzins, A. Korjakins, O. Ozolinsh, K. Kalnins, T. Green, Surrogate models for optimum design of stiffened composite shells, Compos. Struct. 63 (2004) 243–251. https://doi.org/10.1016/S0263-8223(03)00171-5.

[62] P.L. Roe, J.J. Chattot, Innovative Methods for Numerical Solutions of Partial Differential Equations, World Scientific, 2002.

[63] A. Saltelli, M. Ratto, S. Tarantola, F. Campolongo, Sensitivity analysis practices: strategies for model-based inference, Reliab. Eng. Syst. Saf. 91 (2006) 1109–1125. The Fourth International Conference on Sensitivity Analysis of Model Output (SAMO 2004), https://doi.org/10.1016/j.ress.2005.11.014.

[64] J. Sjöberg, Q. Zhang, L. Ljung, A. Benveniste, B. Delyon, P.-Y. Glorennec, H. Hjalmarsson, A. Juditsky, Nonlinear black-box modeling in system identification: a unified overview, Automatica, Trends in System Identification 31 (1995) 1691–1724. https://doi.org/10.1016/0005-1098(95)00120-8.

[65] A. Smolianski, Numerical Modeling of Two-Fluid Interfacial Flows, Ph.D. Thesis, University of Jyväskylä, 2001.

[66] A. Sokolichin, G. Eigenberger, A. Lapin, A. Lübert, Dynamic numerical simulation of gas-liquid two-phase flows Euler/Euler versus Euler/Lagrange, Chem. Eng. Sci. 52 (1997) 611–626. https://doi.org/10.1016/S0009-2509(96)00425-3.

[67] M. Sussman, E. Fatemi, P. Smereka, S. Osher, An improved level set method for incompressible two-phase flows, Comput. Fluids 27 (1998) 663–680. https://doi.org/10.1016/S0045-7930(97)00053-4.

[68] M. Sussman, E.G. Puckett, A coupled level set and volume-of-fluid method for computing 3D and axisymmetric incompressible two-phase flows, J. Comput. Phys. 162 (2000) 301–337. https://doi.org/10.1006/jcph.2000.6537.

[69] T. Tosco, M. Petrangeli Papini, C. Cruz Viggi, R. Sethi, Nanoscale zerovalent iron particles for groundwater remediation: a review, J. Clean. Prod. 77 (2014) 10–21. Emerging industrial processes for water management, https://doi.org/10.1016/j.jclepro.2013.12.026.

[70] T. Tosco, R. Sethi, Transport of non-Newtonian suspensions of highly concentrated micro- and nanoscale iron particles in porous media: a modeling approach, Environ. Sci. Technol. 44 (2010) 9062–9068. https://doi.org/10.1021/es100868n.

[71] G. Tryggvason, B. Bunner, A. Esmaeeli, D. Juric, N. Al-Rawahi, W. Tauber, J. Han, S. Nas, Y.-J. Jan, A front-tracking method for the computations of multiphase flow, J. Comput. Phys. 169 (2001) 708–759. https://doi.org/10.1006/jcph.2001.6726.

[72] G. Tryggvason, R. Scardovelli, S. Zaleski, Direct Numerical Simulations of Gas-Liquid Multiphase Flows, Cambridge University Press, Cambridge, 2011.

[73] S.O. Unverdi, G. Tryggvason, A front-tracking method for viscous, incompressible, multi-fluid flows, J. Comput. Phys. 100 (1992) 25–37. https://doi.org/10.1016/0021-9991(92)90307-K.

[74] E. Walter, L. Pronzato, Qualitative and quantitative experiment design for phenomenological models—a survey, Automatica 26 (1990) 195–213. https://doi.org/10.1016/0005-1098(90)90116-Y.

[75] E. Weinan, B. Engquist, X. Li, W. Ren, E. Vanden-Eijnden, Heterogeneous multiscale methods: a review, Commun. Comput. Phys. 2 (2007) 367–450.

[76] A. Yang, W. Marquardt, An ontological conceptualization of multiscale models, Comput. Chem. Eng. 33 (2009) 822–837. https://doi.org/10.1016/j.compchemeng.2008.11.015.

[77] Z. Yao, M. Le Maguer, Mathematical modelling and simulation of mass transfer in osmotic dehydration processes. Part I: conceptual and mathematical models, J. Food Eng. 29 (1996) 349–360. https://doi.org/10.1016/0260-8774(95)00045-3.

[78] P. Young, Data-based mechanistic modelling, generalised sensitivity and dominant mode analysis, Comput. Phys. Commun. 117 (1999) 113–129. https://doi.org/10.1016/S0010-4655(98)00168-4.

[79] P. Young, Data-based mechanistic modelling of environmental, ecological, economic and engineering systems, Environ. Model. Softw 13 (1998) 105–122. https://doi.org/10.1016/S1364-8152(98)00011-5.

[80] P. Young, M.J. Lees, The Active Mixing Volume (AMV): a new concept in modelling environmental systems, in: V. Barnett, K.F. Turkman (Eds.), Statistics for the Environment, J. Wiley, Chichester, UK, 1993, pp. 3–44.

[81] P.C. Young, Data-based mechanistic modelling and validation of rainfall-flow processes, in: M.G. Anderson, P.D. Bates (Eds.), Model Validation: Perspectives in Hydrological Science, J. Wiley, Chichester, UK, 2001, pp. 117–161.

[82] P.C. Young, K.J. Beven, Data-based mechanistic modelling and the rainfall-flow non-linearity, Environmetrics 5 (1994) 335–363. https://doi.org/10.1002/env.3170050311.

[83] P.C. Young, H. Garnier, Identification and estimation of continuous-time, data-based mechanistic (DBM) models for environmental systems, Environ. Model. Softw 21 (2006) 1055–1072. https://doi.org/10.1016/j.envsoft.2005.05.007.

[84] J. Yvonnet, H.L. Quang, Q.-C. He, An XFEM/level set approach to modelling surface/interface effects and to computing the size-dependent effective properties of nanocomposites, Comput. Mech. 42 (2008) 119–131. https://doi.org/10.1007/s00466-008-0241-y.

Chapter 12

Advanced oil spill decontamination techniques

Saeid Azizian* and Maryam Khosravi
Department of Physical Chemistry, Faculty of Chemistry, Bu-Ali Sina University, Hamedan, Iran
**Corresponding author*

1. Introduction

Oil is one of the most important energy sources and raw material for production of many commodities. Risk of oil spills and its spreading is a serious problem during exploration, transportation, storage, and utilization. Light petroleum, light diesel oil, and heavy crude oil are responsible for a considerable part of pollution by petroleum hydrocarbons (HCs) all over the world. Pollution of oil spill in seawater is especially an important issue nowadays. Spreading of oil spills creates many negative effects and consequences and undesirable environmental problems such as, death of marine animals, inappropriate marine food for human consumption, and reduction of flight abilities of seabirds. Refinery equipment on seacoasts, offshore oil platforms, and collisions involving oil tankers are among the most important sources of oil spill pollution in seawater. Expense to clean up oil spill from seawater depends on various parameters, for example, type, amount, and intensity of petroleum pollution; time of collision; site of collision (site coordinates and also geographic, political, economic, physical, and biological features of collision site); weather of the polluted site; and sea conditions. Sea conditions depend on collision time (whether collision happens at night or during day) and season of the polluted area. Up to now, various methods have been used to clean up oil slicks from water surface and also to prevent pollution access to coast and environmentally sensitive areas. However, the available technologies are difficult, expensive, and time-consuming to use. Therefore, researchers are trying to find more efficient, economic, and fast methods for removal of oil spill from water surface.

Advanced Low-Cost Separation Techniques in Interface Science. https://doi.org/10.1016/B978-0-12-814178-6.00012-1

2. Crude oil

Crude oil is a relatively viscous and dark liquid (from light brown to black), which mainly consists of HCs. This HC-rich material contains small amounts of sulfur, oxygen, nitrogen, and salts [1].

Crude oil contains a broad spectrum of compounds, mainly organic materials, with different molecular weights, so that, it is not possible to describe completely all components of even one type of crude oil [2].

Oil, as a heterogeneous mixture of many types of compounds, can be classified as follows:

- HCs (materials containing only carbon and hydrogen)
- Sulfur-containing compounds
- Nitrogen-containing compounds
- Oxygen-containing compounds
- Organometallic compounds

Water and salts are not considered as oil components, as it is possible to separate them easily from petroleum in oil fields because of density differences. Table 12.1 shows the present elements and their related weight percent in crude oil.

2.1 Hydrocarbons

HCs, consisting carbon and hydrogen, form the main component of any type of crude oil and contain the following subgroups:

- Paraffin compounds or alkanes

  Alkanes are saturated HCs with the general formula C_nH_{2n+2}. They may be branched or without side chains, as a relatively inactive

TABLE 12.1 Elements in crude oil and their related weight percent.

Element	Weight percent
Carbon	83–87
Hydrogen	11–14
Sulfur	0.05–6
Nitrogen	0.1–2
Oxygen	0–2
Metals	Less than 1000 ppm

compound, compared with the other subgroups. The range of carbon atoms in paraffins of crude oil can change between one (methane or CH_4, n = 1) and about 70 carbon atoms ($C_{70}H_{142}$, n = 70). These paraffin compounds often contain many possible isomers in oil (there are 75 isomers for n = 10 and about 100,000 isomers for n = 20). One can imagine how many paraffin compounds can be found in oil, which includes alkanes with 7–70 carbon atoms. Therefore, oil compounds cannot be completely separated. It is very important to know petroleum is not a pure compound but rather a complex mixture of thousands or millions of substances.

- Naphthenic compounds or cycloalkanes

  Cyclic saturated HCs (naphthenics or cycloalkanes) with the general formula C_nH_{2n} are usually found with different percentages in oil. These HCs include ring structures in one part or in all of their skeletons with different carbon number. The boiling point and density values of these compounds are higher than those values of alkenes with equal carbon numbers.

- Olefinic compounds or alkenes

  Olefins or alkenes are unsaturated HCs with the general formula C_nH_{2n}. They almost are known as negligible compounds in oil, but anyway they are produced during processing at high temperatures. Olefins are among the undesirable and unwanted materials present in the final products. Because they are unstable, are easily oxidized, and/or enter into polymerization reactions and produce resins (heavy, sticky polymers). Double bonds in their structures make them very active chemicals. Acetylene HCs are not found in crude oil.

- Aromatic or cyclic compounds

  Aromatic compounds are usually found in crude oil. Unsaturated cyclic compounds with six to eight carbon atoms (benzene, toluene, and xylene) are monocyclic compounds and the most abundant aromatic compounds in crude oil, whereas polycyclic aromatic compounds are present, too [1].

2.2 Sulfur-containing compounds

Sulfur-containing compounds are found with different concentrations in the most types of oil. So far, numerous sulfur-containing compounds have been identified in oil. Apart from gaseous sulfur-containing compounds (mainly as H_2S, which has acidic properties) in crude oil, there are many liquid sulfur–containing compounds such as organosulfur, which are essentially nonacidic.

Most organic sulfur compounds are thermally unstable, and they produce hydrogen sulfide during oil processing. Sulfur-containing crude oils are in lower demand due to their high processing costs and environmental problems (formation of acid rain, etc.). Sulfur oxides (SO_x), which are considered as environmental pollutants, are produced when sulfur-containing compounds are burned.

2.3 Nitrogen-containing compounds

Crude oil usually has low nitrogen content (0.1%–0.9%), but some types of oils may include up to 2% nitrogen. Crude oil without nitrogen-containing compounds is not commonly found. Compared with sulfur-containing compounds in oil, nitrogen-containing compounds are usually more stable. Nitrogen in oils is in the form of heterocyclic compounds such as indole and quinoline.

2.4 Oxygen-containing compounds

Oxygen content of oils is usually less than 2% by weight. Oxygen-containing compounds in oil are generally in the form of phenolic substances, cyclic carboxylic acids, aliphatic acids, and naphthenic acids.

2.5 Organometallic compounds

Several metals including sodium, calcium, manganese, iron, copper, vanadium, and nickel are found in all types of crude oils. The contents of some of these metals, such as vanadium and nickel, are higher than the metal content of others. These metals are usually in the form of water-soluble mineral salts (including sodium chloride) and/or are found as organic substances, containing iron, vanadium, and nickel. Some metals present in oil are in the form of oil-soluble organometallic compounds, organometallic soaps with surface-active properties, and/or colloidal suspensions. Metals are mostly present in the nonvolatile and heavier portion of oil.

2.6 Porphyrins

Porphyrins, as another group of organic compounds in crude oil, usually form the nonbasic part of nitrogen-containing substances. They are regarded both as usual nitrogen-containing substances in oil and as organometallic compounds, found in some types of oil. Porphyrins compose of four pyrrole subunits, interconnected via methylene bonds. The (−NH−) functional group in porphyrin systems allows metals to interact with other molecules and form a complex.

Nickel and vanadium porphyrins have been found in almost all types of crude oils and tar, but they are present in a greater extent in heavy oils.

2.7 Asphaltenes and resins

Asphaltenes are defined as the nonpolar and nonvolatile components of crude oil. They are insoluble in alkanes (normal pentane and normal heptane) but soluble in aromatics (benzene and toluene). Chemical properties of asphaltenes are not well known. In general, asphaltenes are known as a complex organic matter, containing carbon, oxygen, nitrogen, and sulfur. The black color of some crude oils and residuals is related to the presence of asphaltenes, which have not flocculated or precipitated. Asphaltenes have polar molecules with very high molecular weights.

Resins are defined as the polar and nonvolatile components of crude oil. They are soluble in normal alkanes and aromatic solvents but not in liquid propane. Their chemical structures resemble those of asphaltenes but with a lower molecular weight, higher hydrogen to carbon ratio, and fewer heteroatoms. Resins are strongly attracted by asphaltenes, and it may not be possible to separate them from asphaltenes. Asphaltenes are brittle, shiny, dark brown to black, amorphous and solid powders, whereas resins are shiny, adhesive, and dark brown. Table 12.2 presents the compounds found in oil.

3. Petroleum products

The American Petroleum Institute has reported more than 2000 petroleum products are produced globally every year, but only a small number of the main products are considered in designing refineries.

Petroleum products are usually given different names such as crude oil cuts. The main products of oil refineries in ascending order of molecular weight are as follows [3]:

- Light fuel gases, consisting hydrogen and light HCs, used as industrial fuel and/or raw materials in petrochemical industries.
- Liquefied petroleum gas, including propane and butane and/or a mixture of them, used as domestic fuel or industrial fuel.
- Various types of gasoline, used as fuel for automobiles.
- Various types of naphtha, utilized as solvents and used in paints and as industrial solvents in oil and perfume extraction.
- Jet fuel and turbine fuel, used in reaction engines, gas turbines, and rockets.
- Kerosene, used in various lamps for lighting purposes and/or for heat generation.
- Diesel oil and light diesel oil, utilized as fuel in domestic heating facilities or in low-power industrial facilities.
- Light oils, used both as base oils in producing motor oils and as a lubricant in sewing machines, bicycles, etc.
- Heavy oils used as base oils with high viscosity in producing engine oils.
- Cylinder oils, utilized in steam machines and heavy industries.

TABLE 12.2 Chemical compounds in petroleum products.

Classifications	Example classes, names, and compound			
	Chemical class	**Alternate name**	**Description**	**Example compound**
Structures	Alkanes	Paraffins		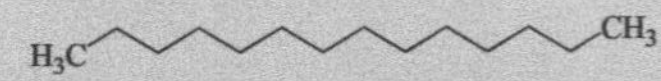 Tetradecane
	Cycloalkanes	Naphthenes		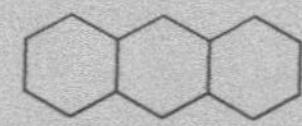 Tetradecahydroanthracene
	Waxes		N-alkanes C_{18}—H_{80}	
Aromatics	BTEX		Benzene, toluene, ethylbenzene, xylenes	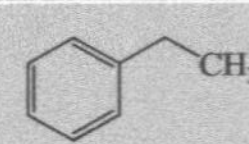 Ethylbenzene
	Polycyclic aromatic hydrocarbons			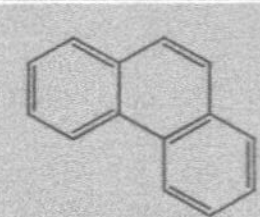 Phenanthrene

	Naphthene aromatics		Combinations of aromatics and cycloalkanes	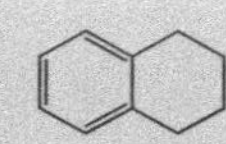 1,2,3,4-Tetrahydronaphthalene
Resins	Class of mostly aromalous polar compounds sometimes containing oxygen, nitrogen, sulfur, or metal			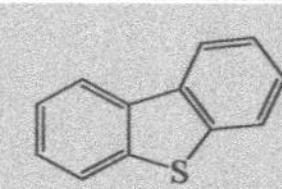 Dibenzo[*b*,*d*]thiophene
Asphaltenes	Class of large anomalous compounds sometimes containing oxygen nitrogen, metals, or sulfur			Structures unknown

- ➢ Paraffins and waxes used in electrical insulation and in preserving food materials.
- ➢ Heavy fuel oils, utilized as fuel in high-power heating facilities and in large diesel engines.
- ➢ Asphalt and tar, used in road building, building impenetrability, etc.
- ➢ Coke, used as an industrial fuel and also in producing electrodes.

4. Seawater oil pollution

According to the study conducted by the American Academy of Arts and Sciences, every year more than 3 million tons of oils on average enter the waters in the world, which oil transport pipelines contribute to produce half of the total. The largest and the most damaging pollution of seawater by oil takes place during wartime. For example, in 1991, when Iraqi troops invaded Kuwait, about 910 million liters of oil entered the sky blue waters of the Persian Gulf by damaging oil tankers and oil exploration platforms. When oil spills get close to coasts, greater damages are inflicted. Close to the coastline, the extent of pollution and also the damages, which oil spills cause, are greater because of the presence of people in the related areas.

In many regions, deep sea oil drilling platforms can be found where special drills are used for drilling, extracting, and transporting of oil. Care must be taken so that the slightest damage is not inflicted on marine organisms such as fish. These platforms must be strong because they may topple in sea storms and releasing millions liters of oil into water [4].

Pipes transporting oil from the seabed are one of the main sources of oil spill pollution. The smallest collision with these pipes causes oil spills, and it may take many years to compensate for the damages. Oil spills, mostly from worn out oil transporting pipes, also cause pollution. Fig. 12.1 shows an instance of seawater pollution caused by oil spills [5].

As it was mentioned above, oil spills in seas (from ships, oil wells, coastal facilities, etc.) cause considerable environmental and economic damages. Results of studies on the fate of seawater oil pollution and on the related self-remediation process indicate the following:

- ➢ The light oil components evaporate
- ➢ The aromatics somewhat dissolve in water
- ➢ The heavy constituents turn into tar balls and move to the shoreline or precipitate on the sea floor
- ➢ The other compounds are mixed with seawater as suspended particles

Oil spill toxicity thus decreases spreading over time through its physical and chemical changes. Oil spills in water are either stable or unstable. In the unstable state, oil is inclined to rapid dispersion in surface water, whereas this process does not happen in the stable state.

FIGURE 12.1 Sea surface and coastline pollution caused by oil spill [5]. *Reprinted with permission from Elsevier.*

When oil enters into an aqueous environment, it undergoes physical, chemical, and biological changes, and this phenomenon influences the environment. As soon as an aqueous environment is polluted by oil, the physical and chemical changes such as evaporation, expansion/spreading, emulsification, degradation and decay, air–sea exchange, and sedimentation will start.

Chemical oxidation of some oil compounds often takes place with the help of sunlight. The decomposed products of the processes include floating tarlike masses and substances precipitated on the seabed.

Biological process also takes place slowly with the physical and chemical processes. A large number of microorganisms are capable of degrading petroleum compounds. HCs metabolized by microorganisms are generally converted to an oxidized compound, which may be further degraded, may be soluble, or may accumulate in the remaining oil.

Knowledge of wind effects and local currents is valuable in determining the spreading rate of an oil spill. In warm waters such as those in the Sea of Oman and the Persian Gulf, the lighter constituents in oil spills evaporate [5,6].

5. Coastal oil pollution

Oil slicks in water are rarely cleaned up completely, and eventually the remaining oil after cleaning up operations moves toward the shore. The fate of coastal oil pollution depends on many factors including type of oil, features of the coastal region, weather conditions, and water waves, which have influence on precipitation of petroleum compounds on the coast. Type and amount of oil and the extent of weathering are among the influential factors before oil reaches the coastlines. When oil reaches coastlines, the ambient temperature, tidal conditions, underlying coastal layers (the composition of materials, type and sensitivity of the regional ecosystem, and the slope of the coastline) are among the influential factors. The extent of oil infiltration and its spread, adhesiveness and the extent of mixing of oil, and other materials present on the coasts are some of the factors determining the cleaning up conditions. Cleaning up operations will be very difficult if oil infiltrates deep into the coastal area. Oil infiltration depends on factors such as oil type and the types of materials present on the coastlines. For example, oil cannot penetrate well into sand and clay beaches, but extensively spreads in hard and rocky shores. Diesel fuels infiltrate several meters into cobblestone beaches, and their cleaning up and removal are very difficult [7].

Weathered oil spills on sand beaches can be cleaned up by using mechanical equipment. The degree of adhesiveness of petroleum products completely depends on oil type and its degree of weathering. Oil spills on surfaces have lower degrees of adhesiveness compared with those that have weathered. Diesel fuel and gasoline have relatively lower adhesiveness. If oil is not adhesive when it reaches the coasts, it will be washed by waves and/or will be finally removed partially from the beaches by tidal cycles. The extent of deposition of petroleum compounds in the coastal region depends on the tide phenomenon. The smallest amount of sedimentation takes place at low tide. It is a less frequent phenomenon because the waves distance themselves from the coastlines. The nature of the intertidal zone (in other words, the intertidal soils and the angle of inclination in the intertidal zone) and energy of the waves influence the fate of the oil slick. A very strong wave can remove oil pollution from the coastal area within a few days. However, in the remote coastal areas, where the energy of the waves is very low, the oil population may persist for tens of years. Fig. 12.1 shows coastline oil pollution on the surface water of sea caused by an oil spill, which water waves have brought to the coast [5].

5.1 Environmental effects of coastal oil pollution

Oil pollution can damage all plants and animals, and consequently, oil pollution destroys the ecosystem and the food chain in the region. Presence of petroleum compounds in coastal regions cannot be ignored because people are living there [8].

Life forms in the intertidal zone are very sensitive and vulnerable due to the presence of motile and nonmotile plants and animals, and therefore, reestablishment of animal and plant populations require months and even years.

Cleaning up operations in this zone, which includes relocation of people and equipment and application of very hot and pressurized water, will endanger the organisms living in the intertidal zone. Cleaning up operations must be undertaken by taking environmental effects into consideration. Cleaning up operations should not be carried out at any cost. In these operations, resuspension of oil and its infiltration into other coasts must be prevented. Sometimes, all plants and animals in a region are killed due to oil pollution, but their populations will be reestablished after cleaning up operations [8].

6. Oil pollution on land

Two types of oil pollution often happen on land. One occurs on ground surface and the other one underground. Many oil pollution events, that take place on ground surface, are in production operations and involve pipelines and power plants, whereas the majority of underground oil slicks are caused by oil pipeline leaks.

Maintaining human security and health are among the important points in cleaning up oil pollution on ground surface, and this issue also should be considered to clean up gasoline spills on land. Reducing long-term environmental effects and preserving agricultural lands are some of the most important challenges involved in oil slicks happening on land. Agricultural lands are preserved through their reclamation. Spread of oil spills on the land surface and their penetrating in the ground make it difficult to determine their extent of spread and also their positions. Oil movements differ greatly based on the oil type, the extent of oil penetration into soil layers (which depends on the various layers in the ground), moisture conditions, land slope, and groundwater flow conditions. Other factors, which may differ in various habitats, include presence of vegetation in the region, plant growth stage, ambient temperature, presence of snow or ice in the region, and form and structure of the regions (such as rocky areas).

Various types of soil including sand and gravel, peat, clay, and silt soils may be exposed to oil pollution. Soil is defined as the loose unconsolidated material located near the surface, whereas rock is the hard-consolidated material and usually exists beneath the soil. Soil particles generally consist of smaller constituents and when these particles join each other, pores and crevices are formed on soil surface. If these particles are large enough and joined, soil is called impenetrable [4].

Since there are different compounds in soils and also soils are not uniform, rates of oil infiltration and spread in various soil layers differ based on the various parts. Soil layers include penetrable ones, having peat, and underlying impenetrable ones, consisting of clay, rocky, and impenetrable beds. If there are crevices in the rocks, oil will easily infiltrate into their layers. Oil adhesiveness is one of the features that influence infiltration of petroleum compounds into soil layers. High-adhesive oils such as bunker oils and tar balls move slowly on the ground when spilled on it, especially when the ambient temperature is below their condensation point. Petroleum compounds with low viscosity, for instance, gasoline are spilled on the ground every year. Precautionary measurements for lightweight compounds must be taken immediately when these compounds are spilled on the ground [8,9].

Crude oil has intermediate adhesiveness. In agricultural lands covered with peat, spilled crude oil saturates soil to a depth of 10–20 cm and rarely reaches the depth of 60 cm.

It only reaches this depth when it is accumulated in dry puddles. If these puddles contain water, oil cannot readily penetrate. Fig. 12.1 shows an instance of oil spread on land [5].

7. Various types of spread oil

Since the fate of oil spills in aqueous environments depends on their compounds, knowledge of materials and various compounds present in oil is necessary to determine the fate of petroleum products spread in water [4].

- The normal (usual) classification:

  Thirty percent gas-fuel or gasoline, 10% kerosene, 15% light oil extract, 25% heavy oil extract, more than 20% oil residues.
- Molecular classification:

  Paraffins (alkanes) 30%; naphthalene (cycloalkanes) 50%; aromatics 15%; nitrogen-, sulfur-, and oxygen-containing compounds 5%.

8. Hazards resulting from oil spills

- Development of diseases in workers, engaged in cleaning up of oil slick

  Oil spills are usually accompanied by emission of oil vapors, and heat, generated by combustion. Inhalation of these toxic materials endangers health of the cleaning up team more than other people. Sometimes, special and scarce chemicals are used to clean up oil slicks. In many cases, these materials may cause skin burns and disorders and/or (at higher concentrations) the development of cancers [10].
- Endangerment of marine life

  Oil slicks, transported under water surface, are in greater contact with aquatic organisms. Therefore, this type of pollution has damaging effects in a short time on marine life [11].
- Intensification of existing diseases

  People, suffering from diseases, for example, asthma and/or other similar pulmonary diseases, are exposed to greater hazards from oil pollution and their chemical compositions compared with other people. Therefore, they should be taken away immediately from the polluted site [10].

- Hazards for pregnant women

  Oil includes many volatile and toxic chemicals, and the inhalation of some of them by pregnant women may cause premature birth, low-birth weight infants, and/or abortion. Consequently, polluted areas by oil are very dangerous for pregnant women, and they must rapidly leave such areas [11].
- Neurological disorders

  Experience on oil spills showed some fishermen, exposed to oil pollution for a long time, develop neurological problems [10].
- Pulmonary diseases

  As toxic and hazardous chemicals spread in water during oil spills, corexit can cause serious pulmonary problems for people engaged in cleanup and for workers that inhale emitted vapors and particles without wearing intact respiratory protection (standard and reliable masks) [10].
- Smoke

  In petroleum fires, oil spills spread on ocean surface, and the particles trapped in smoke move downwards toward land and reach human eyes and lungs. This can cause health problems for some people. In addition to smoke, toxic and hazardous materials are emitted in petroleum fires, and sometimes their harmful effects are tangible for long distances. The extent of damage suffering people by toxic vapors depends on the amount and intensity of inhalation of the released vapors. In petroleum fires, toxic gases such as sulfur dioxide, nitrogen dioxide, carbon monoxide, and polycyclic aromatic hydrocarbons enter the environment in addition to carbon dioxide, and particulate matter [10,11].
- Food contamination

  The movement of oil contamination and pollutants into the food chain is not clear yet, but some researchers believe that oil pollution in one place will be a legacy of pollution in it. Since water pollution always worries governments and communities, it is necessary to put restricted standards at national and international levels to combat it. Destruction of marine systems and surface waters cause irreparable damage to the environment. Nowadays, the self-remediation ability of seas has decreased because of high use of seas and variety and high rate of waste discharged into them. Consequently, seas can hardly neutralize the effects caused by entrance of such materials into them.

  Industries, located near the coastline, cause a major part of seawater pollution by drilling seabed, sea transportation (shipping), natural oil spills, atmospheric precipitation, direct contact between surface water and the surrounding air, and deliberate spills into seas form other sources of marine ecosystem pollution.

  Marine transportations including the ports (commercial ports, oil terminals, dry docks, fishing docks, and ships) produce waste, which can be considered a potential source of sea pollution. Collection, recovery, cleaning up, processing, and correct removal of these materials can considerably reduce seawater pollution, resulting by marine transportations.

Rising environmental problems in most countries becomes very important topic and attracts public attention. It encourages people worldwide to try to remove the existing problems with some kind of common sympathetic feeling. Now prevention of water pollution is considered as one of the most important environmental problems because water of seas and oceans provide about 90% of the oxygen on Earth and form part of the global food chain. Therefore, water pollution means damages inflicted on our planet Earth [11].

9. Change of oil pollution process in seawater

When petroleum compounds are spilled on seawater, their physical and chemical properties change under the influence of the prevailing winds, temperature, and direction of water circulation. Volatile materials separate from oil at 30–40°C and the rest floats on water and moves in direction of water flow. Therefore, when the oil spills are light HCs, most of them evaporate from surface water. In general, the fate of oil spills into seas is a complex process, as shown in Fig. 12.2 [12].

9.1 Evaporation of volatile materials

As long as oil is in the form of a thin layer, its volatile materials evaporate. The extent of evaporation depends on temperature, wind, and vapor pressure. Petroleum compounds with boiling points below 200°C evaporate within a

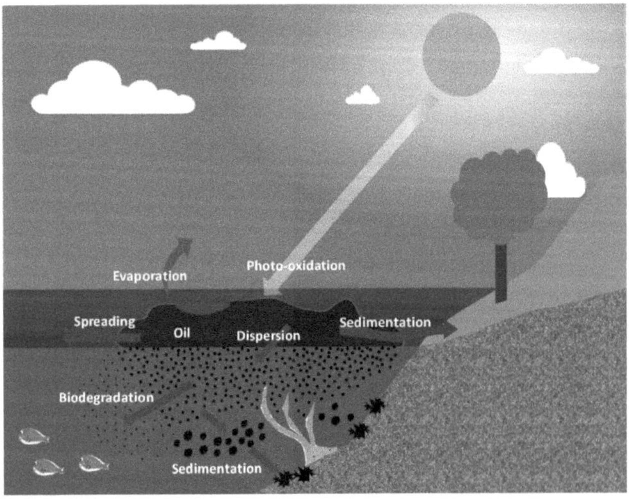

FIGURE 12.2 The variation, transformation, alteration process of oil after spilled into the marine.

few hours, and those with boiling points below 270°C evaporate after several days.

9.2 Partial dissolution of petroleum compounds in water

Some petroleum compounds, consisting of small molecules, especially cyclic substances such as benzene and xylene, are partially water-soluble. These compounds are toxic for marine organisms and destroy them even if small quantities dissolve in water.

9.3 Oxidation of some petroleum products by sunlight

Solar UV radiation causes oxidation of some petroleum compounds and produces various compounds.

9.4 Biological degradation of petroleum products

Seawater contains a range of microorganisms or microbes, which can partially or completely degrade oil to the water-soluble compounds and eventually to carbon dioxide and water. Many types of microbes exist and each of them tends to degrade a particular group of compounds in crude oil. However, some compounds of oil are very stable and may not be degraded easily. The main factors, affecting efficiency of biodegradation, are levels of nutrients (nitrogen and phosphorus) in water, temperature, and level of oxygen. Biodegradation requires oxygen, and it can only take place at the oil–water interface as no oxygen is available within oil. Creation of oil droplets, either by natural or chemical dispersion, increases the oil–water interfacial area and consequently available area for biodegradation process.

9.5 Sedimentation/Sinking

Density values of some heavy-refined products are higher than one, and so they will sink in water. Oil may be denser than water, or it may have enough incorporated sediment to become denser than water. Sinking usually occurs due to adhesion of particles from sediment or organic materials to oil. Shallow waters are often laden with suspended solids, providing favorable conditions for sedimentation. Oil, stranded on sandy shorelines, often mixes with sand and other sediments. If this mixture is subsequently washed off beach back into sea, it may then sink. If oil catches fire after it has been spilled the residues sometimes form, which is sufficiently dense to sink.

9.6 Emulsification

An emulsion forms when a liquid is dispersed in another one. Emulsification of crude oils refers to the process whereby seawater droplets become suspended in oil. This phenomenon occurs by physical mixing promoted by turbulence at the sea surface. The formed emulsion is usually very viscous and more persistent than the original oil. Formation of these emulsions causes the volume of pollutant to increase by factor between three and four. This reaction slows and delays other processes, which would allow oil to dissipate. Oils, containing asphaltene higher than 0.5%, tend to form stable emulsions. The formed emulsions may persist for several months after the initial spill has occurred. Those oils, containing a lower percentage of asphaltenes, are less likely to form emulsions and are more likely to disperse. Emulsions may separate into oil and water again if they are heated by sunlight under calm conditions or when they are stranded on shorelines.

9.7 Natural dispersion

Sea waves break up oil particles into smaller droplets with sizes varying from 1 to 1000 μm. These droplets rise to the surface and tidal waves push them under the surface water. The rate of natural dispersion in large oil spills is 0.5–2 per hour.

10. Containment of oil pollution in water

Containment of oil pollution in water includes preventing the oil spill from spreading and dispersing, diversion to other places, and guiding to especial locations (where it is possible to collect and separate its constituents from seawater). In some of the steps in this separation process, concentration of the oil slick is increased for the collection and in situ burning of the petroleum products [4].

11. Collection of oil products from the surface water

Following containment of oil pollution, collection of oil products from surface water is the next step in the cleaning up operations of the oil slick. This step is usually the most important one in removing oil pollution from surface water. The important point in containment of oil pollution is to raise the concentration of the oil slick in order to increase thickness of the oil layer and facilitate collection of the oil compounds [4].

12. Methods of coping with oil pollution

The common methods in removing oil pollution from seawater include the following:

- ➢ Letting the oil slicks reach the coast
- ➢ Using chemical dispersant
- ➢ In situ burning of oil products
- ➢ Mechanical recovery of oil products

12.1 Letting the oil slicks to reach coast

Oil spilled on water is seldom completely recovered, and some eventually reaches coast. The fate and behavior of oil on coast are influenced by many factors. Some of these factors relate to type of oil spilled, geology of coast and conditions of oil deposition on coast such as weather and waves. There are many important factors to clean up oil such as extent of oil penetrating and spreading, its adhesiveness, and also the amount and type of material mixed with oil. Cleaning up is more difficult work if oil penetrates deeply into coast. Penetration depends on the type of oil and coat materials [13].

Some methods for coast cleaning up:

- natural process
- physical methods

Natural process occurs as a result of the removal of oil from the natural environment by evaporation, oxidation, and biodegradation. Methods, used to physically cleaning up oil from coast, include wiping with absorbent materials, pressure washing, raking or bulldozing.

Collected oil during cleaning up activities must be reused or disposed properly incineration or landfilling.

12.2 Letting oil slicks for natural degradation

Like the previous method, this one is not suitable for ports either. It is used for damaged marine oil pipelines or in cases of mid-sea tanker collisions. Oil slick movement is constantly monitored using predetermined models. If any deviations happen from the adopted model, other methods are immediately used to prevent the oil slicks reaching sensitive areas [4].

12.3 Using chemical dispersants

This is another method of coping with oil pollution. A diverse collection of chemicals is used for cleaning up and removing oil pollution. Utilization of these chemicals is not always effective, and they have toxic effects on aquatic organisms and on wildlife [14].

A typical commercial dispersant contains solvents and surfactants which are sprayed on to the spilled oil. Solvents are used to dissolve the surfactants (some surfactants are solids) and to reduce the viscosity (some surfactants are viscose liquids). The surfactant molecules migrate to the oil−water interface and reduce surface tension. The oil slick breaks into many smaller oil droplets, and the droplets are ultimately degraded by naturally occurring process by microorganisms such as bacteria and fungi. This step may require days or weeks to complete (Fig. 12.3) [14,15].

Advantages of using dispersing agents in coping with oil pollution:

- These agents have applications in wave flows and high turbulent seas.

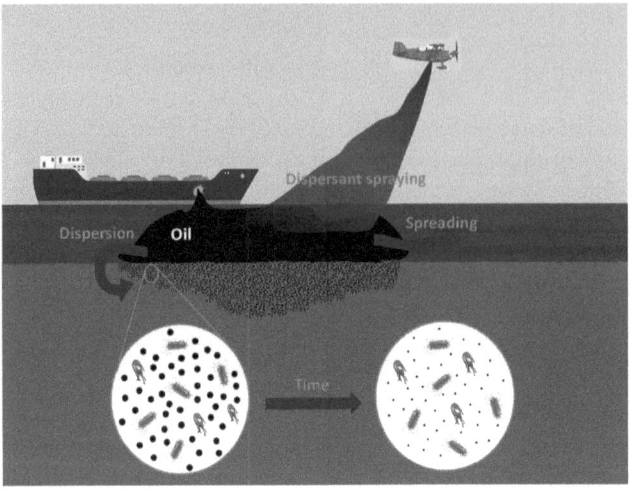

FIGURE 12.3 The chemical dispersion processes for oil spill cleaning up.

- Use of airplanes to spray dispersing agents is the quickest method of coping with oil pollution in seawater.
- Dispersing agents destroy oil slicks on water surface and stop the prevalent winds from moving them. Therefore, the remaining oil slicks become easy to detect and recover.
- This method somewhat prevents marine mammals and birds to be contaminated with petroleum compounds.
- Use of dispersive agents prevents formation of oil floccules in surface water [14].

Disadvantages of using dispersing agents in coping with oil pollution:

- Penetration of petroleum products into deep water damages aquatic organisms.
- This method cannot be used for all petroleum products and can be applied under special conditions.
- If this method is used in coastal waters, it will increase penetration of oil into the sediments on the sea floor.
- Overuse of dispersing agents is toxic for marine organisms.
- There is time limitation in using dispersing agents: they must be sprayed immediately after happening oil spill.

13. In situ burning of oil pollutants

This is one of the cleaning up methods, used in coping with oil pollution. It includes controlled burning of oil at or near the oil pollution site (Fig. 12.4) and has advantages over other cleaning up methods of oil pollution [4].

FIGURE 12.4 Controlled burning to remove oil from the water surface after an oil spill [16]. *Reprinted with permission from Elsevier.*

This method removes large amounts of oil in a wide area very quickly compared with the other methods. This method prevents spreading of oil spills to other regions. From 1970 up to now, extensive research has been conducted on burning methods at pollution site. Under particular conditions, when oil is spilled over ice and mixed with it, burning is the only method that can be used [16].

Some countries approved the in situ burning method, but others have not applied it yet.

13.1 Features of in situ burning

In situ burning has advantages over other cleaning up methods of oil slicks. As a remarkable advantage, it removes large amounts of oil in a wide area very quickly compared with the other methods and prevents spreading of oil spills to other regions. Of course, it can be effective when oil is not exposed to the weathering process to lose its flammable constituents. This method is one of the effective and rapid methods in removing large volumes of oil pollutants, especially in water. In situ burning stops oil spreading to other regions and also prevents pollution of coastlines and ecosystems in the related region. Burning of oil is a final and single-step solution and leaves little residue, which can be collected and/or repeatedly burned to reduce its volume.

Under ideal conditions, in situ burning process requires less equipment and is more efficient than the other methods. Based on experiments, conducted on the coast of Newfoundland in 1993, its efficiency was about 98%–99%[4].

13.2 Disadvantages of in situ burning

The most obvious disadvantage of burning of oil is emission of toxic vapors via huge plumes of thick black smoke. The second disadvantage is the oil layer will only catch fire and burn if it is thick enough. Most oil compounds disperse rapidly in water and the oil layer becomes very thin and cannot be burned. Fire-resistant oil booms are applied to increase thickness of oil slicks in order to facilitate burning [4].

One disadvantage of oil burning is emission of toxic gases into air or in water. This is the noticeable obstacle of the widespread usage and adoption of this method in coping with oil pollution. Atmospheric pollution results from suspended particles in the smoke produced in oil burning, from gases produced in combustion, and from residuals of incomplete combustion remaining in the operational region. Soot particles, which formed during oil burning, consist of carbon particles and a group of absorbed chemicals. Residuals resulting from burning of oil compounds and soluble organic materials penetrate into water through immersion in it or by floating in water. The second problem of oil burning is emission of hazardous gases during combustion. Many studies, which have been conducted on gases emitted when oil is burned, showed during oil combustion different compounds form such as carbon dioxide, carbon monoxide, sulfur dioxide, and some substances in acid form. Production of sulfur oxides is directly related to the sulfur content of oil. When oil burns, volatile organic constituents are evaporated and scattered. Research results indicated that benzene, toluene, xylene, and other volatile compounds are scattered on water when petroleum products burn. When oil burns, oxygen-containing compounds (carbonyl compounds such as aldehydes and ketones) form, which are hazardous for human health [16].

13.3 Collecting and recycling of the oil materials by mechanical methods

Mechanical methods, used to control the oil spill and prevent its spreading to a specific area, include:

- Boom
- Skimmer
- Absorbent packages

13.4 Boom

There are different booms with different applications including [4,17]:

- Fence boom for low depth water and placid seas
- Curtain boom for the use in harbor and relatively placid seas
- Specific inflatable boom for a turbulent sea

≫ Improvised booms made from materials like wood, plastic tubes, tires, and empty barrels
≫ Fire-resistant booms to encircle the oil spill and burn it

The main applications of booms are enclosing the oil spill and diverting the path of oil spill. Fig. 12.5 shows the use of boom with various configurations for collecting of oil spill.

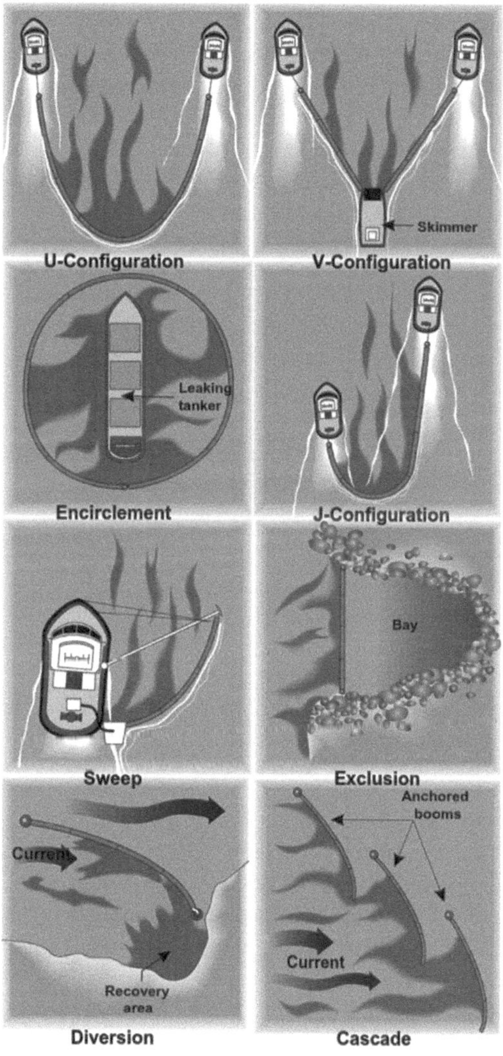

FIGURE 12.5 Illustration of several configurations of booms for collecting of oil spill [17]. *Reprinted with permission from Elsevier.*

The main problems during utilization of booms include the following:

- ➢ In waves with a height above the freeboard, oil spills are spilled out over the boom.
- ➢ In flows more than 1 Knot, oil spills escape from beneath of the boom.

13.5 Skimmers

Skimmers are mechanical tools for collecting oil pollution from the water surface. Due to their large sizes and high capacities, these devices have high efficiencies. The skimmer type depends on the region in which it is used, for example, onshore or offshore, low-depth water, and the lakes. In addition, their efficiencies depend on the viscosity value of different oils. Skimmers came in different shapes and are composed of independent parts, which can be loaded onto a ship. Some skimmers have reservoirs to collect oil, and some others have facilities to purify the collected oil [4,17].

There are different skimmers, including the following:

- ➢ Oleophilic surface skimmers: this type of skimmer is available in a variety models, e.g., brush drum skimmer, brush belt skimmer, adsorbent belt skimmer, drum, discs skimmer, inverted belt skimmer, ropes skimmer (Fig. 12.6). It is used for oil spills with any diameter as it is flexible and resistant.
- ➢ Weir skimmer: oil on the surface of water with a little amount of water is guided into the chamber and then is pumped.
- ➢ Suction skimmer or vacuum devices: this kind of skimmer works like a vacuum cleaner and sucks continuously the oil spill located on the water surface.
- ➢ Elevating skimmers: these devices use conveyors to lift oil from the water surface into a recovery area. They cannot operate in rough waters or in waters with large pieces of debris.
- ➢ Submersion skimmers: this kind of skimmer uses a belt or inclined plane to force the water beneath the surface. The belt forces oil downward toward a collection.

Fig. 12.7 shows the use of various skimmers for collecting of oil spill.

Performance of skimmers is determined based on the amount of the collected oil and also the amount of water, which comes with oil. Isolation of water from the collected oil by skimmer is as difficult as collection of the oil. Efficiency of this method depends on the type, layer thickness and viscosity of oil, sea conditions, wind speed, ambient weather, presence of ice, and other particles in the water. The best performance of skimmers is for thick oil layers, and they usually do not result acceptable efficiency for thin oil layers. Skimmers are placed in parallel to the direction of wind and booms, thus, the wind blow moves the oil spill toward the skimmers [17,18].

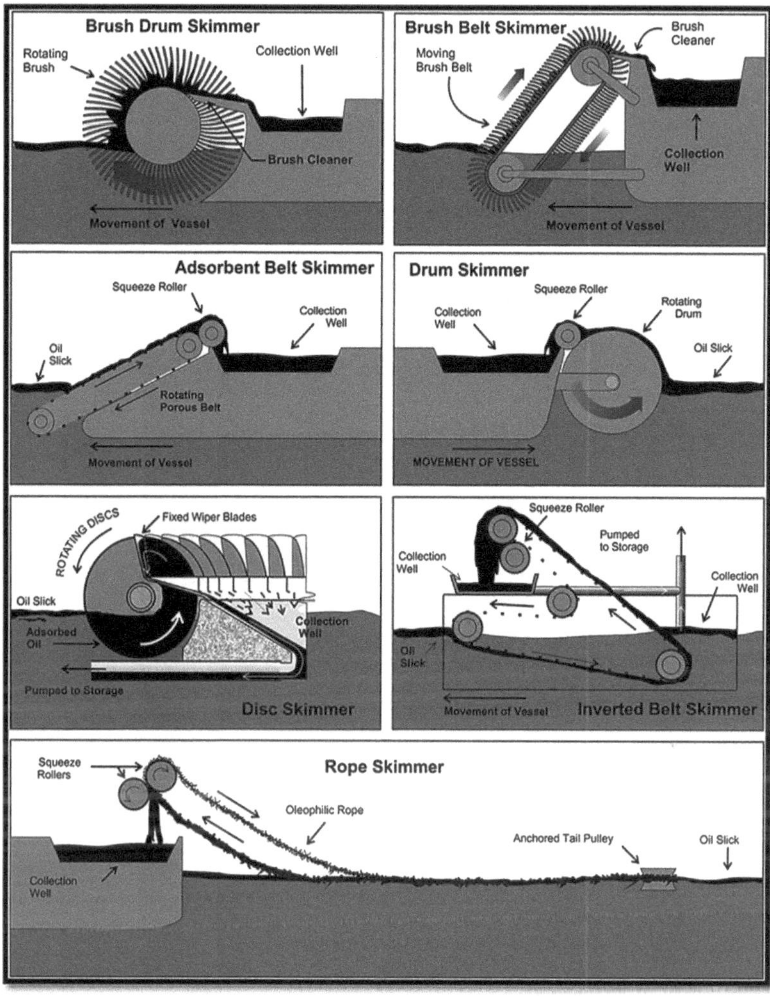

FIGURE 12.6 Illustration of the oleophilic surface skimmers [17]. *Reprinted with permission from Elsevier.*

The weather conditions at the oil spill site affect the performance of skimmers. Most skimmers in calm waters have a higher performance. Most skimmers in water with floating foliage and marine weeds and solids do not work effectively. Some skimmers are surrounded by networks to prevent entrance of ice and other materials into the machine.

Very heavy oils, tarball, and solid particles of oil obstruct skimmers and interrupt oil pumping during oil collecting. A common assumption is that the use of mechanical devices for recycling oil spills from the sea surface is very useful, but these methods have some limitations. For example, the use of

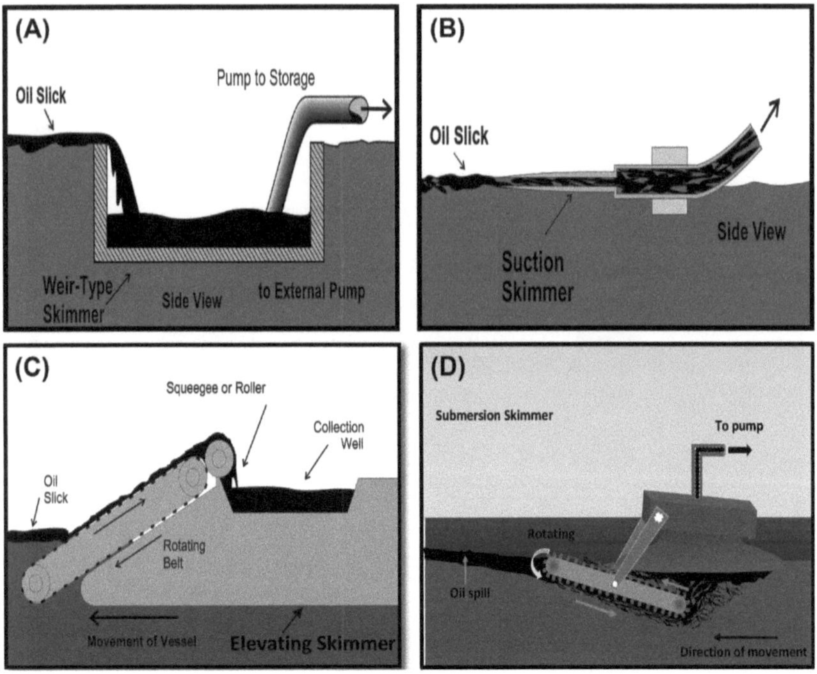

FIGURE 12.7 Illustration of the (A) weir skimmer, (B) suction skimmer, (C) elevating skimmer, and (D) submersion skimmer [17]. *Reprinted with permission from Elsevier.*

inflatable skimmers during severe wind and waves lose its efficiency and usually just a small amount of oil spills can be collected. Because if the spill movement speed is more than 0.5 m/s, the amount of oil spill collection is minimized [17,18].

Skimmer's disadvantages include the following:

- ➢ Skimmer's performance depends on weather conditions
- ➢ In unfavorable weather conditions, it collects water more than oil spill
- ➢ Using a skimmer requires special skills

13.6 Absorbents

One way to deal with pollution of sea and preventing contamination of the marine environment is the use of absorbent materials for collecting oil from the water surface. Absorbent materials are used usually to remove small oil pollution and when oil spills collecting by skimmer from the water surface has limitations. Absorbent materials are used to absorb oil pollution in places having low depth and limited access, as well as in environments that collecting oil spills from the water surface is difficult.

Classification of oil absorbents are natural organics, minerals, synthetic polymers, and hybrid materials.

➢ Natural organic absorbents

Natural organic absorbents include straw, corn wood, peanut peel, wood fiber, and peats made from rotten moss. Natural materials are fairly affordable and abundant, and they have acceptable absorption rate. Wood and lignocellulosic compounds have great potential for absorbing oil pollution. Due to high rate of absorption, the use of these materials is very effective in preventing rapid expansion of pollution. In French Canyon Contamination, about 20,000 pieces of wood were poured over oil spills, and then they were taken off from coast. Natural absorbents are cheap, affordable, renewable, and consistent with the environment. Thus, there is greater tendency to use bio and natural absorbents. Among agricultural wastes, sugar bagasse is one of the most widely used natural absorbents.

The use of natural organic materials such as straw, corn wood, rice peel, feathers, wood, and wool particles during the oil pollution cleaning is popular because of biodegradability, renewability, low cost, and less impact on the marine environment. However, they have also disadvantages, e.g., flammability during storage, absorption of insects by wood, creating dust, dispersion of unpackaged materials, and less absorption efficiencies than polypropylene absorbents.

Many natural absorbents are processed to improve efficiency of oil uptake, but this procedure increases price of the absorbent material. One of the processing methods is acetylation of the free hydroxylase group existing in various absorbents, such as straw, rice, wood, and bagasse by using acetic anhydride. Since the acetylated straw has low water absorption capacity, it can be used several times to collect oil pollution. Consequently, acetylation of the rice straw and other materials with biodegradability feature such as bagasse and wood waste are among convenient and economical methods of removing oil pollution from the water surface [19].

➢ Mineral absorbents

Mineral absorbents include zeolites, clay minerals, silica, etc. Minerals are considered as very efficient and low-cost sorbents due to chemical and mechanical stability, availability, high surface area, and structural properties. Mineral absorbents are raw materials of natural origin, used in a powder or granular form. Pellets of agglomerated particles of mineral sorbent tend to reduce to an impalpable simply upon being rubbed together or being squeezed in the hand. Mineral adsorbents are generally noncombustible and resistant to acids and bases. Their primary disadvantage is a risk of dust formation during application in open spaces. The absorption capacity of the mineral absorbents for straight chain HCs is only about to 15% of the dry weight of the sorbent particles. They are not generally preferred for the removal of oil spills from water surfaces because of their low floating and low oil absorbability compared with polymers or natural organic sorbents [19]. However, the modified mineral absorbents by organic compounds are used as suitable absorbents for cleaning up of contaminated water with oil.

➢ Synthetic polymeric absorbents

Synthetic absorbents including polymer materials such as polypropylene, polyethylene, polyurethane (PU), polyester, and polystyrene are widely used due to hydrophobic and oleophilic characteristics. Other synthetic materials, used in recent years for the removal of oil spills include rubber waste and rubber compounds, inherently absorb liquid and gas HCs. So, the use of rubber compounds in absorbing oil spots can be one of the right and the cheapest methods of recycling these materials. On the other hand, the synthetic polymers have very slow degradability, which is a major disadvantage for them [19].

➢ Hybrid absorbents

Hybrid devices exploit absorbents of both natural and synthetic origin with comparable contributions to sorption, resulting in a structure with properties superior to either of its contributing components. A typical hybrid sorbent may comprise synthetic polymeric foam, synthetic polymer fibers, natural fibers, wood shavings, treated cellulose, exfoliated graphite, or vermiculite, and all clubbed together into a single structure. The oil-absorbing component was made up of fibers of natural or synthetic origin. The hybrid devices have a large specific surface area and oleophilic and hydrophobic nature [19].

14. Methods of using absorbent materials

Collection of oil contaminations from the water surface by using absorbent materials is done in the following ways:

14.1 The massive use of absorbent materials

Massive absorbents are products without a shape of their own, made up of particles without any links between them. These materials may include short strands or minerals (pearlite and vermiculite), herbal-based materials (straw and peat), or artificial materials like polymer materials (PU, polypropylene, and polystyrene). These materials are spread widely on water and due to directly contact with contamination, they can penetrate among thin layers of the pollution [20].

14.2 The use of absorbent materials in sheet or roll shape

➢ Absorbent sheets are usually in square form or in strip form. Polypropylene absorbent sheets are an example of these types of sheets. Rolls are usually made up of nonwoven textile with inside strings to strengthen them. In order to use absorbent material in the form of absorbent sheet or roll, the following points should be taken into consideration [20].

- Sheet- or roll-shaped absorbent materials are more effective if, for a long period of time, remain on the water. Scattering of absorbent sheets by water or wind makes it difficult to collect them from the water surface.
- In cases of low area, in order to increase the absorption rate, these absorbents are floated indefinitely. One can also use inflatable booms to surround contamination and make better use of the absorbent material. With the help of this method, the collecting process of these absorbents from the water surface becomes easier.
- For oil spill collecting, if condition is not suitable for using the inhibitor booms, absorbent rolls should be used. Maritime flows and waves cause dispersion and movement of absorbent rolls on the water. To use absorbent rolls, ropes must be used to prevent dispersion of rolls in the operation area.

14.3 The use of absorbent materials in the form of pads

Absorbent pads are thin, flexible products, and capable for absorption of petroleum materials. The length of these pads is usually less than 1 m. They are sufficiently resistant to be handled (e.g., felts and polypropylene sheets).

The absorbent pad is placed on the pollutant oil surface, and then recovered after impregnation by either hand or using forks, rakes, or other tools [20].

14.4 The use of absorbent materials in the form of boom

The oil absorbent boom is a highly versatile containment product that frequently is used in the cleaning up of small water spills. The length is far greater than the other dimensions and is more than 1 m. Particularity of certain model for use on water is a "skirt," as a flexible ballasted band, which improves their ability to contain a floating pollutant. These products combine the properties of "physical" booms (containment) and sorbent booms (trapping and sorption).

Given their relatively compact form, these products are more suitable for fluid pollutants (e.g., light refined products) which can be soaked up rapidly and completely. To increase absorption and preventing the release of contamination, the booms must be connected to each other at their end [20].

14.5 The use of absorbent material in the form of long string fibers

The absorbent fibers such as long rope strings are usually made of polypropylene fibers. Such absorbents with obvious advantages of low cost, tunable hydrophobic and oleophilic properties, and available usage in large scale show their great potentials for oil–water separation. These fibers are used to absorb low amounts of petroleum materials with a medium to high viscosity as well as to quickly absorb oil contaminants that have turned to emulsion in water [4].

14.6 The use of absorbent materials depending on the area location

Sometimes, to deal with pollution, absorbent materials are used with regard to the environmental situation and abundance of absorbent materials at that point. Examples of temporarily using of absorbent substances are the use of straw, sweeping grass, dried ferns, bagasse (cane waste), and other available and abundant materials in the region. In operations against oil pollution, identification and the use of indigenous plants have a higher priority than the use of industrial absorbent materials because they have less impact on the marine environment [4].

15. Manual cleaning

Clearing small oil spills in remote areas is done usually manually. Isolation of heavy oils in comparison with light petroleum in this method is much faster. Sometimes cleaning oil spills close to coastlines is done by mechanical excavator, by rakes, or by cutting off plants contaminated with petroleum materials.

Regarding all the facts listed above, these conventional isolation systems are expensive and need a lot of space and energy, and also they have low separation efficiency [4].

16. New methods for cleaning up of oil spills

Mechanical extraction by using absorbents is one of the most efficient and economical methods to clean up the oil spills, involving the use of absorbents for condensation and transformation of these materials in solid or semisolid states. Hence, ideal absorbents must be hydrophobe and oleophile, resulting in features such as high absorption rate, high floating property, reusability, and oil keeping capacity.

Separation by using absorbents, which have selective tendency to water or oil, is a relatively new field in industrial development with a very bright future. Using chemical composition and proper surface structure, one can produce special wetting levels with selective superwetting characteristic with respect to the water for separating various mixtures of water, oily materials, and their emulsions [21,22].

17. The wetting phenomenon

Wettability is the ability of a liquid to contact and spread on a solid surface, resulting from intermolecular forces. Degree of wettability is determined based on the balance between cohesive and adhesive forces. In other words, wettability means tendency for the maximum contact surface of liquid with a solid surface [23].

Tendency of a fluid to spread on a solid surface specifies the wetting properties of a liquid relative to a solid surface. This tendency for spreading is usually expressed by measuring the contact angle, θ. By reducing the contact angle, the wetting property of liquid increases. The zero-degree contact angle represents the full wetting, and the contact angle of 180 degrees implies absolute nonwetting. The median wetting property has different definitions, but usually refers to angles from 60 to 90 degrees[24].

17.1 Adhesion force

The gravitational force between liquid molecules and the solid surface is called the adhesion force. Note that this force is between two molecules of different kinds of substances, causing them (liquid and solid) to adhere to each other [23].

17.2 Cohesive force

The gravity force between liquid (or solid) molecules is called cohesive force. In fact, this force exists between molecules of the same type of materials and causes them to shape. For example, the cohesive force between water molecules causes that water look evenly integrated into a container.

Fig. 12.8 shows the adhesive and cohesive forces in a system containing liquid and solid [23,25].

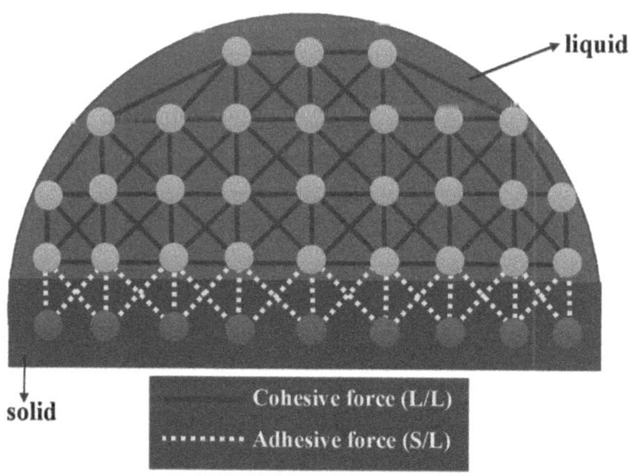

FIGURE 12.8 Schematic presentation of cohesive and adhesive forces.

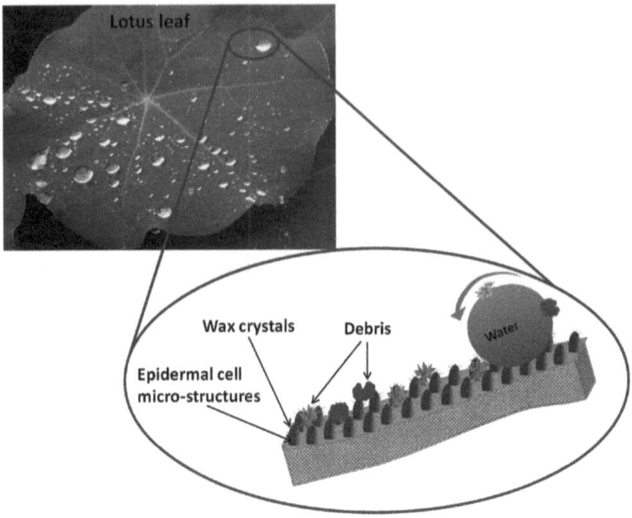

FIGURE 12.9 Structure of the lotus leaf (micrometers and nanometers lint) and picture of the water drop on the lotus leaf; the debris particles stick to the water droplet and detach from the surface.

As shown in Fig. 12.9, the adhesive forces cause the dew drops to be placed on the lotus leaf, and the cohesive forces causes the water droplet to look as a sphere.

18. Defining superhydrophobic surfaces

When liquid droplet is placed on the solid surface, the lack of its spread leads to a fixed shape for it as well as the balance contact angle. The balance contact angle shows the wettability of the solid surface by the liquid [23,25]. From a geometric point of view, the contact angle is formed by liquid at the boundary of the three phases of liquid, solid, and gas. If the balance contact angle of the water droplet is less than 90 degrees, the surface is hydrophilic, and if it is more than 90 degrees, the surface is called hydrophobic.

When the effective contact angle of the water droplet is equal to or bigger than 150 degree, the solid surface is called superhydrophobic. At present, by using advanced methods in the field of nanotechnology, solid surfaces with contact angles close to 180 degrees have also been reported [23,25].

18.1 Natural superhydrophobic surfaces

There are several examples of natural superhydrophobic surfaces, e.g., plants and insects. The most remarkable natural superhydrophobic surface is the lotus leaf. The high hydrophobic degree by lotus leaf inspired researchers to do initial research on the hydrophobic phenomenon.

Ward et al. observed for the first time that the lotus leaf is clean in marshes and there is no pollution on it. In the 1960s, studies done by scanning electron microscopy showed that surfaces that look quite flat in macroscopic scale actually are rough in the microscopic scale. When a surface has a texture on the micrometer or nanometer scale, the interface between air and water droplet, located on the solid surface, increases and the capillary force between the droplet and the surface severely decreases. So, the water droplet takes a spherical shape and easily slides and therefore pollution and dust particles are absorbed to it. High hydrophobic nature of these leaves is attributed to the simultaneous existence of micro- and nanostructures as well as hydrophobe waxes on the leaf surface (Fig. 12.9) [26–28].

18.2 Superhydrophobic surfaces

When a surface has a rough texture on a nanometer scale, the interface between air and water droplet, which is located on the surface, increases and the capillary force between the drop and the surface becomes very low. Therefore, the water droplet takes a spherical shape and fails to stand on the surface. Generally, to achieve superhydrophobicity at a surface, superhydrophobic coatings can be used. Surface free energy and surface roughness are two important factors for preparation of superhydrophobic surfaces.

Surface roughness causes to trap a high amount of air between the water droplet and the surface. The maximum wettability angle obtained for water on a flat surface by reducing the surface energy is about 120 degrees. Fabrication of uneven structure is necessary to reach higher contact angles and by taking various approaches, one can fabricate rough surfaces with low surface energy with contact angle of up to 170 degrees [27].

➢ Surface energy

To achieve the superhydrophobic surface, the decrease in surface energy is very common. Surfactants are used for lowering the surface energy. Of course, to have a superhydrophobic surface it is necessary to have a surface with very low surface tension along with appropriate nano- to microstructures [29].

➢ Surface roughness

Surface roughness is an important factor to achieve superhydrophobic property. Thin coatings of materials with low surface energy on rough surfaces lead to formation of superhydrophobic surfaces. Since, in such surfaces, air bubbles are caught in the ups and downs positions of the surface, when water is put on the surface, it does not contact with all surface points. Due to low energy of surface and the presence of air, water does not penetrate into the valley, and consequently, the surface area reduces causing a reduction of friction, and droplets slip on the surface easily [29].

For a surface to be superhydrophobic, contact angle must be enough large ($\theta \geq 150$). This means that trapped air between the water droplet and the surface on which it is located must be existed. A high-quality superhydrophobic surface must have a uniform layer of trapped air that leads to very high contact angle (close to 180 degrees), very small rolling angle (close to zero), and a low contact angle hysteresis (close to zero) [23,30].

If a water droplet is placed on a flat and uniform surface, its localized contact angle is a function of the surface tension of the liquid and the surface chemistry (surface energy). However, by increasing surface roughness, effective contact angles will also increase. A surface that is both hydrophobic and rough can become superhydrophobic; the roughness magnifies the effect of the hydrophobicity and can lead surfaces being very hydrophobic that water droplets on them appear by spherical shape [31].

19. Advancing and receding contact angles

It has recently been argued that measuring the contact angle of a fixed droplet is not enough and it is better to use also the other angles (advancing and receding contact angles) to ensure of the superhydrophobicity of surface. The difference between these two contact angles (hysteresis) reaches the maximum when the liquid drop begins to slide down the tilted surface. The contact angles of forefront and trailing edges of the liquid drop just prior to movement of its contact line are called the advancing (θ_A) and receding (θ_R) contact angles, respectively (Fig. 12.10A). At the superhydrophobic surfaces, these two angles are very close to each other.

If the drop moves on the tilted surface or in other words moves at the boundary of three solid–liquid–gas phases, the angle created in front of the droplet is called a dynamic advancing angle and the angle created behind the droplet is called a dynamic receding angle (Fig. 12.10B) [32–34].

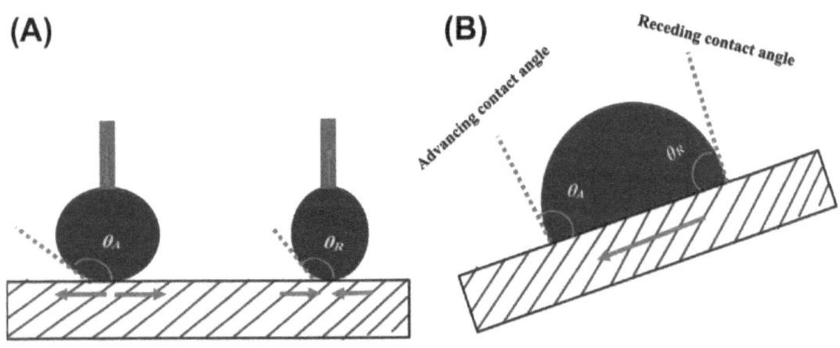

FIGURE 12.10 (A) The liquid droplet in contact with the surface (advancing and receding angles are shown with θ_A and θ_R, respectively). (B) dynamic advancing and receding angles.

20. Contact angle hysteresis

The difference between the advancing and the receding angles is called contact angle hysteresis. Roughness of solid surface is examined with the roughness factor, which is effective in appearance of contact angle hysteresis. Contact angle hysteresis may be observed due to the change in solid surface energy because of absorption of water or liquid [35]. In determining the surface hydrophobicity, reporting the contact angle hysteresis is important as contact angle.

According to Eq. (12.1), the force required to start moving the contact line of droplet on the surface is proportional to the difference between the advancing and receding angles or the hysteresis contact angle. In the other words, lower hysteresis contact angle is the result of a higher surface hydrophobicity [36].

$$F = \gamma_{lv}(\cos\theta_R - \cos\theta_A) \tag{12.1}$$

F is the critical force for contact line movement, γ_{lv} is the liquid surface tension, θ_R and θ_A are the advancing and receding contact angles, respectively [36]. Water droplets on a surface without contact angle hysteresis easily slide.

21. The sliding angle of drop

By setting a hydrophobe surface horizontally and putting a water droplet with a specific volume on it, and then increasing the surface sliding angle gradually, the droplet starts moving on the sloping surface. The angle obtained in this case is called the sliding angle of droplet. Contact angle shows static properties in thermodynamic balance state, while the droplet sliding angle shows dynamic properties of solid–liquid contact surface [37].

Many empirical equations are stated to obtain the drop sliding angle and its relation with other angles [37]. For example, we can refer to Eq. (12.2), which states the relation between contact angle hysteresis and the drop sliding angle.

$$\frac{mg\sin\alpha}{w} = \gamma_{lv}(\cos\theta_R - \cos\theta_A) \tag{12.2}$$

where α is the drop sliding angle, m is the water drop mass, and w is the solid–liquid contact width. This equation predicts that for a given mass of water droplet, a smaller contact angle hysteresis will result in a smaller sliding angle and easier roll-off. According to Eq. (12.2), it is clear that the same hysteresis levels do not show always the same drop sliding angles. This can be due to differences in value of m/w for different contact angles. So, to specify the sliding properties of the surfaces with different contact angles, direct use of the drop sliding angle is preferred [37]. A simple geometric analysis predicts that it is difficult to roll off a liquid droplet from the surface if the static apparent contact angle is low (<140 degrees) unless the contact angle hysteresis is very small (<5 degrees) [38]. It is not easy to obtain low-contact angle hysteresis for surfaces with low contact angles. On the other hand, a liquid droplet will easily roll off if the apparent contact angle is larger than ~150 degrees, even if the contact angle hysteresis is larger than 5 degrees [38]. This is due to the reduction of the solid–liquid contact (small w).

When water droplets roll over the surface, they can easily trap and remove dust particles from the surface.

22. Young–Dupre equation

When the liquid droplet is in contact with the flat surface of solid and its vapor, in its equilibrium state, the contact angle can be obtained from the Young–Dupre equation (Eq. 12.3):

$$\cos\theta_Y = (\gamma_{sv} - \gamma_{sl})/\gamma_{lv} \quad (12.3)$$

where γ_{lv} is the liquid–vapor interfacial tension, γ_{sl} is the liquid–solid interfacial tension, γ_{sv} is the solid–vapor interfacial tension, and θ_Y is the equilibrium contact angle. Fig. 12.11 schematically presents the contact of a water drop with a flat surface [39,40].

Typically, surfaces with small interfacial tension (γ_{sl}) have small amounts of polar groups on the surface or even have no polar group at the surface. Usually, these surfaces show high contact angles. To calculate the balance contact angle of rough surfaces, the Young equation cannot be used. For rough surfaces in addition to the surface tension, the surface roughness affects contact angle. For this reason, other theories are used to calculate the balance contact angle, which will be referred in the next section [40].

An ideal solid surface is flat, rigid, and chemically homogeneous, for which the Young equation is applicable. In other words, all the surfaces are nonideal, unless neglecting effect of roughness, chemical heterogeneity, and surface reconstruction [40]. A specific example for nonideal state is a surface

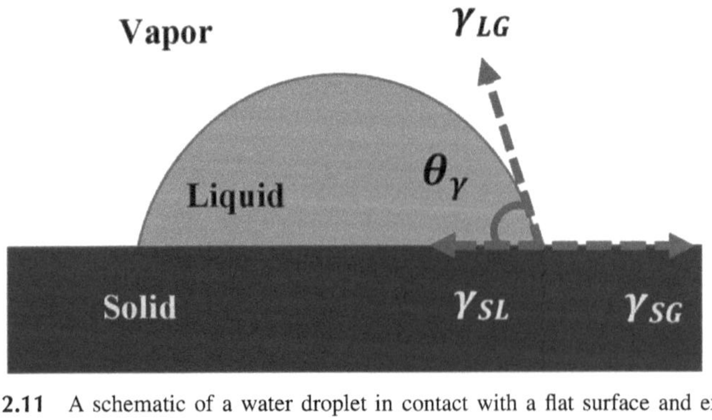

FIGURE 12.11 A schematic of a water droplet in contact with a flat surface and existing parameters in the Young–Dupre equation.

with rough texture. In the homogeneous wettability condition, liquid fills the bumps of the rough surface. On the other hand, the nonhomogeneous wettability occurs when the surface is actually a composite of two materials or phases. An important example of such a composite surface is a surface composed of air and solid. Such surfaces have different effects on the wettability angle. The Wenzel and Cassie–Baxter models are two main models explaining wettability of the textured surfaces. However, these models can be used only when the droplet size is large enough in comparison with the roughness of the surface [41].

23. Wenzel theory

At the rough surfaces, droplets confront two different conditions. In the first mode, the water droplet contacts with the rough surface placed under the drop, resulting in wetting of all the grooves below the drop surface, which is named in memory of Robert N. Wenzel, who outlined the principles of this state in 1936 (the Wenzel state) [42]. Wenzel explained that at the rough surface, the actual area of the solid–liquid contact under the drop was greater than the flat surface. If solid–liquid energy at the interface is less than that value in the solid–vapor interface, wetting of the rough surface decreases the system energy and therefore the rough surface becomes wet. However, when the solid–liquid energy at the interface is more than solid–vapor interface, the surface inherently rejects the water and its wetness is unfavorable [31,42]. Modified Wenzel's equation is:

$$\cos \theta_W = r \cos \theta_Y \tag{12.4}$$

where θ_W is the apparent contact angle in Wenzel's theory and r is the surface roughness ratio, which is obtained in accordance with Eq.(12.5). In fact, r is the ratio of the actual surface, A_r, to the geometric surface, A_O [31,42]:

$$r = \frac{A_r}{A_0} = \frac{\cos \theta_W}{\cos \theta_Y} \tag{12.5}$$

A schematic of the Wenzel surface is shown in Fig. 12.12A.

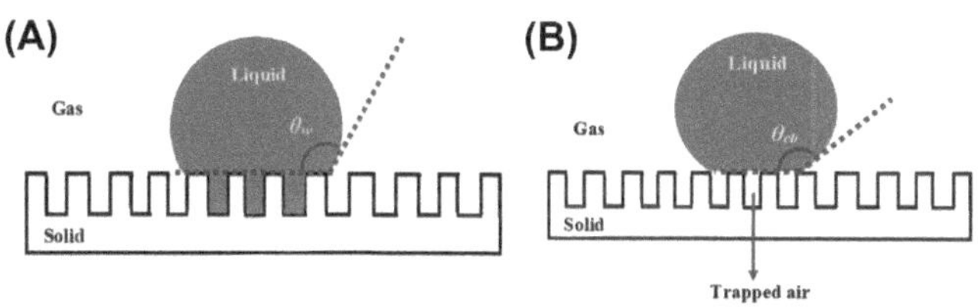

FIGURE 12.12 A schematic presentation of (A) Wenzel surface and (B) Cassie–Baxter surface.

Wenzel equation represents that roughness modifies wettability and anti-wettability features, depending on the surface nature. If the flat surface is hydrophile ($\theta_w < 90$), roughness makes the surface more hydrophile, and if the surface is hydrophobe ($\theta_w > 90$), the surface roughness increases the surface hydrophobicity [42].

Nevertheless, for surfaces with high roughness or for those with porous surfaces, the right side of the Wenzel's equation may be larger than 1, which for such cases the Wenzel's equation is not valid and the Cassie–Baxter model is used [42].

24. Cassie–Baxter theory

The second possible mode for a droplet on a rough surface according to Fig. 12.12B is suspension of a water droplet on the surface ridges, which in this case, air is trapped between the surface ridges and water droplet. This condition is called Cassie–Baxter state. In other words, Cassie–Baxter expressing an equation examined the liquid contact angle on a solid–vapor surface. Actually, on these surfaces, air is trapped between the surface grooves, which indicates its difference with the Wenzel theory [43]. Eq. (12.6) is known as the Cassie–Baxter equation:

$$\cos\theta_{cb} = f_{sl}\cos\theta_Y - (1 - f_{sl}) \tag{12.6}$$

where f_{sl} is the solid–liquid surface area ratio and θ_{cb} is the contact angle in Cassie–Baxter theory. The Cassie–Baxter model is a more general model that can be used to predict entire wetting regimes from low-extreme to high-extreme conditions, whereas the Wenzel model can predict only moderate homogenous wetting regimes. From the Cassie–Baxter model, it can be noticed that a reduction in the solid fraction and an increase in the air fraction would enhance the water repellency of a surface regardless of whether the surface is hydrophobic or hydrophilic [43].

25. Solid–air composite surfaces

Investigation of hydrophobicity mechanism by changing surface roughness is complex. For most hydrophobic surfaces, formation of a solid–air composite surface under the water droplet is very important. The composite surface reduces the contact area between solid and liquid, and thus reduces both adhesion of liquid droplet to the solid surfaces and hysteresis contact angle. Materials with superhydrophobic properties are widely used in anticorrosion coatings, antifreezing coatings, liquid-repellent textiles, and separation of oil from water [44].

In most of these applications, presence of the trapped air between surface irregularities can reduce penetration of liquid (for separation of water and oil, antifog), ionic solutions (for anticorrosion, water sweetening, and batteries), and heat (for antifreeze).

These findings provide a way to develop methods for preparation of superhydrophobic surfaces by mimicking the natural surfaces. Many scholars are working around the world to produce surfaces, which have special surface morphologies at micro- and nanoscales besides having very low surface energy.

To create suitable roughness on the surface, different methods were used such as nanoparticles deposition, chemical vapor deposition, laser patterning, electrospinning, sol–gel, hydrothermal, and chemical corrosion. These methods along with a hydrophobe layer such as silane, fluorocarbon, and fatty acids lead to the superhydrophobicity of the surface [45]. All the mentioned methods can be used for hard surfaces such as metal and glass and also for textiles and sponges.

For example, TiO_2/single-walled carbon nanotube (SWNT) composite, as coating material, was prepared through a modified liquid phase deposition process [46]. Then, the TiO_2/SWNT composite was modified with 1*H*,1*H*,2*H*,2*H*-perfluorodecyltrichlorosilane (FDTS) to lower the surface energy chemically. The treated rough composite displayed superoleophobicity with an oil contact angle of 160.4 degrees and an oil-sliding angle of 5.9 degrees to a silicon oil droplet in air. When the surface was irradiated with UV light for 13 h, the composite character thoroughly converted from superoleophobic to superoleophilic. This conversion was described to the decomposition of FDTS monolayer by the photocatalytic action of the TiO_2 component [47].

26. Superhydrophobic and superoleophilic absorbents

Several hydrophobic products are presented based on the nano- and microscale materials to absorb oil pollutants because of environmental concerns due to oil pollution and oily sewages. New materials are designed to absorb oil from water surface. Some of these absorbents include the following:

- Synthetic absorbents such as polypropylene and PU are the most common organic synthetic materials for removal of oil pollution because they have lightweight and oil-friendly feature. As the ultralight and open-cell PU foams have low density and high porosity, they are considered as a good oil sorbent. However, PU forms due to the reaction of di-/poly-functional hydroxyl (polyol) or amino-containing compounds with di-/poly-isocyanates, they have lots of ether-ester-, carbamate-, and amide-groups, causing PU to absorb both oil and water. To increase hydrophobicity of adsorbent, nano- or micromaterials are used for coating of the PU foam. The other advantages of the PU are low price and nonbiodegradability, which make PU a potential absorbent for oil spills. Absorbent materials may be mixed with hydrophobic and oil-friendly compounds for improved performance. Oil-friendly compounds tend to absorb oil and repel water. Creation of organic friendliness and hydrophobicity properties sounds useful because these properties help to increase suspension of absorbents on the water surface. Sponges, used during the separation process, never break, and they can be used several times to clean the oil spill. However, the recycled oil can be reused, and this is an effective incentive for the oil companies to use this substance in cleaning up the oil pollution (Fig. 12.13) [48,49].

➢ The use of hydrophobic textiles such as fibrous substrates is a common technology for separation of water−oil mixture. Textile fibers have great features such as high area to volume ratio and the structures with various porosities. A drawback of absorbents, made by textile fiber, is the reduced strength leading inefficient use of them. An effective way to improve mechanical properties and reusability of superhydrophobic textile fibers for oil−water separation is preparation of their composites with other materials. In selecting the reinforcing components, it should be noted that the reinforcing part should be durable and flexible to endure the mechanical stresses involved during refinery of adsorbents. The reinforcing component needs to be oil friendly and hydrophobe to assist the absorption of petroleum and oily materials. To meet these criteria and to modify the porous fibers, different polymers are used such as polystyrene, polyvinylidene fluoride, PU, and polyvinyl chloride as the reinforcing part. Prepared superhydrophobic composite packages have good absorbance performances and maintaining high oil absorption capacities after several cycles, indicating excellent reuse feature of these composites as absorbents (Fig. 12.14) [50].

➢ Aerogel is a very porous material which can absorb a large amount of fluids within its empty space. Since a large volume of aerogel is air, it is known as the lightest solid substance in the world. Sometimes aerogels

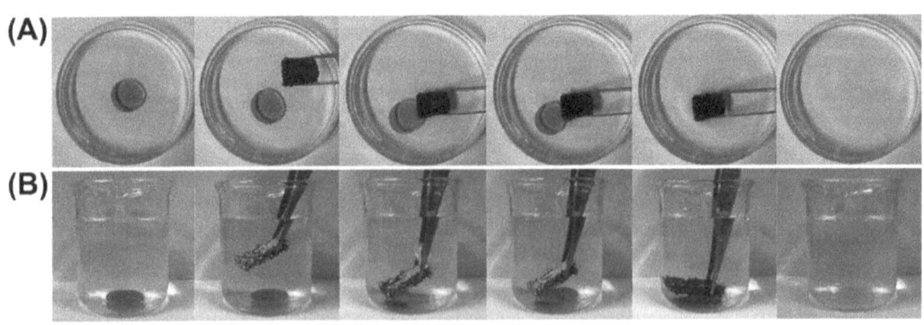

FIGURE 12.13 Process of removing (A) hexane oil and (B) dichloromethane from the surface of water by the as-prepared sponge [49]. *Reprinted with permission from Elsevier.*

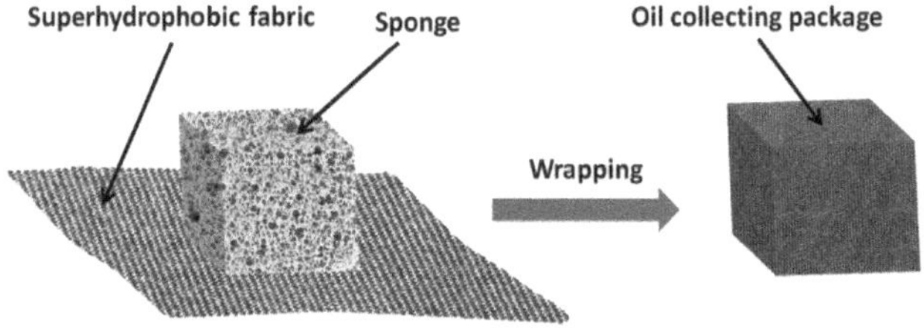

FIGURE 12.14 Fabrication of the superhydrophobic oil collector package using polyurethane sponge and superhydrophobic carbon cloth.

are called “frozen smoke.” Aerogels can be made from silicate or carbonic materials. Hydrophobic aerogels can absorb organic compounds more than 100 times of their initial weight and without absorbing any water. These adsorbents can also be used several times with the same initial efficiency. Various aerogels can be used to remove petroleum and oil contaminants such as hydrophobic silica-aerogels (a combination of nanogel and aerogel), aeroclay, maerogels (aerogels made from plant wastes), and the carbon aerogels. Since aerogels are very light, large volumes of them can be carried easily. Hydrophobic aerogels float on the water surface and in a few seconds absorb a lot of oily and petroleum materials. Some oily materials are present in the waterbed and carbon aerogels can absorb them under the water easily. The absorption rate for oily petroleum materials varies depending on the type of the constituting structure, pore size, and functionalized carbon aerogel type. Therefore, carbon aerogel type can be selected depending on the pollutant type. For recycling of the carbon aerogel, usually, burning and distillation methods are used. If oil or pollutant, absorbed by carbon aerogels, is of little value, it can be burned to clean up the used aerogel without being slightly damaged. Then, aerogel can be reused with the same absorption capacity. But, if the oily or petroleum materials are enough worthy, distillation can be used easily to extract the oil from aerogel and use again the carbon aerogels to absorb oil contaminants (Fig. 12.15) [51].

- Hydrophobe and oil-friendly meshes are used as an important material in cleaning up the oil spills, recently. Superhydrophobic meshes with their lattice structure can be used as an important tool in the oil spill cleaning up industries. High absorption capacity and high structure stability are among the advantages of these materials for oil spill cleaning up. For better performance, meshes are designed in different forms, for example, as floating boat on water surface or as oil collector container (Fig. 12.16) [52].

In addition, by deposition of hydrophobic material with nano-/microstructures, it is possible to prepare superhydrophobic meshes, Fig. 12.17, to trap even very small amount of oil spills and remove the pollution completely [52].

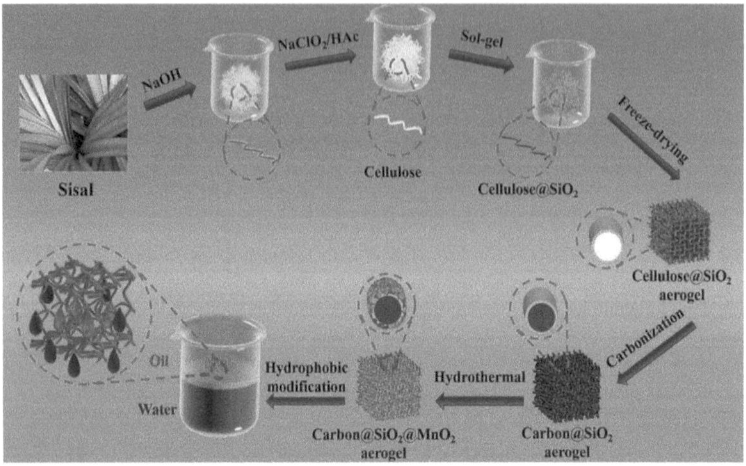

FIGURE 12.15 Schematic illustration of the facile fabrication of superhydrophobic aerogels for fast and selective oil–water separation [51]. *Reprinted with permission from Elsevier.*

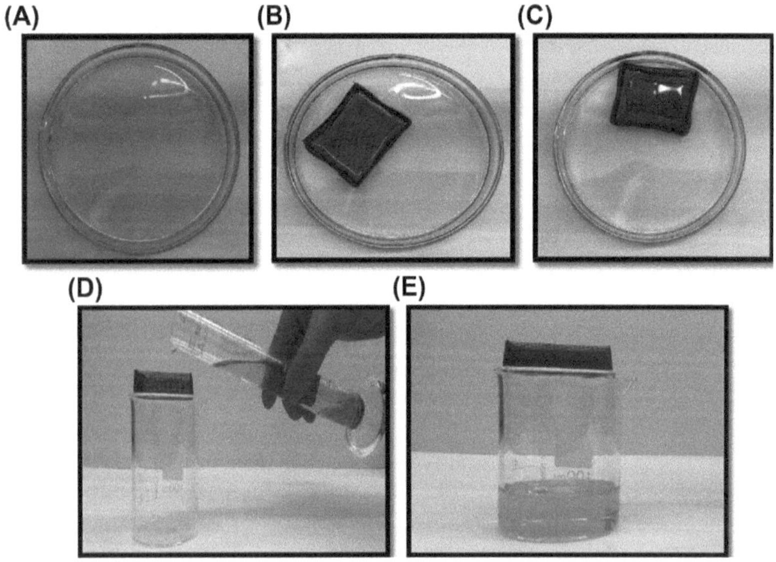

FIGURE 12.16 (A) Photographs of the separation process of kerosene from water by using (A–C) superhydrophobic miniature boat and (D and E) superhydrophobic oil collector container [52]. *Reprinted with permission from Elsevier.*

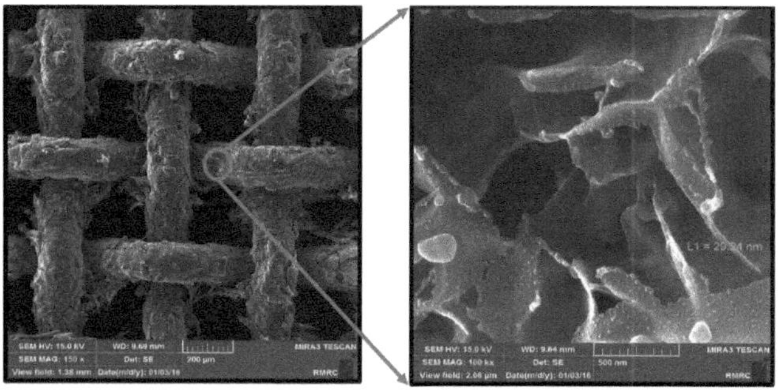

FIGURE 12.17 Scanning electron microscopy images of superhydrophobic and superoleophilic nanostructured deposited layer on steel mesh [52]. *Reprinted with permission from Elsevier.*

27. Design of systems for cleaning up marine oil spill in the future

The researchers and oil companies are trying to take some precaution to solve the problem of oil spill in sea, in river, or on ground. All of the previously mentioned methods and techniques concern removing of the oil spill from water only. But it is not an easy task to do as the spill usually gets stretched and spreads wider by passage of time.

Recently, the collection systems based on a simple, flexible, and intelligent design are used for oil separation and regeneration of sorbents. Researchers have designed new systems to collect oil from the water surface/water bed and to recycle the collected oil at the end of cleaning up procedure. Some of the new systems include the following:

➢ An example of the oil−water separating device is shown in Fig. 12.18. It mainly consists of compressive roller wheel, running roller wheel, propeller, and so on. A superoleophilic sponge is fixed on the crawler belt.

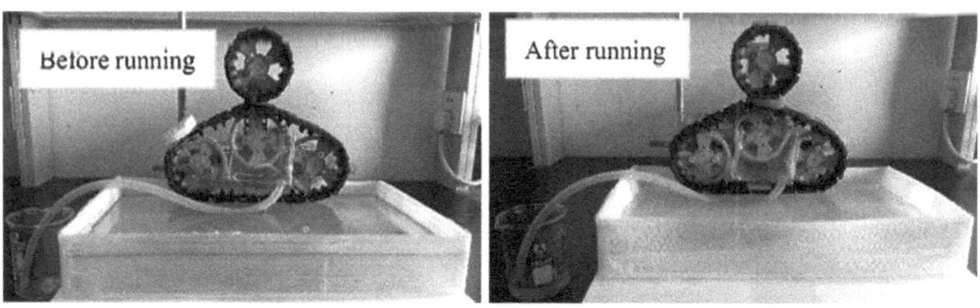

FIGURE 12.18 Photograph of an oil−water separator [53]. *Reprinted with permission from B. Liu, L. Zhang, H. Wang and Z. Bian, Preparation of MCC/MC silica sponge and its oil/water separation apparatus application, Ind. Eng. Chem. Res.* ***56*** *(20), 2017, 5795−5801.*

By moving the device, the sponge could absorb oil and other organic solvents. Then absorbed sponge was transmitted to the top of the device to be compressed. As sponges have a good elasticity and mechanical property, it is easy to release the collected oil and other organic solvents under compression. Interestingly, the presented oil−water separation device can be operated easily, and it is possible also to make a routine plan in the oil and chemical spills/leaks area by the help of GPS and the wireless control system [53].

- An oil collection apparatus, based on a simple combination of porous hydrophobic and oleophilic materials (PHOMs) with pipes and a self-priming pump (Fig. 12.19), is introduced for consecutive collection of oil from the water surface [54]. By using PHOMs, only the floating oil is absorbed by the PHOMs, and it flows along the pipes to the collecting tank. This oil collection system not only will use small amount of PHOM material but also makes the collection of oil spills easier and faster. It has been shown that PHOMs with a suction force can efficiently separate oil and water from the water surface with simulated waves. Therefore, this external pumping as a novel technique will bring PHOMs a step closer to practical application in oil spill remediation [54].
- A similar device for oil spill collection is "oil gathering package." This package is made by wrapping a pristine polymeric sponge with the superhydrophobic/superoleophilic fabric connected to a suction pump (Fig. 12.19) [50,55]. By putting the package into a mixture of oil/water, due to the superoleophobicity of the outer fabric, oil permeates through

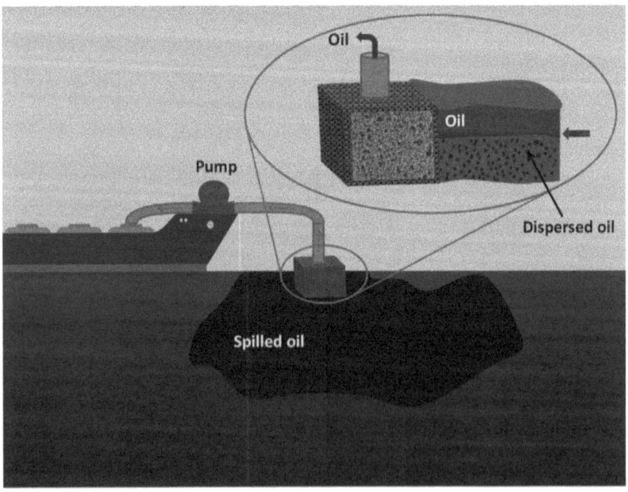

FIGURE 12.19 Schematic of the sponge wrapped with superhydrophobic/superoleophilic fabric for preparation of "oil gathering package" and a suction pump connected to oil gathering package for oil spill collection.

the fabric and absorbs by the inner sponge, forming an oil storeroom. Then the absorbed oil is sucked by a pump to the collecting tank. This approach of absorption using superhydrophobic package combining with pump suction could successfully remove dispersed and floating oil from water. When the system is working, oil is sucked and collected continuously, while water is repelled, indicating a great potential for practical application [55].

➢ To overcome the difficulty of existing oil–water separation methods, a new and simple filtering device is introduced. By this method, floating oil can be collected. A small-scale oil collection device is shown in Fig. 12.20A, consisting of a superhydrophobic steel mesh mounted in a leak-proof manner on the open end of a container. The container is partly immersed in a tilted fashion into a floating oil–water mixture as shown in Fig. 12.20A. Part of the mesh is submerged and remained in contact with both the floating oil and the water underneath. Because of the affinity of the oil toward the oleophilic mesh, the floating oil film is wetted and spread over the mesh. The oil phase passed through the partly submerged mesh (which has negligible penetration resistance against the oil), flowed down along the inclined inner wall of the container, and collected by gravity at the bottom, while the water phase was not transmitted through the mesh and remained outside the collection vessel [56].

➢ A new setup was constructed for oil–water separation, which uses two antagonistic meshes, hydrophilic poly sodium methacrylate (PNaMA) mesh and hydrophobic poly stearyl methacrylate (PStMA) mesh (Fig. 12.20B). These meshes were first primed with small amount of water for the PNaMA mesh and small amount of oil for the PStMA mesh. When a mixture of oil/water is pumped into this apparatus, water passes through the PNaMA mesh, and oil through the PStMA mesh. Oil–water mixtures with different volume fractions were also tested and passed through the meshes easily. The oil–water separation device can be used to run

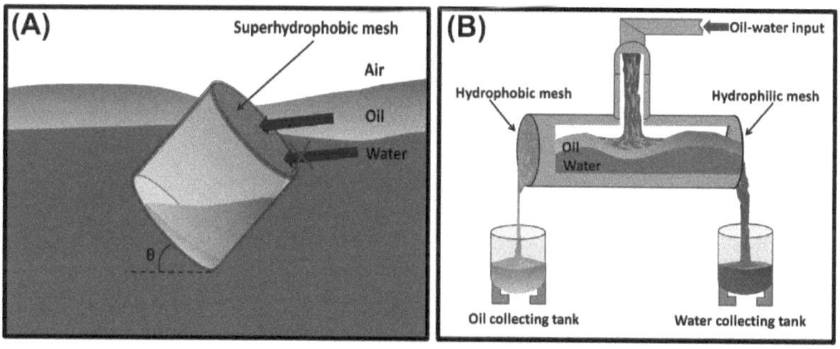

FIGURE 12.20 (A) Schematic of oil collection procedure, by superhydrophobic mesh mounted on a container. (B) Oil–water separation system using two hydrophobic and hydrophilic meshes.

continuously and separate large volumes of oily water at high flow rates [57].

- More interestingly, due to the good mechanical flexibility of the metallic meshes, the superwetting mesh film (SMF) can be simply folded or bent to prepare closed 3D-SMF for large-scale transportation or selective absorption of nonpolar (or weakly polar) organic compounds or oils from water (Fig. 12.21) [58]. In this case, both oil distributed on water surface and underwater can be quickly absorbed and directly stored inside the 3D-SMF without absorption of water because of the excellent superhydrophobic and superoleophilic properties of the SMF. In addition, no leakage of oil from 3D-SMF was observed even when moving the oil-loaded 3D-SMF under water, thus making 3D-SMF an ideal candidate for use as a porous three-dimensional superwetting tank for the capture or transportation of oils or organic compounds in water. In this way, the absorbed organic compounds or oils can be simply removed and recovered from water only by extracting the absorbates from the 3D-SMF. Oils and organic compounds can also be continuously separated and recovered from water by applying the columnar 3D-SMF to the oil–water system, which would be of technological significance and may have potential for a wide range of applications such as water treatment and oil spill cleaning up [58].
- Fig. 12.22 demonstrates an efficient approach for the cleaning up of oil spill by using an oil collector, fabricated by a superhydrophobic metallic

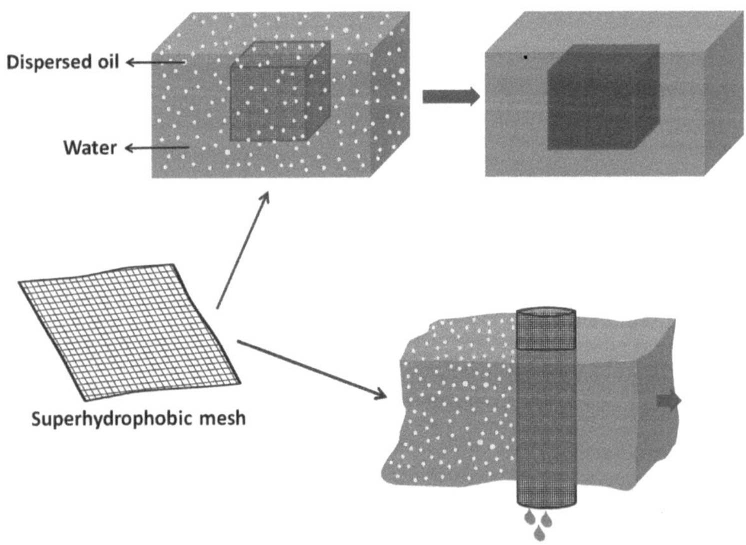

FIGURE 12.21 Scheme illustrating a cubic and columnar 3D-SMF prepared from the SMF for selective oil absorption from oil–water mixture. *SMF*, superwetting mesh film.

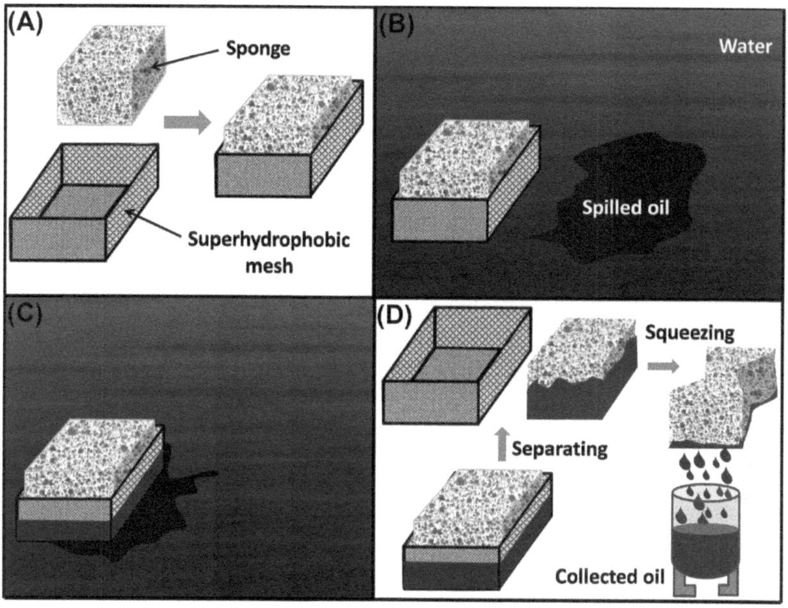

FIGURE 12.22 (A)–(D) A schematic illustration of the oil separating and collecting from water surface by using sponge–mesh package.

mesh combined with a superoleophilic polymeric sponge. Sponge is a cheap and commercially available superoleophilic porous material. However, it usually absorbs both oils and water, which may reduce the oil separation efficiency and selectivity. The combination of superhydrophobic mesh and superoleophilic sponge provides a good solution. When the oil collector floats on the surface of oil–water mixtures, oil penetrates through the walls of mesh and then diffuses into the pores of the sponge rapidly due to the superoleophilicity of the two materials, whereas water is completely repelled because of the superhydrophobicity of the mesh. All of the starting materials are commercially available, and the fabrication process is convenient and efficient. The oil collector can selectively absorb various oils while repelling water completely. Additionally, the absorbed oil in the sponge can be quickly released by squeezing. Furthermore, the collector exhibits good stability and repeatability in oil–water separation cycles. Therefore, this oil collector may provide a useful method for oil spill cleaning up from the surface of water [59].

➢ Fig. 12.23 schematically depicts the superhydrophilic mesh for spilled oil collection. The as-prepared mesh, tilted to certain degrees, is installed on the tail of the ship. The two sides of the as-prepared mesh are fixed on the holder and storage container. The oil–water mixtures are driven to the top

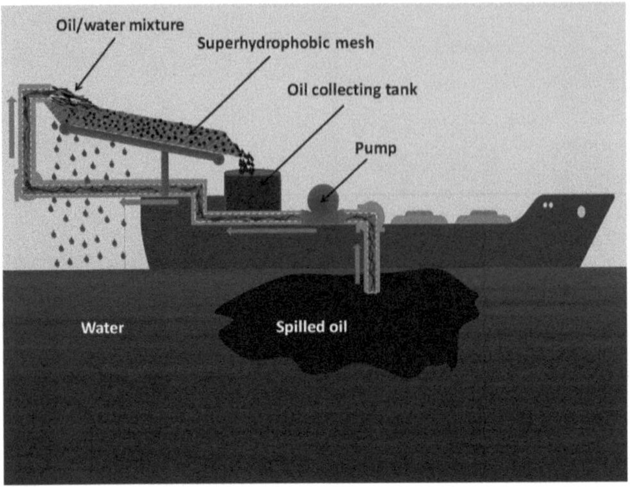

FIGURE 12.23 Illustration of the oil spill collector using superhydrophilic metallic mesh.

of the as-prepared mesh by a pump and then flowed on a parallel direction along the mesh surface by gravity during period when water gradually passes through the superhydrophilic mesh, while the oil flows over the mesh and finally pours into the storage container. [60]

28. Conclusions

Production and consumption of oil and petroleum products are increasing worldwide, and consequently, demand to treat polluted water by oil is increasing accordingly. Obviously, protection of the environment is very important by decreasing oil spills as much as possible. There are many barriers to clean up oil spills such as lack of government budget and high cost of doing the required process. Therefore, discovering of new and low-cost methods/materials for oil spill cleaning up is very important from practical and environmental points of view.

There are different techniques for removing of oil pollution from seawater, which depend on various factors, for example, type, amount, location, and characteristics of oil spill.

Among the introduced approaches for cleaning up of oil spill, the use of superhydrophobic and superoleophilic materials as absorbent have recently drawn a great deal of attention from both fundamental and practical points of view. The main advantages of superhydrophobic and superoleophilic absorbents are possibility of synthesis in large quantities, fast cleaning process for different oil pollution from water surface, and also easy oil separation from water under static and turbulent conditions. These advantages make it possible to apply superhydrophobic absorbents in real systems such as oil spill cleaning up from sea and river surface water.

As well as it is possible to design a simple, flexible, and intelligent collection of systems for oil separation process and to regenerate and reuse superhydrophobic absorbents.

References

[1] J.P. Wauquier, Crude Oil—Petroleum Products—Process Flow Sheets, Editions Technip, 1995.

[2] B.K.B. Rao, Modern Petroleum Refining Processes, forth ed., Oxford &Ibh Publishing Co. Pvt. Ltd, 2001.

[3] C. Song, X. Ma, New design approaches to ultra-clean diesel fuels by deep desulfurization and deep dearomatization, Appl. Catal. B 41 (2003) 207—238.

[4] M. Fingas, The Basics of Oil Spill Cleanup, CRC Press, Taylor & Francis Group, 2012.

[5] Y. Bai, Q. Bai, Oil Spill Response Plan, Subsea Pipeline Integrity and Risk Management, Elsevier, 2014, pp. 377—390, chapter 12.

[6] I. Leifer, W.J. Lehr, D.S. Beatty, E. Bradley, R. Clark, P. Dennison, Y. Hu, S. Matheson, C.E. Jones, B. Holt, M. Reif, D.A. Roberts, J. Svejkovsky, G. Swayze, J. Wozencraft, State of the art satellite and airborne marine oil spill remote sensing: application to the BP deepwater horizon oil spill, Remote Sens. Environ. 124 (2012) 185—209.

[7] L.B. Corti, T. Hfpner, Geo-biological aspects of coastal oil pollution, Palaeogeogr. Palaeoclimatol. Palaeoecol. 219 (2005) 171—189.

[8] R.G. Prasad, M.V.V.S. Anuprakash, Pollution due to oil spills in marine environment and control measures, IOSR J. Environ. Sci. Toxicol. Food Technol. 10 (2016) 1—8.

[9] I.A. Mendelssohn, G.L. Andersen, D.M. Baltz, R.H. Caffey, K.R. Carman, J.W. Fleeger, S.B. Joye, Q. Lin, E. Maltby, E.B. Overton, L.P. Rozas, Oil impacts on coastal wetlands: implications for the Mississippi river delta ecosystem after the deepwater horizon oil spill, Bioscience 62 (2012) 562—574.

[10] A.D. Mark, R.G. Kesava, Crude oil spill exposure and human health risks, J. Occup. Environ. Med. 56 (2014) 1029—1041.

[11] J. Chen, M.S. Denison, The deepwater horizon oil spill: environmental fate of the oil and the toxicological effects on marine organisms, J. Young. Investig. 21 (2011) 84—95.

[12] T.J. McGenity, B.D. Folwell, B.A. McKew, G.O. Sanni, Marine crude oil biodegradation: a central role for interspecies interactions, Aquat. Biosyst. 8 (1) (2012) 1—19.

[13] A. Olita, A. Cucco, S. Simeone, A. Ribotti, L. Fazioli, B. Sorgente, R. Sorgente, Oil spill hazard and risk assessment for the shorelines of a Mediterranean coastal archipelago, Ocean Coast Manag. 57 (2012) 44—52.

[14] S. Rahsepar, M.P.J. Smit, A.J. Murk, H.H.M. Rijnaarts, A.A.M. Langenhoff, Chemical dispersants: oil biodegradation friend or foe? Mar. Pollut. Bull. 108 (2016) 113—119.

[15] R. Almeda, S. Cosgrove, E.J. Buskey, Oil spills and dispersants can cause the initiation of potentially harmful dinoflagellate blooms ("red tides"), Environ. Sci. Technol. 52 (2018) 5718—5724.

[16] M. Fingas, In Situ Burning: An Update. Oil Spill Science and Technology, second ed., Elsevier, 2017, pp. 483—676. Chapter 10.

[17] M. Fingas, Physical Spill Countermeasures. Oil Spill Science and Technology, Elsevier, 2011, pp. 303—337, chapter 12.

[18] A.T. Hoang, V.V. Pham, D.N. Nguyen, A report of oil spill recovery technologies, Int. J. Appl. Eng. Res. 13 (2018) 4915–4928.

[19] N. Bhardwaj, A.N. Bhaskarwar, A review on sorbent devices for oil-spill control, Environ. Pollut. 243 (2018) 1758–1771.

[20] G. Wang, Q. Sun, Y. Zhang, J. Fan, L. Ma, Sorption and regeneration of magnetic exfoliated graphite as a new sorbent for oil pollution, Desalination 263 (2010) 183–188.

[21] F.X. Merlin, P.L. Guerroué, Use of Sorbents for Spill Response. Operational Guide, Research and Experimentation on Accidental Water Pollution. Translated by Alba Traduction, CEDRE, Centre of Documentation, 2009.

[22] M.O. Adebajo, R.L. Frost, J.T. Kloprogge, O. Carmody, S. Kokot, Porous materials for oil spill cleanup: a review of synthesis and absorbing properties, J. Porous Mater. 10 (2003) 159–170.

[23] A. Carré, K.L. Mittal, Superhydrophobic Surfaces, CRC Press, Taylor & Francis Group, Leiden. Boston, Boca Raton, 2009. ISBN:978900419333.

[24] S.S. Latthe, A.B. Gurav, C.S. Maruti, R.S. Vhatkar, Recent progress in preparation of superhydrophobic surfaces: a review, J. Surf. Eng. Mater. Adv. Technol. 2 (2012) 76–94.

[25] R.J. Crawford, E.P. Ivanova, Superhydrophobic Surfaces, Elsevier, Amsterdam, 2015. ISBN:9780128011096.

[26] Z. Guo, W. Liu, Biomimic from the superhydrophobic plant leaves in nature: binary structure and unitary structure, Plant Sci. 172 (2007) 1103–1112.

[27] T. Darmanin, F. Guittard, Superhydrophobic and superoleophobic properties in nature, Mater. Today 18 (5) (2015) 273–285.

[28] H.Y. Guo, Q. Li, H.P. Zhao, K. Zhou, X.Q. Feng, Functional map of biological and biomimetic materials with hierarchical surface structures, RSC Adv. 5 (2015) 66901–66926.

[29] G.B. Darband, M. Aliofkhazraei, S. Khorsand, S. Sokhanvar, A. Kaboli, Science and engineering of superhydrophobic surfaces: review of corrosion resistance, chemical and mechanical stability, Arabian J. Chem. (2018) in press, Corrected Proof.

[30] S. Li, J. Huang, Z. Chen, G. Chena, Y. Lai, A review on special wettability textiles: theoretical models, fabrication technologies and multifunctional applications, J. Mater. Chem. A. 5 (2017) 31–55.

[31] R.N. Wenzel, Surface roughness and contact angle, J. Phys. Chem. 53 (1949) 1466–1467.

[32] K. Kawasaki, Study of wettability of polymers by sliding of water drop, J. Colloid Sci. 15 (1960) 402–407.

[33] C.G.L. Furmidge, Studies at phase interfaces. I. The sliding of liquid drops on solid surfaces and a theory for spray retention, J. Colloid Sci. 17 (1962) 309–324.

[34] G. McHale, N.J. Shirtcliffe, M.I. Newton, Contact-angle hysteresis on super-hydrophobic surfaces, Langmuir 20 (2004) 10146–10149.

[35] A.J. Hall, Textile Finishing, Heywood Books, American Elsevier Publishing Company, Inc, New York, 1966.

[36] W. Chen, A.Y. Fadeev, M.C. Hsieh, D. Öner, J. Youngblood, T.J. McCarthy, Ultra-hydrophobic and ultra-lyophobic surfaces: some comments and examples, Langmuir 15 (1999) 3395–3399.

[37] H. Murase, T. Fujibayashi, Characterization of molecular interfaces in hydrophobic systems, Prog. Org. Coat. 31 (1997) 97–104.

[38] D. Quéré, Non-sticking drops, Rep. Prog. Phys. 26 (2005) 2495.

[39] A. Patnaik, R.S. Rengasamy, V.K. Kothari, A. Ghosh, Wetting and wicking in fibrous materials, Text. Prog. 38 (2006) 1–105.

[40] K. Liu, X. Yao, L. Jiang, Recent developments in bio-inspired special wettability, Chem. Soc. Rev. 39 (2010) 3240–3255.

[41] A. Marmur, Wetting on hydrophobic rough surfaces: to be heterogeneous or not to be, Langmuir 19 (2003) 8343–8348.

[42] R.N. Wenzel, Resistance of solid surfaces to wetting by water, Ind. Eng. Chem. 28 (8) (1936) 988–994.

[43] A.B.D. Cassie, S. Baxter, Wettability of porous surfaces, Trans. Faraday Soc. 40 (1944) 546–551.

[44] J. Bico, U. Thiele, D. Quere, Wetting of textured surfaces, Colloids Surf. A. 206 (2002) 41–46.

[45] B. Bhushan, Y.C. Jung, Wetting, adhesion and friction of superhydrophobic and hydrophilic leaves and fabricated micro/nanopatterned surfaces, J. Phys. Condens. Matter 20 (2008) 225010–225034.

[46] M. Zhang, T. Zhang, T. Cui, Wettability conversion from superoleophobic to superhydrophilic on titania/single-walled carbon nanotube composite coatings, Langmuir 27 (2011) 9295–9301.

[47] X. Chen, A. Selloni, Introduction: titanium dioxide (TiO_2) nanomaterials, Chem. Rev. 114 (19) (2014) 9281–9282.

[48] M. Khosravi, S. Azizian, Synthesis of a novel highly oleophilic and highly hydrophobic sponge for rapid oil spill cleanup, ACS Appl. Mater. Interfaces 7 (45) (2015) 25326–25333.

[49] H. Liu, Y. Kang, Superhydrophobic and superoleophilic modified EPDM foam rubber fabricated by a facile approach for oil/water separation, Appl. Surf. Sci. 451 (2018) 223–231.

[50] M. Khosravi, S. Azizian, Fabrication of an oil spill collector package by using polyurethane foam wrapped with superhydrophobic ZnO microrods/carbon cloth, ChemPlusChem 83 (2018) 455–462.

[51] D. Yuan, T. Zhang, Q. Guo, F. Qiu, D. Yang, Z. Ou, Recyclable biomass carbon@-SiO_2@MnO_2 aerogel with hierarchical architectures for fast and selective oil-water separation, Chem. Eng. J. 351 (2018) 622–630.

[52] M. Khosravi, S. Azizian, Preparation of superhydrophobic and superoleophilic nanostructured layer on steel mesh for oil-water separation, Separ. Purif. Technol. 172 (2017) 366–373.

[53] B. Liu, L. Zhang, H. Wang, Z. Bian, Preparation of MCC/MC silica sponge and its oil/water separation apparatus application, Ind. Eng. Chem. Res. 56 (20) (2017) 5795–5801.

[54] J. Ge, Y.D. Ye, H.B. Yao, X. Zhu, X. Wang, L. Wu, J.L. Wang, H. Ding, N. Yong, L.H. He, S.H. Yu, Pumping through porous hydrophobic/oleophilic materials: an alternative technology for oil spill remediation, Angew. Chem. 126 (2014) 3686–3690.

[55] C. Zhou, Z. Chen, H. Yang, K. Hou, X. Zeng, Y. Zheng, J. Cheng, Nature-inspired strategy toward superhydrophobic fabrics for versatile oil/water separation, ACS Appl. Mater. Interfaces 9 (10) (2017) 9184–9194.

[56] J. Song, S. Huang, Y. Lu, X. Bu, J.E. Mates, A. Ghosh, R. Ganguly, C.J. Carmalt, I.P. Parkin, W. Xu, C.M. Megaridis, Self-driven one-step oil removal from oil spill on water via selective-wettability steel mesh, ACS Appl. Mater. Interfaces 6 (22) (2014) 19858–19865.

[57] G.J. Dunderdale, C. Urata, T. Sato, M.W. England, A. Hozumi, Continuous, high-speed, and efficient oil/water separation using meshes with antagonistic wetting properties, ACS Appl. Mater. Interfaces 7 (34) (2015) 18915–18919.

[58] H. Sun, A. Li, X. Qin, Z. Zhu, W. Liang, J. An, P. La, W. Deng, Three-dimensional superwetting mesh film based on graphene assembly for liquid transportation and selective absorption, ChemSusChem 6 (2013) 2377–2381.

[59] F. Wang, S. Lei, C. Li, J. Ou, M. Xue, W. Li, Superhydrophobic Cu mesh combined with a superoleophilic polyurethane sponge for oil spill adsorption and collection, Ind. Eng. Chem. Res. 53 (2014) 7141–7148.

[60] K. Yin, D. Chu, X. Dong, C. Wang, J.A. Duan, J. He, Femtosecond laser induced robust periodic nanoripples structured mesh for highly efficient oil-water separation, Nanoscale 9 (2017) 14229–14235.

Index

'*Note:* Page numbers followed by "f" indicate figures, "t" indicate tables and "b" indicate boxes.'

B

M

N

O

P

Q

R

X

Y

Z

FSC
www.fsc.org
MIX
Papier | Fördert gute Waldnutzung
FSC® C083411

Kolibri 360 GmbH
Ferdinand-Jühlke-Straße 7
99095 Erfurt, Deutschland
produktsicherheit@kolibri360.de